Management of Natural Resources—Institutions for Sustainable Livelihood:

The Case of Rajasthan

Published by Academic Foundation
in association with the
Institute of Development Studies,
Jaipur, Rajasthan

Sunil Ray is a professor at the Institute of Development Studies, Jaipur. He has earlier been on the faculty of the National Institute of Rural Development, Hyderabad. He has also served at the Institute for Social and Economic Change, Bangalore. He has been a Visiting Fellow at the Institute of Development Studies, Sussex (UK). Prof. Ray has conducted several research studies sponsored by both national and international organisations. He has been in a Review Mission Team of the European Commission. Besides having published extensively in national and International journals, Prof. Ray has authored several books including *Protection and Industrial Growth in India* and *Natural Resources, Organization and Technology Linkages*. He has also co-authored *Agricultural Services and the Poor: Case of Livestock Health and Breeding Services in India.*

Management of Natural Resources—Institutions for Sustainable Livelihood:

The Case of Rajasthan

Sunil Ray

ACADEMIC FOUNDATION
NEW DELHI

First published in 2008
by

ACADEMIC FOUNDATION
4772-73 / 23 Bharat Ram Road, (23 Ansari Road),
Darya Ganj, New Delhi - 110 002 (India).
Phones : 23245001 / 02 / 03 / 04.
Fax : +91-11-23245005.
E-mail : academic@vsnl.com
www. academicfoundation.com

in association with:

Institute of Development Studies, Jaipur
www.idsj.org

Cataloging in Publication Data--DK
Courtesy: D.K. Agencies (P) Ltd. <docinfo@dkagencies.com>

Ray, Sunil.
Management of natural resources : institutions for sustainable livelihood : the case of Rajasthan / Sunil Ray.
p. cm. Includes bibliographical references (p.)
Includes index.
ISBN 13: 9788171886272
ISBN 10: 8171886272

1. Agricultural resources--India--Rajasthan--Management. 2. Forest management--India--Rajasthan. 3. Water resources development--India--Rajasthan. 4. Natural resources--India--Rajasthan--Management. 5. Sustainable agriculture--India--Rajasthan. I. Title.

DDC 333.709 544 22

Typeset by Italics India, New Delhi.
Printed and bound in India.

CONTENTS

List of Tables, Figures, Annexures and Maps

TABLES

FIGURES

ANNEXURES

MAPS

Foreword

With burgeoning population and changing life style the pressure on natural resources is becoming unsustainable. This concern for deteriorating population-resources nexus is now shared by a large number of scholars, policy makers and, fortunately, by public at large. However, there have not been many instances of restoring such balance. In this context, the book by Dr. Sunil Ray is a most welcome contribution. The value of the book is enhanced as it gives sound theoretical reasoning and illustrates the underlying principles by carefully conducted case studies in different ecological zones.

The main thesis of the book is that the state by itself cannot preserve environment or rejuvenate natural resources. An active involvement of community is the precondition for any measure of success. The community's involvement, Dr. Ray rightly emphasises, is predicated upon several factors. Complexity of these factors, such as power relations, extent of benefits expected and sharing of these benefits, has to be properly appreciated. Also, the scale of operations has to be properly defined. An activity, which needs to be undertaken at the basin level, has scant chance of success if confined to a village. And if the objective is to institutionalise such community-led efforts the 'rules of game' should be clear and enforceable. Such rules will have to specify the arrangements for contributions by the participating households, sharing of benefits and, procedures for conflict resolution.

A point rightly emphasised by Dr. Ray, which is often forgotten, is the role of the supportive institutions in ensuring success or failure of the community efforts. For example, access to credit and terms on which it is made available for any project can prove to be a critical facilitating, or inhibiting, factor. While the initiative can come from the community, the support has to be extended by several external agencies.

With all the handicaps in enhancing the role of communities in development and management of natural resources, Dr. Ray is not pessimistic about the future prospects. As he rightly maintains, as the quantum of resources become inadequate, and their quality deteriorates conditions are created for collective action by all those who are affected. Scarcity becomes a great leveler.

Dr. Ray's work is an important contribution in understanding the logic of collective action and forging right institutional arrangements in diverse social and ecological conditions.

V.S. Vyas

IDS, Jaipur

Preface

An already limited renewable resource base in a complex ecosystem is exhibiting signs of fragility in almost all agro-climatic parts of the state of Rajasthan. Institutional renovation at the initiative of the local community gained considerable momentum in the recent past as an approach to sustain regeneration of the resource base and its conservation. While there is no reason here to debate whether community management can be substituted for private management of common-pool renewable natural resources, one might be inclined to verify the objective conditions under which community institutional approach produces desirable results in the context of the profound social heterogeneity. This gives new direction to the discourse on the management of the natural resource base against the backdrop of increasingly competing claims on it. The rise of competing claims, as known, is increasingly shrinking the base of livelihood generation for millions. The issue is how the rural community in a state like Rajasthan could achieve an optimum level of regeneration of the resources and sustain it within the institutional framework. The present study is an effort to examine this issue.

We are grateful to the Ford Foundation for providing financial support for this study. I express my deep sense of gratitude to Prof. Shabd Swarup Acharya, former director, IDSJ, for his support and cooperation I received at every stage of my work during his tenure. Prof. Sarthi Acharya, director, IDSJ, has been a constant source of inspiration for me. I am thankful to him. I am indebted to Narayan Belbase and Anirban Ganguly, programme officer and programme associate, respectively of the Ford Foundation, New Delhi, when I began to work on this project. Their comments on my research proposal were rich and invaluable. I thank Vasant Saberwal, programme officer of the Ford Foundation for his valuable cooperation.

I am extremely grateful to the eminent academicians and practitioners, who were members of the Review Committee, for attending meetings and giving their insightful comments. The members of the committee included Prof. V.S. Vyas, Prof. S.S. Acharya, Prof. Katar Singh, Prof. Sarthi Acharya, Dr. Suresh Kumar, Neelima Khetan, Ajay Mehta and others, who were invited to attend the meetings. I am indebted to several scholars who went through some parts of the report at drafting stage. They included Prof. V.S. Vyas, Prof. Jean Dreaze, Prof. Czech Conroy, Dr. B.D. Tripathi, Dr. Ganesh Pangare and Prof. M.S. Rathore. I express my gratitude to the hydrologists and geo-hydrologists of the state groundwater department, Government of Rajasthan and the Central Groundwater Board, Jaipur for their help and cooperation in estimating groundwater recharge. Their comments on the draft were extremely encouraging. The draft report was presented in the final workshop in which several eminent researchers and practitioners made insightful comments on the draft. I thank them all. Special thanks are due to Prof. Katar Singh, Dr. Suresh Kumar and Shri Ajay Mehta for taking pains for going through the entire draft and making valuable comments. I am thankful to Prof. K.N. Joshi for his valuable cooperation at the initial stage of the project. I also thank Mr. Roop Singh, Herbarium Assistant, Department of Botany, University of Rajasthan for identification of plants. I take this opportunity to thank my colleagues working in the accounts and administration sections and the library of the Institute for their cooperation.

Cooperation and the help received from the intervening agencies, the NGOs, were immense. I thank them all profusely. Without the cooperation of the village community, it would have been impossible to conduct the study. I would like to place on record my appreciation for their selfless cooperation that we received during our field survey.

Finally, I do not know how to express my gratitude to my colleagues, who worked with me on the project and whose selfless cooperation made it possible for me to complete the project. Jagdish Sharma took pains in organising the field survey meticulously, initiated investigation of each research problem proficiently and documented field observations systematically. His contribution was immense. I am extremely grateful to him.

Help received from Alka Upadhay and later Ankita Rastogi for ecological analysis, Mahendra Bijarnia for statistical analysis, Sunil

Pareek for preparation of maps and geographical analysis and Vinita Agarwal for social analysis were invaluable. All of them were equally helpful in the field investigation. I thank them all. Ajay Garg remained all along with me until the project was over. His deep understanding of the Rajasthan village and the communities chosen for the present study and careful field investigation were extremely helpful in each step of the analysis. I am thankful to him. I am also thankful to others who worked for the project. They included Dr. Purushotham Sharma, Anindita Kali, Kanika Panwar, Rashmi Verma, Swati Krishnatrey, Sagar Meena and Kavita Mishra. I am also grateful to Sanjeev Mathur, who displayed his commitment with full sincerity in providing me computer assistance at each stage of the research work. Finally, I thank Nina, my wife and Sanjukta and Archita, my daughters for being understanding and patient.

Sunil Ray

1

Introduction

Background

The continued depletion of the natural resource base in Rajasthan is reflected in the growing vulnerability of the people to livelihood crisis. There is a growing mismatch between the claims of the burgeoning population on the natural resource stock and the resource regeneration. Recurrent droughts in the recent past have aggravated the problem. The certainty of livelihood generation, gained through a process of 'natural adjustment' between the claims of the people on resources and their regeneration, has ceased to exist. The communitarian approach towards resource conservation has lost its appeal.

Nor has intervention by the state made any substantive dent on it. With its narrow development perspective, the state has failed to recognise the interface that exists, or should exist, between livelihood generation and the local natural resource base. It now grapples with continued occurrence of dark zones in poor recharge of groundwater juxtaposed against its uses and deterioration of its quality, decline of forest cover, growing desertification, denuded village pastures, and finally, the shrinking of the life support system, the ecological biodiversity. On top of it, there is continuous colonisation of common land by private individuals (Ray, 1997; Mohnot, 2002; Bhise, 2004). All these together reinforce the criticality of livelihood generation in rural Rajasthan that seeks restoration of its local natural resource base as the only possible solution.[1]

---

1. Villagers especially the poor collect fuelwood, crop wastes, cow dung, weeds, fodder, organic manure (dry leaves, forest litter), building materials (poles, thatch and silt), fruits and vegetables, herbs, fibres, etc. from the local common natural resource base. Water is collected by all from community ponds, lakes, rivers, etc. Raw materials that are collected for traditional occupation including art and crafts from the common resource base provide employment and income to the rural poor (Singh, 1994a).

In the absence of adequate public action, villagers evolve coping strategies locally and fall back upon them for their survival. These strategies are distressful more often than not, including as they do the costs of human and livestock migration, working as wage labour in distant places, sale of livestock at low prices, shifting to petty trade, cutting down on food and clothing, and, of course, on children's education, etc. (Rathore, 2004). While they provide dependable routes to many for bare survival in the short run, they can never define terms of sustainable development for rural Rajasthan. Besides, they are riddled with a host of conflicts to which the villagers are often vulnerable. Such compelling circumstances tend to narrow down the scale of choices of survival strategies for the poor. In turn, this escalates transaction cost of their livelihood generation to a point of distress.

Antithetical to it is the state action that fails to recognise that terms of sustainable development of rural Rajasthan are rooted in conservation and growth of renewable natural resources. The obscurant approach as followed by the state for resolving resource crisis lends credibility to 'shallow management' of the resource base, wherever it intervened (Saxena *et al.*, 1997; Jairath, 1999). While this must not render inconsequential the role of the state, probing of development links that are intrinsically missing in its perspective is definitely a trying proposition. Initiatives taken by the officials individually within the administrative contours of the government in Sukhomajri in Haryana and Arabari in West Bengal are some instances that strike a positive note on the possible collaboration between state and community for better management of natural resources (Menon *et al.*, 2005).[2] However, these examples are limited and failed to yield community management in the real sense when the same is driven to other parts of the country by the government (Saxena *et al.*, 1997; Jairath, 1999).

It is not operational inefficiency alone that causes failure of the state. The significant aspect of state failure must be seen in the light

2. The Sukhomajri experiment in Haryana was led by a technocrat, who got constructed soil conservation structures on the slopes of the catchment followed by regulation of grazing activities on the slopes. Initiative was then taken up for setting up a sharing mechanism to manage water-harvesting structures. Subsequently, many 'hill resource management societies' came into existence with government support. Similarly, a forest officer in Arabari, West Bengal, struck a deal with the local communities in order to save the plantations done by the forest department. This finally led to what is popularly known as 'joint forest management'.

of its flawed ideological spectrum that has failed to bring 'environmental rationality' to the centre of its development discourse (Leff, 1993). Its performance, as a result, remains illusory. Nor can the market act as a dependable source to resolve environmental crisis (Platteau and Abraham, 2002; Chopra, 2005), for it is incomplete due to its inability to value environmental resources fully. Even contingent valuation, which is claimed to be an alternative methodology for valuing environmental resources outside the market, cannot serve the purpose because environmental goods are non-additive (Perrings, 2005; Chopra, 2005). Besides, absence of commitment to intergenerational transfer of resources with respect to private discount rate and incomplete information about the future market make the market-driven argument for natural resource management untenable. The arguments of non-linear processes that govern the ecosystem do not match with the market that functions well only if the processes that govern transformation of goods and services are 'linear' (Dasgupta, 2001a). The non-commensurability argument further de-legitimises the instrumental value of market as an institution in managing natural resources (Espeland, 2001).

It was this state-market dilemma that gave an entry to the institutional approach and generated a new development discourse in the field of sustainable management of natural resources. It brought the community to the centre of the discourse in that collective action was underlined as a strategy for improving public good. While the approach influenced civil society organisations, donor agencies and policy makers profoundly, the advocacy of 'getting institutions right' gained considerable momentum for sustainable management of 'nature capital'. Rajasthan is one of the early states in India to have seen considerable potential in it for strategic intervention. Several initiatives, as a result, were taken by the civil society organisations and the government to renovate indigenous institutions at the local level. The objective was to strike a balance between ecological imperatives and development compulsions and sustain such a balance through local institutional renovation. The present investigation is about the verification of this communitarian institutional approach in rural Rajasthan in terms of achieving this goal.

I

SEEKING RESOLUTION: AN INSTITUTIONAL APPROACH

Institutional approach, in the present context, suggests controlling human behaviour by the humans themselves when they interact with a degraded ecosystem for its restoration. Endless subsidisation by nature seems to have been threatening 'co-evolution' of both humans and the ecosystems and, hence, human survival (Costanza *et al.*, 2001; Norgaard, 1984). No system exists that does not involve control. While physical and behavioural laws control many processes of ecosystem and govern conditions of survival and reproduction of all living components of the ecosystem, human controls include physical and behavioural laws, selection mechanism and rules in use (Costanza *et al.*, 2001).

What does human control represent in the present context? It is control over behaviour that manifests in a variety of ways including habit and routinised activities, emulation, etc. It is sought only to see that nature capital regenerates, the ecosystem regains its resilience and co-evolution continues to exist uninterruptedly. In other words, humans desire a transition from old habits and practices that have failed to embody a system of control and facilitate co-evolution. Yet, they are socially acceptable and continue to disturb co-evolution. However, when humans find that co-evolution is threatened due to the practices of such old habits, they are forced to seek transition to new habit and practices that call for a system to control their old habits. It is a shift from 'habitual rationality' in Weberian sense to 'problem-solving rationality' (Redmond, 2004).

This is what one calls 'dual mode rationality' in that 'man is as much a rule-following animal as a purpose-seeking one' (Hayek, 1973). Contrary to the single mode rationality (neoclassical variety of rational choice theory or maximisation principle), individuals here make a conscious attempt to adopt a new course of action, which is collective, for a common cause. It is an evolutionary adaptation for survival in that it suggests better methods to cope with the changing circumstances, threats and opportunities, not to maximise utility (Nozick, 1993). In other words, human systems evolve through generating adoptive capacity with a wide-ranging coping ability. The rule-following behaviour creates space in social sphere of the community by means of bringing in a shift of the previously internalised habits and norms. By doing so, they enhance their coping ability while they interact with the degraded ecosystem and attempt to restore it. The reformed habits and practices result in institutional renovation or reforms that bring resilience to the human system, on the one hand, and the ecosystem, on the other. This makes co-evolution to continue (Munasinghe and Reid, 2005).[3] However, the challenge that lies before the rational utility maximising humans is how to sustain institutional renovation through their collective mode of action. Although it is argued by some that it may sustain as long as perceived cost of the member of collective action is more than perceived benefit, it is too simplistic to approximate cost in reality (Singh, 1994a). If other factors including social and economic heterogeneity, diverse stake in the resource base, etc. are incorporated into the cost functions, the argument may loose its rigour. What is therefore, suggested is to recognise the importance of equity that plays a significant role in sustainable collective action (Oakerson, 1986). But, then, one must link it to the power structure within which institutional renovation is sought.

Criticality of Nature Capital

It is argued that nature capital is critical since no human or manufactured substitute currently exists. Besides, it contributes to the life-support system (Ekins *et al.*, 2003). Some have tried to understand its criticality in two different perspectives—ecocentric (most important to maintain environmental health) and anthropocentric (ecosystem services that are important to our survival and well-being and can never be substituted) (Groot *et al.*, 2003). The dichotomy in the perspectives may be essential at one level in order to gain insight into the differences in the nature of functions. However, at another level, such an understanding may hide the complementarity that exists between the two and reduce the integrative character of the nature

3. Preservation of biodiversity strengthens resilience of the ecosystem. It is like preservation of capital stock that protects the economic systems from external shocks. In social systems, resilience depends on the capacity of human societies to adopt and function despite stresses and shocks. The long-term concept of co-evolution of social, economic and ecological subsystems within a larger more complex adoptive system provides useful insights into the harmonious integration of various elements of sustainable development (Hanna and Munasinghe, 1995).

capital. In any case, these perspectives are central to any study of human interaction with thc ecosystem, and therefore, it is important to contextualise them to draw lessons (English Nature, 1994).

In the present context of research, it is a set of renewable environmental resources such as groundwater, forest and forest produces, village pastures, land, etc. whose immediate relevance to human survival and well-being can hardly be questioned, especially in a state like Rajasthan. One is rather more inclined to develop a strong perspective on the criticality aspect of these nature capitals, which are passing through crises owing to their degeneration. It is pointless to overemphasise how survival conditions of humans in rural Rajasthan are shaped by these environmental resources (Jodha, 1992). Here, human lives are so interwoven with the local ecosystem that one can hardly talk about the use of substitutes of the environmental resources, as mentioned above, and seek solution to the mismatch between increasing consumption requirement and their availability within the framework of the market. While one may not have difficulty in finding substitutes for fuelwood, grass, fodder, timber, etc. one does not know how their costing will be socially uneconomic (Dasgupta, 2001a). Not to talk about economic calculation as some economists do in the case of non-renewable resources (Myers and Simon, 1994; Simon and Khan, 1984). The substitution argument in favour of environmental resources in the context of the present discourse is misdirection for it reduces the usefulness of biodiversity.[4]

Achieving biodiversity is, therefore, an incontestable proposition, which is yet to be an integral part of the development discourse of rural Rajasthan. The argument, therefore, is how to bring back resilience in the degraded ecosystem by means of expanding biodiversity, while the latter continues to provide environmental services. How prepared are humans to switch over to a new behavioural system to respond to ecological distress? This question has an immense significance for livelihood in the present context of globalisation that intensifies uses of resources and over extracts niche resources with little concern for their environmental and socio-economic consequences (Jodha, 2001).

---

4. The currently dominant literature on economic growth ignores these issues and overemphasises the growth of ideas. According to these literature, growth of ideas can circumvent any constraint the natural environment may impose on the ability of economies to grow indefinitely (Dasgupta, 2001a).

II

REMAKING OF INSTITUTIONS: AN OVERVIEW

While collective action of the community as opposed to individual action is seen as an irrefutable approach in the development literature, it seeks to reform old habits and practices as mentioned earlier. These sets of reformed habits and practices that the community seeks to achieve through self-enforcing control of its behaviour are, as one may call, institutional renovation or reforms within the context of 'methodological collectivism' (Samuels, 1990). A potentially strong route is discernible in sustainable management of natural resources in these theoretical constructions. For, it justifies rebuilding of the community institution and put in place self-enforcing mechanism of controlled human behaviour. However, difficulties arise when such institutions fail to sustain in practice due to freeriding that does not allow collective action to produce collective benefits (Ostrom, 1990).

If social actors of the community, with diverse interests under common norms of behavioural pattern, bring institutional resilience, they can bring back ecological resilience by means of expanding biodiversity. Hence, what is inescapably true is the emergence of a system of cooperation among diverse interest groups of people in that the degree of freedom available to the individual member of the same system is restricted (Sengupta, 2001). While this gives rise to the emergence of the institution, the underlying condition for it being able to gain resilience is the complete absence of freeriding of any individual member of the system of cooperation that the community has itself evolved. However, development praxis in the field of sustainable management of natural resources seems to have insinuated its culpability and resultant loss of institutional resilience because of freeriding. One is then drawn into reworking of the fundamental principles of institutional development that helps to achieve the objective. Theoretical edifice, as it increasingly received intellectual legitimacy for remoulding the communitarian approach for natural resource management, is yet to address the loss of institutional resilience mainly due to it that power structure permeates.

Institutional discourse, as it evolved in the recent past with its great propensity for empirical verification, has eschewed such an important political economy question. It seems to have been carried away more by democratic decentralisation. While development literature recognises how powerful the latter could be as opposed to bureaucratic centralisation, it remains rhetoric and fails to address the reality where power relations never allow decentralisation to take root in the socioeconomic and political milieu of the community. The limited appeal of the game theoretic approach that concentrates on the conditions under which stable solutions can exist, fails to capture this, and by doing so, it fails to capture evolutionary process of the institution (Chopra, 2005).

Understanding Community

The ability of remaking of community institutions lies more in its homogeneity assumption that leads one to remain uncritical about the use of the term 'community'. Many have pointed out that such an uncritical view about the community obscures power relations that disempower people and render them voiceless (Mansuri and Rao, 2004). Even those who have tried to understand community from the viewpoint of geographical and conceptual boundaries are not straightforward and much is left to be desired (Mansuri and Rao, 2004). The 'they do not belong to us' argument may take one nearer to reality while one attempts to conceptualise the same (Elias and Scotson, 1965), but it is too difficult to obtain a definite grasp of social and economic differentiation that social relations reproduce continuously within a community. It may be puzzling and seek empirical significance of community institution if one goes by differentiation, draws boundaries and conceives institutional remaking. This then suggests parallel existence of many communities within the same community and leads one nowhere in locating community institutional approach for redress of the common problem that each one is facing. To contextualise it, we may find different social groups in terms of caste living in a given geographical boundary, say a village. Alternatively, we may have another configuration of the same population in terms of the economically rich and poor people across all caste groups. The issue is how does one draw boundary and define community, and seek empirical significance of community institutional approach for, say, sustainable management of natural resources.

While one sees a somewhat weak theoretical assertion in this regard, one may find it comfortable to draw a boundary based on shared values that give a sense of collective wisdom and make a distinction between outside and inside (Falk, 1994). Once it is done and the 'they do not belong to us' argument is juxtaposed with this, one may find it less problematic to draw boundary at all levels of analysis, including the macro ones. It may then produce and reproduce internal unity (Casson and Wadesen, 1998). The central question is how does one underline the essence of the community in relation to the shared values. To the extent values are commonly shared, homogeneity of mind prevails and creates what Veblen calls common institutional 'furniture' (Dolfsma and Dannreuther, 2003). However, if boundaries are made flexible to the point where they are inconsequential, as viewed by the neo-liberals, no scope is left for analysis of power relation's *vis-à-vis* social relations or social structural explanation of economic life. Dolfsma and Dannreuther argue that, "...the concept of boundary allows for an analysis of social relations, and their effects on polis and economy, in a way that is able to conceptually come to grips with the 'threshold effects' endemic in society" (Dolfsma and Dannreuther, 2003).

One must not ignore the analytical importance of boundary or demarcation issue to the study of long-term cooperation and institutional sustainability. This is what is held by Ostrom as one of the design principles of cooperation (Ostrom, 1990). However, what is equally important is to examine the role of power relations in shaping the nature and direction of cooperation and its implications for renovated community institution even within the boundary. This is seemingly missing in the design principles of cooperation as constructed by Ostrom. One may, however, find Galbraith's variety of 'condign power' useful to explain how it plays an instrumental role perpetuating unequal power relations (Galbraith, 1983). For, it is a case of submission of will of the individual to the powerful one. The powerful one may exercise its power through organisation and reproduce same relations of unequal power and authority (Leach *et al.*, 1997a; 1997b and 1997c).

To wriggle out development discourse from the conceptual confusion of the community, Plateau and Abraham go by the economic concept of the community (Platteau and Abraham, 2002). The key characteristics namely, perfect information, continuous interaction and trust, that they bank upon to explain community, has an implicit recognition of the need to understand the community *vis-à-vis* its social relations in an imperfect world. They perceive institution remaking as a transition from status-based to norm-based society. It is but a radical transformation of the deep-rooted values, beliefs, and the practices that have to do with the mode of social relations, power structure, sources of legitimacy and perception of the outside world (Platteau and Abraham, 2002).

One must, therefore, recognise that socioeconomic hierarchies undermine potential for collective action for the common good within the community (boundary) and increase potential for 'local captures' of benefits (Dasgupta *et al.*, 2004). It suggests that one must bring power and power relations to the centre of the development process that results from a complex interaction of market and institutional forces. 'Getting the power process right' is still a fundamental challenge in the field of development (Cameron, 1995).

Social Capital, Power Relations and Institutions

One may, therefore, argue in favour of getting the power process right within the institutional framework at the community level through collective action. It is in this context that institutionalised behaviour based on the concept of social capital (social networking, norms and mutual trust) gained considerable currency in the development literature (Bourdieu, 1983; Coleman, 1988; Putnam *et al.*, 1993). However, the central issue is how does one conceive the framework of social capital that allows institutional development through collective action and get the power process right. Bonding and bridging social capital are two significant elements that are given instrumental roles in Putnam's work for sustenance of community institution. But, it hardly recognises the existence of power structure and exclusion of the poor from development. Besides, it undermines the role of the state (Harriss, 2001; Staveren, 2003). Some have gone further to observe that this framework falls very much in line with the 'post-Washington consensus' that has given space to non-market interventions to resolve market imperfections that have social origins (Fine, 2001; Bebbington *et al.*, 2004).

It may not be wrong to emphasise the role of non-market institutions such as community institutions in the event of the failure of both market and state for sustainable management of natural resources. But one fails to understand why the development state is kept out of the conceptual framework of social capital in that contributions of political action, role of law, civil and political liberties, etc. may be instrumental to create an enabling environment for improving social capital and facilitating collective action to sustain (Bhattacharya *et al.*, 2004). The World Bank's notion of social capital as the 'missing link' in development is understandable. However, the framework within which the notion is sought to be legitimised following Putnam's work is misleading.

There are several other critiques levelled against its integration into economic theory. Apart from the problems associated with its measurement (for example, how to measure trust in a society or weigh its differences, etc.), its instrumental and functionalist views are criticised (Staveren, 2003). Not all these criticisms, however, reduce the importance of social capital as an instrument to mobilise social actors for collective action outside the power structure. There are cases that show how class-based political action improved social networking and brought about economic transformation through collective action (Heller, 1996). One must recognise that the concept of social capital serves to analyse inequality in the society. Hence, omission of power conflicts from the analysis contributes to the explanations why poverty is so persistent (Durston, 2001).

Several researchers have pointed out how social and economic heterogeneity can have negative consequences on collective action related to management of common-pool resources (Hardin, 1982; Dayton-Johnson and Bardhan, 2002; Menon *et al.*, 2005). Some studies, however, reveal the opposite (Wade, 1988; Jayaraman, 1981; Easter and Palanisami, 1986; Tang, 1991; Ostrom, 1990; Mosse, 1997; Braun and Feldbrügge, 1988). Although each case is distinguishable in terms of its socio-cultural and economic environment, the power structure has a direct and indirect role in impinging upon collective action in the first case. These successful cases suggest how homogeneous material base of the individuals facilitates collective action and making of an institution.

However, there are cases that exemplify how even heterogeneity does not constrain collective action (Baland and Platteau, 1998). This is what happened in the arid areas of western Rajasthan, as Baland and Platteau explain in the same paper (quoting the illustration from Jodha's works (Jodha, 1992), before the modern state was formed by conglomerating the princely states of Rajputana. At that time, common grazing lands were under the effective control of *jagirdars*, the big landlords.[5] This is clearly a case of compliance of the villagers with the rules of conservation imposed by the authoritarian ruler rather than collective action achieved through consolidation of social capital under the condition of heterogeneity. Had this been collective action by the village community in that the regulatory authority was the community itself rather than the landlords, the village commons would not have suffered degradation. The local elite would have been restrained from misuse for illicit personal advantage at the expense of the resource use efficiency (Baland and Platteau, 1998).

This is not to decry the importance of collective risk-sharing that once used to be an institutionally proven stable arrangement and was rooted in high risk and low productive arid Rajasthan (Jodha, 2001). Concern for collective sustenance at the community level was so high that the semblance of its continuity is found even today in the maintenance of village commons such as *oran*, repository of biological diversity, in some places in Rajasthan. However, despite such unwritten codes of conduct and maintaining sanctity, quality and productivity of these village commons have gone down considerably in the recent past (Mohnot, 2003).[6]

While both illustrations show how informal village institutions could conserve village commons, compliance of the community was maintained through two different routes. One was to the authoritarian rulers while the other one was to God or local heroes or martyrs.

---

5. "By virtue of their dominant position, they could appropriate a large share of the benefits accruing from the exploitation of common property resources (the best pastures were indeed earmarked for the animals they owned). It is therefore not surprising that they took upon themselves the task of deciding and implementing 'conservation measures', which ensured considerable stability to these resources. Thus, they charged grazing taxes, organised rotational grazing (evenly scattered) around water points, decreed the periodical closures of part of the commons and periodical restrictions on entry of certain animal species, appointed watchmen to monitor compliance with grazing regulations, imposed penalties on herd owners found guilty of violating those regulations, and used their authority to extract regular contributions for maintenance works from poor users. After Independence, following a land reform that resulted in the privatisation of a large part of the village grazing areas and in the dissolution of the *jagirdar* rule and its replacement by the *panchayat* system, collective maintenance of the commons was discontinued" (Baland and Platteau, 1998).

6. While some are about 150-500 hundred years old, some others have been in existence for 700-800 years. Most of the *orans* support human and livestock population of two or more villages of the area, some 5 or more while a few supports as many as 30 villages. There are two-three *orans* that support as many as 100 villages (Mohnot, 2003).

The common aspect in both was the compliance, which was imposed from above. The status of local heroes or martyrs was raised to a level that commanded tremendous respect from the community. Once the village commons, in the second case, were dedicated to them, the maintenance of their sanctity gained overriding importance in terms of their conservation. A kind of mystical faith associated with the act of dedication was an instant source for creating values that went against indiscriminate human interference and degradation of the village commons. Although there was a concern for collective sustenance that had a definite contribution to this system of compliance, imposition from the above associated with fear and respect in both cases as illustrated above gave credible support to the sustenance of the community institution in the past.

While one may learn from this lesson, it is imperative to read the past carefully to get at the truth. Hence, what was possible to achieve as a stable social arrangement may have less operational significance today in the face of steep rise of claims on resource base as a result of pressure of population on the one hand and social disintegration on the other. The rapacious claims on the resources by the rich and powerful further aggravated the balance between the people and their environment (Vyas, 1991). The questions, therefore, are: Is institutional remaking of the village community possible in Rajasthan for sustainable regeneration of resources under growing market influence, social disintegration and demographic pressure? If so, under what conditions is it possible? How does one see the conceptual framework of social capital in the realm of power structure and suggest institutional renovation for meeting environmental entitlements of all? How does the state define its engagement with the community institution and seek its engagement for sustainable management of natural resources (Dasgupta, 2001; Nadkarni, 2002)?

These questions are, perhaps, not new. However, what is missing is a coherent response to these questions within the complex relationships between ecological change and community response to it. Its implications for sustainable management of natural resources in a power theoretic framework are relevant to the study of institutional sustainability. This provides an interdisciplinary framework for the present study that seeks to empirically examine the development significance of institution for natural resource management with the following objectives.

III

Objectives

- To examine social conditions under which institutional development takes place and how it determines the entitlements of different social actors to improved natural resources base;
- To evaluate the role of interventionists in creating scope for local institutions to become self-sustainable in the long run;
- To examine how efficient interventions are in reducing the impact of environmental cost in: (i) improving village economy, in general, and household economy of the poor, in particular; and in (ii) improving women's well-being;
- To assess how ecological imperatives are respected and built into intervention mechanisms that aim at initiating a process of sustainable development;
- To explore alternative institutional outcomes under different conditions and identify the best scenario under which institutions achieve stability and become more successful in management of natural resources.

Methodology

The case study method was followed to examine the objectives mentioned above. Both primary and secondary data were collected to examine: (1) demographic transition and changing livestock composition of the study area; (2) history of governance and pattern of utilisation of natural resources; (3) historical context of the formation of the content of social capital; and (4) changing natural resource maps, etc. Data and information were collected from both secondary and primary sources on many other aspects as well. Apart from household survey for primary data collection, several focus group discussions were conducted in each village with different groups of people—including the elderly, women of all social groups together, women of only backward communities, executive members of newly formed village institutions, *panchayat* members and NGOs—and all villagers together. The field survey was conducted in different phases over a period of two years from 2002-2004. The vegetation sampling was done in each study area twice in a year, before and after rainfall.

Appropriate tools were applied for analysing resource entitlements that emerged because of development initiatives taken by the local institutions. The study followed the methodology developed by the Ministry of Water Resources, Government of India for estimating groundwater recharge (Annexure 1). Suitable statistical tools were used to examine the ecological impact of institutional intervention and analytical tools were used to examine the economic benefits that accrued to the social actors. The ecological impact was examined on the basis of the estimation of several indices following standard methodology (Annexure 2). The scenario analysis of probable alternative institutional outcomes was carried out on the basis of suitable statistical tools under different conditions. This was done in respect of those study areas where local institutions were found active during field survey (Annexure 3).

Study Areas

Five different cases spread over all agro-climatic zones of Rajasthan were selected for the present study. They included: (1) pastureland development (Ajmer district); (2) revival of traditional rainwater harvesting practices (Alwar district); (3) farm forestry development (Udaipur district); (4) revival of degraded forest (Banswara district); and (5) micro watershed development (Jodhpur district) (Map 1).

It may be noted that the study areas were selected after conducting several rounds of discussions with many civil society organisations, which were engaged in conservation and growth of renewable natural resources in the state. Discussions were also held with the officials of the concerned departments of government of Rajasthan. Trips were then undertaken to the places where interventions were made for resource conservation in order to gain first-hand knowledge. Finally, cases that could fulfil the basic criteria were selected for the present study. The basic criteria that were considered for selection of the study areas were the spread of the cases over all ago-climatic zones, length of duration after the withdrawal of interventionist (the capacity of the village community in managing its resource base independently after the intervention was phased out); and the remoteness of the study area. Once the study areas were selected, meetings were held with the villagers in each village. This was followed by collection of basic information of each household in the village. Appropriate sampling design of the household was taken up in the next stage.

Sampling Design

The sample size was calculated with true proportion of 0.5 with maximum acceptable difference of 0.07 at 95 per cent confidence interval for the total number of households. The total number of household for the entire study was 2,639 of which 635 households were sampled, which was approximately 24.06 per cent of the total households of the villages under study (Table 1.1). The sample size was calculated through PEPI VERSION 4, considered a standard statistical tool for sample size calculation. The acceptable difference was further reduced by different sampling schemes including: (1) stratified sampling; (2) two-phase stratified sampling; (3) two-stage stratified sampling; and (4) cluster stratified sampling. Appropriate sampling schemes, including geographical spread of the village, caste composition, and landholding pattern, were chosen depending upon the characteristics that differed between the selected villages. Apart from sampling of households, several focus group discussions, as mentioned above, were also conducted in each village.

The Plan of the Book

The book is organised into nine chapters. Chapter I critically reviews the ongoing discourse on institutional management of the renewable natural resources and delineates the issues that are examined in respect of cases selected for the study. While the case of development of degraded pastureland is examined in Chapter II, the revival of traditional rainwater harvesting practices and its impact on the local ecosystem is probed in Chapter III. Chapter IV concentrates on examining how farm forestry development through plantation could serve both commercial and environmental interests of the community. Chapter V evaluates the significance of regeneration of a degraded forest. The premise is that, if the conservation programme of renewable resources is appropriately designed and implemented, village community need never be vulnerable to impoverishment even in desert conditions. This is examined in Chapter VI, which shows how micro watershed development can restore a degraded ecosystem under arid conditions. Chapter VII examines how an organised attempt to restore the local ecosystem through new institutional

Map 1

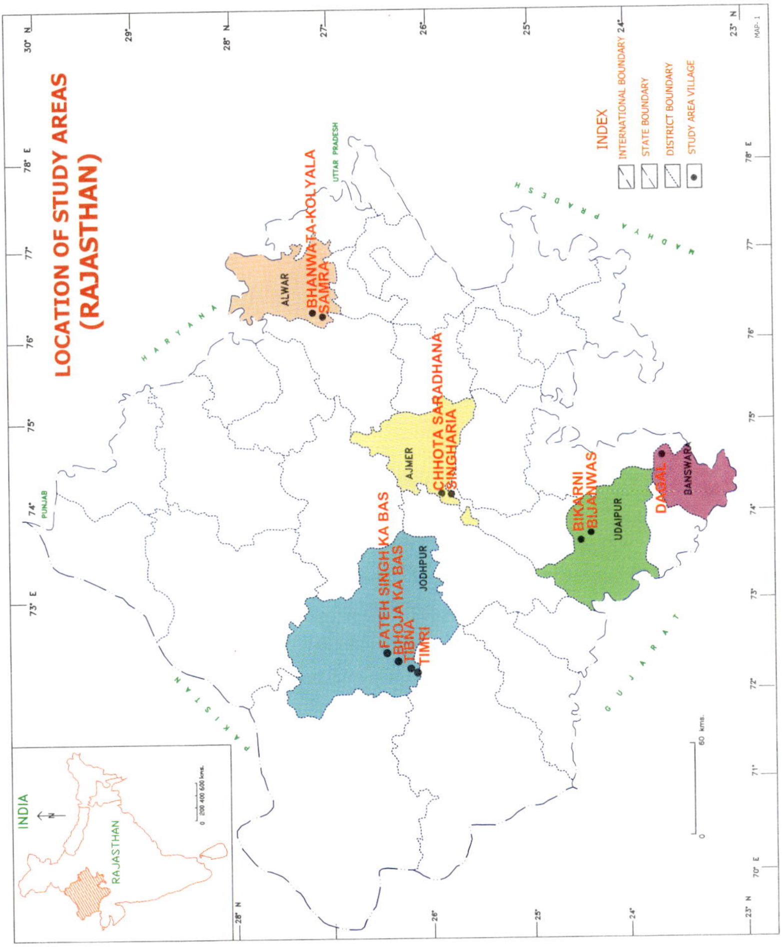
LOCATION OF STUDY AREAS
(RAJASTHAN)
INDIA
RAJASTHAN
0 200 400 600 kms.
PUNJAB
HARYANA
UTTAR PRADESH
MADHYA PRADESH
GUJARAT
PAKISTAN
ALWAR
BHANWATA-KOLYALA
SAMRA
AJMER
CHHOTA SARADHANA
SINGHARIA
JODHPUR
FATEH SINGH KA BAS
BHOJA KA BAS
TIBNA
TIMRI
UDAIPUR
BIKARNI
BIJANWAS
BANSWARA
DAGAL
INDEX
INTERNATIONAL BOUNDARY
STATE BOUNDARY
DISTRICT BOUNDARY
STUDY AREA VILLAGE
0 60 kms.
MAP-1

Table 1.1

Estimation of Sample Size of the Study Areas

Cases	*Selected Villages*	*Total No. of HH*	*Sampled HH*	*Sampling Criteria*
Pastureland development	Chhota Saradhana	83	50	
(Ajmer district)	Singhariya	199	120	Stratified random
Total		282	170	sampling
Rainwater harvesting (Alwar district)	Bhanwata-Koliyala	89	53	(i) Two-stage stratified sampling
	Samra	286	73	(ii) Multi-phase
Total		375	126	sampling
Farm forestry development (Udaipur district)	Bijanwas and Bikarni	817	142	Two-phase stratified random sampling (probability proportional to size)
Revival of degraded forest (Banswara district)	Dagal	351	78	Two-stage stratified random sampling (probability proportional to size)
Micro watershed	Tibna/Timri	471	69	Cluster stratified
development	Bhoja ka Bas	217	29	sampling (probability
(Jodhpur district)	Fatehsingh ka Bas	126	21	proportional to size)
Total		814	119	
Grand total		2,639	635	For all case studies

Note: HH-Household. The number of sample households in Bijanwas was 107 and, for Bikarni, it was 35.

practices provides an important route to women for their empowerment. Chapter VIII summarises main findings of the study and Chapter IX seeks to address the implications of the study and searches for a new perspective for sustainable management of natural resources.

2

Pastureland Development

Introduction

Pastureland development continues to be peripheral in the development discourse of Rajasthan. The incredible contribution pastureland makes to the state's economy, mainly through conversion of grass grown on it to milk, wool and leather, is generally overlooked in what appears to be a classic case of the 'development paradox'. While industrial processing, based on these raw materials, receives attention from the government and policy planners, the nature capital, the grassland that makes the enormous contribution of feeding the industrial development of the state, lacks the consideration that it deserves.

It may be noted that the land available (including barren and uncultivable land, other uncultivable land excluding fallow land, permanent pasture and other grazing land and culturable waste) for grazing and fodder collection for livestock continuously declined over the years. It declined from 11.60 million hectares in 1975-76 to 8.96 million hectares in 2001-02. Besides, there had been a steady encroachment of these lands in different parts of the state over the years (Mohnot, 2003; Bhise, 2004; Ray, 1997).[1] Additionally, the decline of production of grazeable biomass, continuous loss of vegetation cover, falling short of protein due to the lack of mixed type of herbage containing both leguminous and non-leguminous forage and grasses weakened the regenerative capacity of the available grasslands (Ray, 1999). That this steadily declining regenerative capacity of the grasslands is further minimising its potential contribution to livestock as a sustainable source of livelihood to millions of rural Rajasthan is completely overlooked. Even at the risk

1. As a result, carrying capacity of land of livestock declined continuously. For instance, even without taking encroachment into consideration, cattle units per hectare of land increased in the state from 2.31 in 1972 to 4.33 in 2003 (Ray, 2007).

of oversimplification, one can hardly discount the resulting negative implications for the sustainable development of livestock that provides a safety net, particularly to the poor in distressing conditions. Reversing the trend seems to be the only way of stopping the degeneration.

In such a scenario, Chhota Saradhana and Singhariya, two villages located in Ajmer district of Rajasthan (Map 2), show how the trend of degeneration of grassland ecosystem can be reversed and resilience restored through institutional reform. The villagers have been following renovated institutional practices of using their grazing lands and fuelwood collection for more than a decade or so, and largely on their own. Although the Magra Mewar Vikas Sanstha (MMVS), a non-governmental organisation (NGO), had a presence in the villages initially for some years, the village development society (VDS) carried forward the management of the pastureland without any external assistance once MMVS existed in the early 1990s. It embarked upon institutional renovation according to the need of the village community without compromising the growth and conservation of the grassland ecosystem.

The present study seeks to analyse how institutional renovation evolved and helped the village community of Chhota Saradhana and Singhariya villages to grow and conserve their natural resource base and gain from it. This case study has five sections—sections I and II concentrate on institutional development and analyse gains accrued to the village community in both the villages through the restoration of degraded pastureland; sections III and IV take up a comparative analysis of ecological changes and probable alternative scenarios of institutional outcomes; and section V looks at the lessons learnt.

I

CHHOTA SARADHANA

The first part of the study on Chhota Saradhana briefly describes the social context in which institutional reform was sought and underlines the changing social processes. While the second part traces how the village institution reformed itself for the conservation and development of common pastureland, the third segment analyses benefits that accrued to different groups of households, including the poor.

Map 2

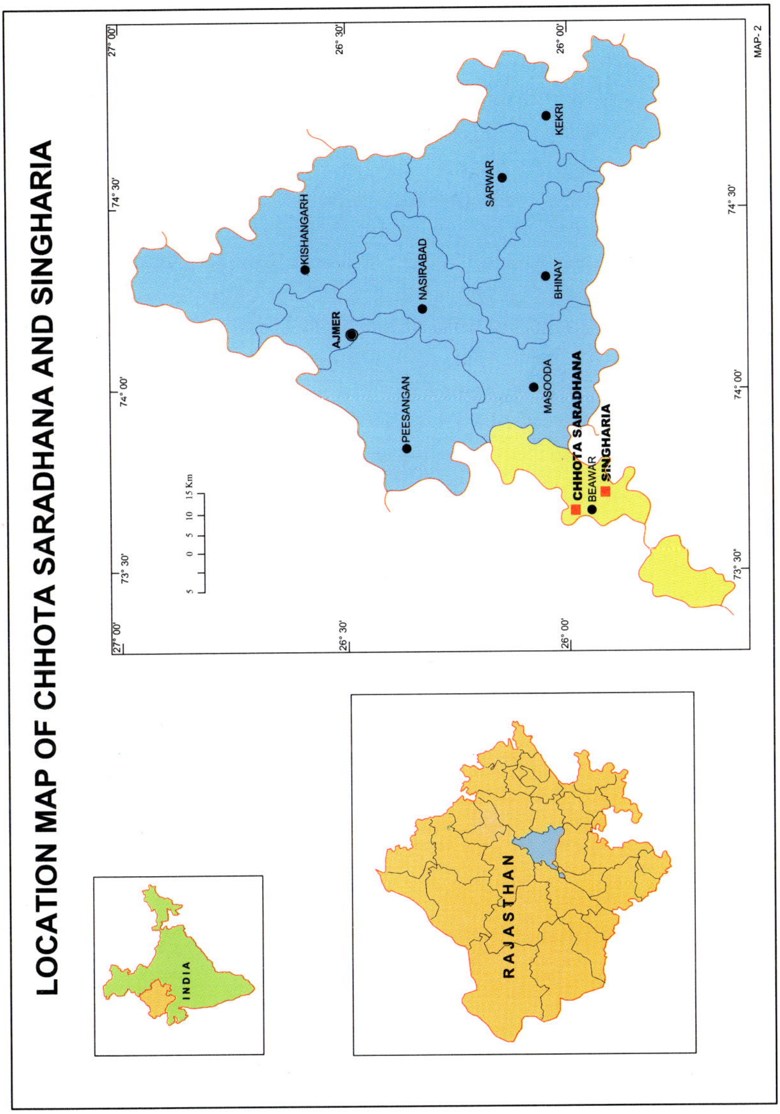
LOCATION MAP OF CHHOTA SARADHANA AND SINGHARIA
INDIA
RAJASTHAN
KISHANGARH
AJMER
NASIRABAD
PEESANGAN
SARWAR
KEKRI
BHINAY
MASOODA
CHHOTA SARADHANA
BEAWAR
SINGHARIA
5 0 5 10 15 Km
27° 00'
26° 30'
26° 00'
73° 30'
74° 00'
74° 30'
MAP- 2

The Social Context

Residents of Chhota Saradhana, a nondescript village in the Jawaja block of Ajmer district, could never have imagined even a decade ago, that it was possible for them to meet their increasing fuel and fodder requirements by developing their own natural resource base. The gradual loss of vegetative cover that had resulted in fodder and fuel scarcity in the village saw its severest manifestation during the drought of 1987. Only in such adverse conditions, did the villagers of Chhota Saradhana take note of the efforts of Dhaneswar Acharya, founder of MMVS. Much before the drought of 1987, Acharya, the force behind natural resource conservation through pastureland development in all eight villages of Jawaja block, had advised Chhota Saradhana to develop its own pastureland then in a state of extreme degradation.

In 1987, Acharya launched an awareness development programme through non-formal education, a significant aspect of which was to bring people together and persuade them to prioritise their problems, work out the means for their redressal along with assessing the need of external assistance. There was consensus in the community with no exception. A consensus for restructuring production conditions of fuel and fodder and a consensus for what one may call 'institutional renovation' sought through social action, which helped villagers gain access to means of reproduction.

The Village Dynamics

The early settlers in Chhota Saradhana were the pastoralist *Gurjars* with their traditional inclination towards conserving grazing lands and vegetation cover. As elsewhere, livestock rearing was their principal occupation with agriculture taking a secondary place. After them came the *Jats* with agriculture as their primary occupation. Though the two communities' occupational engagement differed initially, with varying concerns for production conditions, the village community was never riddled with any conflict on sharing of resources. The abundant availability of natural resources also helped the 'natural economy' of the village retain its stability. As time went by, the *Gurjars* stepped up their engagement with agriculture and the *Jat* community expanded its household activities towards livestock rearing.

Chhota Saradhana witnessed several other changes. Apart from occupational diversification, there was expansion in the range of caste groups. As many as seven caste groups, including *Jat* and *Gurjar*, live in the village now. A notable feature of all caste groups is that livestock rearing became equally indispensable to all. In fact, dependence on it increased over the years due to drought-induced uncertainty in practising agriculture. It led to intense competition for grass for livestock subsistence and for fuelwood for human subsistence. Conditions of scarcity began to disturb the village economy gradually pushing the villagers to a point of no return in 1987—one of the severest drought years in the recent past. The time was ripe for institutional reform of the village community for sustainable management of pasturelands to meet the fodder crisis and for initiating a plantation programme to meet the fuelwood crisis. But before examining how institutional renovation could succeed in establishing a new management regime for natural resource conservation and growth, it is important to understand the social structure of the village and the ownership pattern of land resources. It is also important to examine whether power relations contribute to shaping of institutional renovation.

Social Structure and Landholding Pattern

The number of households in the village almost doubled in three decades—from 47 in 1971, the number had gone up to 83 in 2001. As mentioned earlier, there are seven caste groups of which the largest number of households belonged to the *Gurjar* community (Table 2.1).

Table 2.1

Distribution of Households across Different Caste Groups and Landholdings

Caste Groups	*Landless*	*Marginal*	*Small*	*Medium*	*Large*	*Total*
Gurjar	15	24	3	3	-	45
Jat	-	-	4	17	1	22
Meghwal	5	-	-	-	-	5
Bawri	-	2	-	-	-	2
Kumhar	-	2	-	-	-	2
Rawat	-	-	2	-	-	2
Sadhu	-	5	-	-	-	5
Total	20	33	9	20	1	83
Percentage	24.10	39.75	10.84	20.09	1.20	100.00

Note: Marginal farmer: 0-4 *bighas* of land; small farmer: 4-8 *bighas* of land; semi-medium *farmer*: 8-16 *bighas* of land; medium farmer: 16-40 *bighas* of land; Large farmer: above 40 *bighas*.

Source: Field survey.

As Table 2.1 shows, only 21 per cent of the households had medium and large holdings while the landless, small and marginal landholders constituted the rest. The majority of the *Gurjar* community fell under this lower landholding category. Eighteen, out of the 22 households possessing large and medium landholdings, belonged to the Jat community. Households that belonged to other caste groups were numerically weak and, therefore, traditionally did not have any significant voice in the decision making process of the village. However, at the time of the drought, social forces seemed to have polarised all groups vertically and restored balance between the two dominant communities of *Jats* and *Gurjars*—although the *Gurjars* had the largest numbers they had far less ownership of agricultural land than the *Jats*, but they scored in livestock ownership over all the communities (Table 2.2).

Table 2.2

Ownership Pattern of Livestock across Social Groups

Caste	*Total No. of Households*	*Total Number of Animals*				
		Cow	*Buffalo*	*Sheep*	*Goat*	*Bullock*
Meghwal	5	3	1	-	51	-
Sadhu	5	2	1	-	2	-
Jat	22	12	33	-	35	4
Kumhar	2	1	-	-	1	-
Rawat	2	1	4	80	20	-
Bawri	2	1	2	-	2	-
Gurjar	45	27	49	155	322	5
Total	83	47	90	235	433	9

Source: Field survey.

The significant element that contributed to the social polarisation was the uniform material conditions that all social groups encountered during the drought. This provided a homogeneous ground for the reformation of institutional practices of the consumption and conservation of grassland resources. Consequently, power relations in whichever form they existed could not weaken the norms of mutual trust and reciprocity. On the contrary, the alternative credit delivery system developed by the villagers themselves created a larger social space for deepening them further. The traditional moneylending system, known for its instrumental role in shaping rural power relations, was replaced by this alternative system.

It may be noted that there were seven active women self-help groups (SHGs) in the village with a membership of 115 and an annual collection of Rs. 1,38,000. These SHGs were organised by the villagers themselves. Any member requiring credit paid interest at 2 per cent per annum. Similarly, the men of the village formed three groups with 20 members each with an annual collection of Rs. 4,80,000. These groups themselves evolved norms for delivering credit to their members at minimum interest. In this social context, the awareness development programme lent credible strength to social capital in the village community for institutional renovation for the development and management of pasturelands.

The atmosphere of reciprocity and mutual trust changed the nature of the social embededdness of the village community. All social groups were prepared to respond to the renewal of traditional institutional practices for collection of fuelwood and fodder as suggested by MMVS. Of course, it was a response born out of the sheer necessity for the survival of both livestock and human beings. But one cannot underestimate the outcome of institutional renovation achieved through the strengthening of social capital. It may be noted that this had no correspondence with the capacity building of the village, achievable, as normally understood, through education.

Chhota Saradhana exemplified an extreme form of educational backwardness. A cursory glance at Table 2.3 shows that 80 per cent of the members of the sample households in the age group of above 30 years were illiterate and around 73 per cent of all members in the age group of 21-30 years were illiterate. Unfortunately, of the ones who did begin their education, only a very few could proceed up to secondary school. While the absence of adequate incentive from the village community did affect educational levels, the necessity of eking out a living assumed greater importance in the trade-off at the household level.

with a commodified exchange system that works within the basic model of utility maximisation. While such a system appreciated traditional practices of using grazing land guided mainly by self-goal choice, it never encouraged the village community to improve the conditions of the grazing land or renovate the traditional grazing practices. But, at the same time, the system could also not undermine the basic cultural traits of individuals such as the sense of belonging, mutual trust and reciprocity that constitute social capital and provide instant legitimacy to the village institution to retain its collective character (Mohapatra, 2001; Warren *et al.*, 2001). This seemed to have encouraged the villagers of Chhota Saradhana to jointly internalise external intervention through social mediation and decide to renovate the traditional institutional practices of using grazing lands. Social capital, thus, must be looked upon as input into the process by which institution was created at the local decentralised level for development. In order to ensure sustained development of the common pastureland, its increasing stock formed a necessary condition (Chopra, 2002).

Social Mediation

However, social mediation, that crystallised social capital for an improved outcome, gained considerable momentum in Chhota Saradhana under certain specific material conditions that must not be overlooked. Of course, the intense drought conditions of 1987 made a definite contribution to facilitating social mediation. Factors, external to the village, such as the likely encroachment of the neighbouring villagers and the fear of allotment of the common land to retired army personnel by the government were perceived as threats to control over the land. The instrumental value of all these was enormous for the social mediation that helped the villagers to internalise the need for institutional renovation and accept collaborative intervention of an external development agency such as MMVS. The leadership provided by Devaji Patel, a resident of this village was unique in that it succeeded in mobilising all social groups towards reviving the village pastureland. The first step was to form an informal group. Once the villagers did that, the following steps were taken.

Raising of Village Nursery

In order to meet the fuelwood crisis the immediate task of the informal group was to raise a village nursery with the active support of MMVS and Society for Promotion of Wasteland Development (SPWD). The notable feature of this initiative was that the womenfolk of the village were brought to the centre of this activity for not only plantation but also protection of the saplings. The second most important aspect was that absolute care was taken to protect the local species including those that were gradually disappearing. The once dominant plants in the area, *kumpla, daand, gaigan kheenp* (local names) had disappeared completely while *arunja; khejri, deshi babool and thor* (local names) were on the verge of extinction. Efforts were made to restore these plants in addition to growing new plants including the *Prosopis juliflora* seedlings and Acacia holoceracca and plants locally known as *nibbu, neem, anar, ber, bamboo, subabool, jamun,* etc.

However, due to lack of adequate care and scarcity of water, most of the plants failed to survive. The only exceptions were the *vilayati babool* (Prosopis juliflora), *thor, baradi, khejri, deshi babool,* etc. that could manage to survive even without regular supply of water. The entire land, unfit for cultivation and considered 'wasteland', was now full of these species that helped the villagers meet the fuelwood crisis. Also importantly, it provided respite to the women from walking miles everyday for fuelwood collection. This also boosted the self-confidence of the villagers and renewed their sense of identity gradually eroding in the struggle for survival in times of drought.

Grassland Development

The second most important step taken by the informal group constituted by the villagers was grassland development for meeting the fodder crisis. The following step-by-step approach was adopted.

Land Preparation

Firstly, 106 hectares of land was earmarked and fenced with locally available materials such as stone in place of the option of wire fencing, which had low cost-effectiveness. An anicut was constructed inside the pastureland for conservation of rainwater and retention of soil moisture even as the land was cleared and levelled. This was followed by the construction of staggered trenches and contour bunds for rainwater percolation and soil conservation. Three significant aspects of all these activities need to be underlined. First, the women's participation was overwhelmingly large. Women from all social groups were divided into several groups of four-five. Each group was assigned different jobs. The second aspect was complete use of locally available materials for fencing, as mentioned above, and the third one was the unique combination of local knowledge with the knowledge imparted by the scientific organisations invited for preparation of land and conservation of soil and water.

Organisational Networking

There were several technical aspects related to the revival of grassland ecosystem that needed special attention for ensuring sustainability. The informal group held several discussions with MMVS and invited many organisations to train villagers on various aspects of grazing land management without causing any harm to the natural process of growth. For instance, scientists from Central Arid Zone Research Institute (CAZRI) trained villagers for seed treatment and production of grass. They cultivated grass on an experimental basis in three soil-nutrient conditions in three plots only to examine under what conditions grass could grow better with low investment. The *Dhaman* variety of grass was sown on one plot with fertiliser, on the second without fertiliser and, on the third rocky plot with an abundance of fertiliser and water. The experiment on the second plot showed encouraging results in the context of water scarcity. This prompted the villagers to sow seeds of Dhaman grass in the entire fenced area. The training provided by SPWD for construction of check dams, contour bunds, etc. also proved extremely fruitful. Assistance was also sought from other NGOs including Centre for Community Economics and Development Consultants Society (CECOEDECON) for development of *nadi*, inter-cooperation for seed-testing, Oxford Committee for Famine Relief (OXFAM) for women's empowerment, UNNATI for strategy formulation, etc. All these organisations played their role at different points of time and at various stages of the grassland development. The interesting aspect of the entire exercise was that no assistance was sought from the government, which, however, did play a positive role in allowing the village community to incorporate some part of the revenue land into pastureland. It was in this context that the informal group was formalised as VDS.

Formalisation of VDS

The villagers' group constituted informally had to be registered under the Societies Act when the community approached the revenue department for incorporation of 24 hectares of uncultivable land lying adjacent to the fenced in 106 hectares of pastureland.

However, whether VDS was formal or informal, basically it was a combination of the social forces of the *Gurjars* and *Jats* that determined the activities of VDS. As Table 2.5 shows, these two communities had the maximum members in the VDS, while the representation of the other social groups was marginal. While the lower numerical strength of these social groups most likely precluded greater participation in VDS, there was no evidence during the field survey to suggest that they were taken into confidence in the decision-making processes. For example, though three women members were drawn from numerically small social groups, because it was imperative for VDS to induct women, they were hardly given space and the opportunity to articulate their opinion. However, it is also true that no decision went against them since poor households as theirs also belonged to the *Gurjars* and *Jats*.

Table 2.5

Composition of Members in VDS

No. of Members	*Social Groups*							
	Gurjar	*Jat*	*Rawat*	*Sadhu*	*Bawri*	*Bhambi*	*MMVS*	*Total*
Initially	7	6	1	1	-	1	-	16
At present	7	7	-	2	1	1	1	19

Source: Field survey.

The second most important aspect of VDS was that the community leaders selected the members all along. Since the village community did not exhibit any interest in an election, and primarily depended upon mutual trust and social relational network, the acceptability of the body was high and gained instant legitimacy in the bi-annual meetings that took place with all the men of the village community. The women's role in the decision making process was as peripheral as that of other marginalised social groups. However, their contribution to the growth and conservation of grassland, as we shall see later, cannot be ignored.

Framing of Rules

The rules framed by the village community with the help of MMVS determined the external limits of the individual's conduct. They were mainly meant for regulating the behavioural pattern of every household in favour of growth and conservation of the common pastureland. The rules that were framed at the initial stage were as follows:

- Nobody could cut plants or lop trees from the pastureland for fodder or fuel.
- Grass, fuelwood and tree lopping would be sold to the villagers through the open auction system as instituted by VDS.
- The meetings of VDS with the village community would be held twice a year at Holi and Deepawali. Discussions related to the management of the pastureland would be held in such meetings. VDS would implement the decisions taken therein.
- Anybody found violating the rules would have to pay a penalty of Rs. 500 and Rs. 51 in case of an animal found grazing in the pastureland.
- No common road could be laid inside the pastureland.
- Each household had to provide at least one worker for any development work, for instance, the construction of the wall of the pastureland, failing which it would have to pay Rs. 50 as labour charge.
- Grass produced in the pastureland would be harvested only after dispersal of seed. It would then be distributed (if distribution of grass were to take place free of cost) equally to all households.
- Some portion of the common land would be utilised for cultivation of green grass only and cultivation would be contracted out. Any household could receive such a contract provided it meet the conditions laid down by VDS.

It may be noted that the enforcement of these rules took place without discrimination. The system of penalty was accorded equal importance, so much so that it made everybody in the community cautious about the protection of pastureland. Another important aspect was the built-in flexibility of the rules that gave considerable space to the village institution to switch to new practices, if necessary, as a part of the coping mechanism.

Institutional Flexibility

The open auction system adopted for selling grass created

discontent among many villagers. For, it was encouraging inequity in accessing grass grown on common land. This led the community to replace the open auction system by direct sale of grass when it matured. Under this system, households interested in purchasing grass had to enter into contract with VDS. The contract identified the area within the protected pastureland from where grass might be collected, the price of grass to be paid to VDS (as fixed by VDS) and household's responsibility for cutting the grass. This arrangement continued until 1994 without any interruption. However, in 1994, the demand for green grass increased steeply due to increase in the population of buffaloes in the village (the improved conditions of village pastureland under the new institutional arrangement had provided considerable incentive to villagers to increase buffalo population for more income). In this new context, the earlier system of grass collection was found to be inadequate, which was why the 'head load' system was introduced.

Head Load System

Under this system, first, only one member per household was allowed to cut green fodder to the extent she/he could carry the weight of the grass load. However, if there was good yield, each household was allowed two head loads a day. Second, one had to carry the head load from pastureland directly to the household through the route as specified by VDS. There were cases of imposition of fine of Rs. 51 per person on five persons who violated this rule. All collections had to be entered into the register, failing which a fine of Rs. 51 was imposed upon the offender. Third, nobody was allowed to enter the pastureland after 12 noon and lastly, one had to pay charges per head load as fixed by VDS from time to time on the basis of grass yield. Table 2.6 shows how the rate varied depending upon the yield during 1995-2001.

No rate per head load was, however, fixed after 2001. Persistent drought had reduced grass yields considerably. At the time of the field survey (during drought years), households were allowed to take their animals inside for grazing. It was, as told by the villagers, a part of the coping mechanism followed by the village institution in times of fodder crisis.

Table 2.6

Charges Per Head Load of Grass

Years	*1995*	*1996*	*1997*	*1998*	*1999*	*2000*	*2001*
Rate (in Rs)	2.0	2.0	7.0	3.0	5.0	5.0	5.0
Reason	HY	HY	LY	HY	LY	LY	LY

Note: HY- higher yield; LY- lower yield.
Source: Field survey.

Coping Strategy

Alongside all these initiatives, the village community decided to cultivate green fodder in some parts of the pastureland throughout the year. This decision was taken against the backdrop of the suffering animals had to undergo during the drought of 1987. One may consider this as an innovative coping strategy in the sense that it encouraged private initiative for fodder cultivation while the management and supervision remained under the control of village community or VDS. Again, this exemplifies the institutional flexibility of the village community of Chhota Saradhana. It shows how people could evolve their coping strategy by combining both private initiative and community ownership in grassland management.

Immediately after the decision was taken, a borewell was installed on the protected common land. The purpose was simple. It was exclusively meant for creating an extra facility for watering the plants in the common land, producing fodder and providing drinking water to the villagers when it was required during drought. Besides, it was also meant to provide water for production of fodder during the lean months of January-June. Of course, in drought years this system was operational throughout the year. In view of the capacity of the borewell, it was also decided to bring under fodder cultivation no more than 20 *bighas* of the protected pastureland.

The village community, along with VDS, decided to cultivate the *lucerne* and *kasni* varieties of grasses and oat in the *rabi* season and *jawar* and *chari* in the *kharif* season on that portion of the pastureland that was brought under grass cultivation. Initially in 1989, only three *bighas* of land was brought under *kasni* grass cultivation on an experimental basis. The entire job of crop raising to harvesting was given on lease for a year to one of the villagers. The village community and VDS together decided the terms of contract on the basis of which the job was given. These were as follows:

- The entire cost of cultivation, including electricity, had to be borne by the person who got the contract.
- Selling or auctioning of grass, if necessary, had to be conducted by the same person.
- Villagers had to contact the same person directly for procuring grass through the system of exchange as suggested by VDS.

It is important to note that the area under fodder cultivation was no longer fixed. Initially, it was three *bighas* of protected pastureland that was earmarked for fodder cultivation. Subsequently, around 1995-96, pressure on fodder increased significantly due to the steady increase in buffalo population. This led VDS to extend the area under fodder cultivation to 30 *bighas*. The community took considerable trouble in preparing the rocky and unsuitable land for cultivation of fodder. The cultivation continued until 1998 when the area began to face severe water shortage due to the onset of drought. The society was then forced to reduce the area of the land for fodder cultivation to eight *bighas*. However, it did not completely stop fodder cultivation.

Distribution Mechanism

While the entire land, being brought under cultivation of green fodder (whether it was 30 or 8 *bighas*), was given on contract, fodder produced on one-fourth of the land area was exclusively owned by VDS. Grass produced on rest of the land was meant for selling to the villagers.

The terms of contract with VDS were different from the contract with the villagers for grass distribution. The individual who had taken the job on contract paid charges as fixed by VDS to the latter in four installments for the grass produced on this plot. He could then auction the same to the villagers. It may be noted that charges increased at different rates over the years. Besides, the contractor also changed from time to time as decided by VDS and the village community.

> The system that existed for sharing of grass by the village community was slightly tricky but appeared to be innovative in variety of ways. The entire land was demarcated in terms of what is locally called 'biswa' (a measure of unit of land, 20 biswa=1 bigha of land). Any household could get the contract for grass production from either one biswa or more than

> one depending upon his or her requirements. He had to pay one-time charges fixed for grass collection from each biswa. The rate could vary from the initial Rs. 30 per cutting of grass in one biswa to Rs. 300 for the same. VDS and the contractor depending upon the input costs of both material and wages jointly decided the rate.

The venture was a profitable one even for the contractor. The net income (gross income less input and wages) that he earned per *bigha* of land was estimated to be Rs. 9,335. The size of net income per *bigha* should not necessarily lead to the conclusion that all eight *bighas* of land, which was under cultivation of grass, yielded the same amount of earning. There was variation in the yield of different types of grasses; some varieties of grass, such as the oat, got exhausted after only two cuttings. However, this was a relatively risk-free venture and there was continuous demand for green fodder from the farmers, thus proving an incentive for any individual in the village. It was, however, risk-free only so long as there was no water scarcity, which got mitigated due to the installation of the borewell, and continuous support was received from the village community. Additionally, there was the burden of protecting the produce from animals of other villagers and even from animals of their own village, especially during the drought when animals were free to move inside.

The system, as explained above, shows how institutional reform became instrumental in evolving coping strategies. The unique feature of this renovation was that the village institution allowed market forces to determine production and exchange of grass only to the extent required for facilitating a smooth exchange to take place and remove uncertainty regarding fodder availability.

Institutional Renovation or Failure?

Although institutional renovation made a significant difference to the livestock economy and grassland ecosystem of Chhota Saradhana during the last one decade or so, the state of management of the pastureland at the time of the study seemed to give confusing signals. The common pastureland was now divided into six equal plots without fencing inside and allocated to six groups of households formed through a lottery system designed by the villagers themselves. Each group had its own leader who, along with the members, was responsible for the management and development of the plot of land. The group now had its own norms of management. Only group members could harvest grass and fuelwood from it. Though this may

Figure 2.1

Collection of Grass from Common Protected Pastureland and Other Sources Per Household

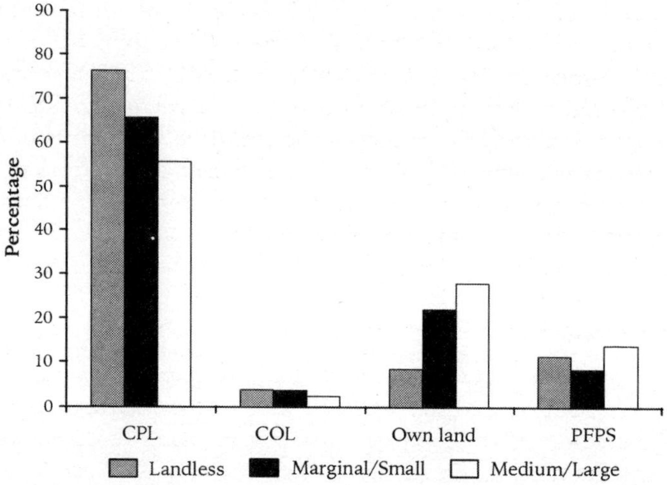

Note: CPL = Common protected land, COL = Common open land, PFPS = Purchase from private sources

Source: Field survey.

be considered as another form of institutional renovation, it may have serious ramifications on the sustainability of grassland in the future. As the field survey revealed, the village community adopted this option to cope with serious financial difficulties; persistent drought for the past three years had continuously depleted the village common funds.

In 1988-1999, the investment in the development and maintenance of the pastureland was Rs. 9.21 lakh, while the share of the villagers and income earned from the pastureland was Rs. 3.60 lakh (39 per cent of investment); the earnings from the pastureland through charges received for dry and green fodder and selling of fuelwood was Rs. 1.89 lakh. The rest was received from external sources. The lack of funds made management of the pastureland difficult, it did not convince many in the village as a valid reason for division of the land. The overriding importance given to 'group rationality' over the collective rationality of the village community was likely to weaken the social capital. In such a situation, it is apprehended institutional renovation that ensured equity and sustainable management of the pastureland might lose its direction in the future.

Gains from Institutional Renovation

One of the significant aspects of the change that came about in the village in the wake of revitalisation of degraded pastureland was that it reduced uncertainty regarding fodder availability. This, in turn, acted as an incentive to the farmers for expansion of their livestock resources. The assured stock of fodder, unseen in recent past, imparted, as the field survey revealed, enormous confidence to the village community for managing the grassland resource base. The question was whether the poor were able to gain equal access to this resource base and improve the status of their livestock resources. Of course, improvement in the grassland ecosystem, which was central to institutional intervention, was a direct gain that one must account for.

Access to Resources

Several studies have indicated that the poor are deprived of institutional outcome due to lack of access to the improved resource base (Ahluwalia, 1997); power relations, that the traditional institutional arrangements nurture, do not allow the poor to gain equal access to the resource. However, Chhota Saradhana turned this premise on its head; power relations, in whatever form they existed, could never affect institutional renovations and make accessibility to the resource base inequitable. Access of the poor to the common protected pastureland can be gauged on the basis of the extent to which they could meet their requirement from the same. There were four sources of fodder available to farmers—the common protected pastureland, common open land, own land and other private sources.

The collection of grass from different sources is shown in Table 2.7 and Figure 2.1. While no payment was required for collecting grass from the common open land, it was mandatory for collection from the common protected grassland and other private sources. The price differential was substantially high at Rs. 25 per quintal from the former and Rs. 100 from the latter. Table 2.7 shows that the landless, marginal and small farmers gained more from the common protected pastureland than other categories of farmers. Each sample landless household was estimated to have collected as much as 20 quintals of grass from the protected pastureland, 77 per cent of the total requirement, on an average during the year of the study.

Table 2.7

Collection of Grass and Green Fodder from Common Pastureland and Other Sources Per Household During 2001

(in quintal)

Category of Landholdings	*Common Protected Pastureland*	*Common Open Land*	*Own Land*	*Purchased from Private Sources*	*Total*
Landless	20.0 (76.57)	1.00 (3.83)	2.20 (8.42)	2.92 (11.18)	26.12 (100.00)
Marginal/small	25.5 (65.70)	1.44 (3.71)	8.59 (22.13)	3.28 (8.45)	38.81 (100.00)
Medium/large	33.6 (55.73)	1.38 (2.29)	16.89 (28.09)	8.40 (13.94)	60.25 (100.00)
Average per household	26.2 (63.54)	1.32 (3.20)	9.21 (22.31)	4.52 (10.95)	41.28 (100.00)

Note: Figures in the parentheses are percentages of the quantity of grass collected to the total grass collected.

Source: Field survey.

Although each sample household of medium and large farmers collected the same from the same land, about 34 quintals, it was around 56 per cent of the total requirement of grass/green fodder. Grass/green fodder purchased from private sources also included the one cultivated in a part of the common protected pastureland contracted out by VDS to individual farmers of the village. It may be noted that landless farmers leased in land from the medium and large farmers and also collected grass grown naturally there. If all sample households are taken into consideration, about 63.5 per cent of the total requirement of grass per household was met by the common protected pastureland on an average.

The dependence of the landless households on common protected pastureland for fuelwood was no less. As it appears in Table 2.8 and Figure 2.2, landless households collected maximum fuelwood from the common protected pastureland as compared to other category of farmers. Small and marginal farmers did not appear to depend on the same to such an extent, for they collected fuelwood from their own lands. Each household of landless category of sample households collected 30 per cent of their total requirement on an average during the year under study. In the case of small and marginal farmers, it was 19 per cent. However, it was 25 per cent in the case of medium and large farmers, who, at 38 per cent, depended comparatively more on their own land for fuelwood collection. Tree lopping made available from the protected pastureland were used as fuelwood once the animals ate the leaves. The dry plants available from the same protected pastureland were also used as fuelwood. However, the villagers paid VDS for this.

Table 2.8

Collection of Fuelwood from Common Pastureland and Other Sources Per Household During 2001

(in quintals)

Category of Landholding	*Common Protected Pastureland*	*Common Open Land*	*Own Land*	*Total*
Landless	3.17 (30.40)	7.25 (69.60)	0.00 (0.00)	10.42 (100.00)
Marginal/small	2.18 (19.12)	7.68 (67.22)	1.56 (13.66)	11.42 (100.00)
Medium/large	2.69 (25.05)	3.98 (37.01)	4.08 (37.94)	10.75 (100.00)
Average per household	2.68 (24.70)	6.30 (58.00)	1.88 (17.13)	10.86 (100.00)

Note: Figures in the parentheses are percentages of quantity of fuelwood collected to the total collection of fuelwood.

Source: Field survey.

Figure 2.2

Collection of Fuelwood from Common Protected Pastureland and Other Sources

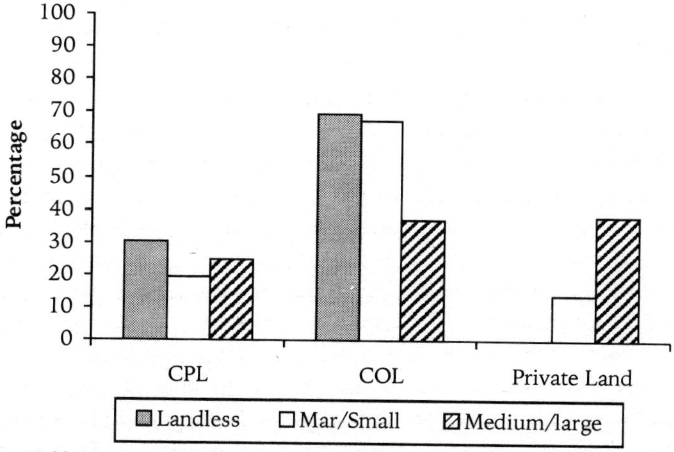

Source: Field survey.

It may be noted that the total requirement of both grass and green fodder of all sample households together was estimated to be 2,064 quintals (not shown in the Table 2.8) during the year. The largest quantity, which was 1310.50 quintals (63.49 per cent), was collected from the common protected land alone. Only 66 quintals of grass (3.20 per cent) was collected from the common open pastureland by all sample households. However, the role of common open land in mitigating the fuelwood crisis was significant. Of the total requirement of fuelwood of the entire sample, estimated to be 543 quintals (not shown in the table) during the year under study, 315 quintals (58 per cent) was collected from the common open land. Fuelwood collected from the protected pastureland was estimated to be around 134 quintals (25 per cent). At the initial stage of the intervention, as mentioned earlier, a plantation programme was initiated for meeting the fuelwood scarcity. Fuelwood, which was made available to all households from both protected and open land, seems to have sufficiently fulfilled the village requirement. The poor were no less benefited from the institutional intervention.

Improvement in Livestock

In order to gain insight into the improvement of status of livestock, especially of the poor, it was necessary to understand the nature and trend of expansion of their animal stock *vis-à-vis* that of relatively affluent farmers. Data was collected from the sample households for three different years under different conditions. Data for 1991 was collected from MMVS only to indicate the status of livestock during the initial years of the intervention. While the year 1997 was normal and farmers reaped the benefits of their grazing land management, the worst-hit drought year was 2002.

Two important aspects of the livestock position can be noted in Table 2.9. First, landless households witnessed a considerable rise in the stock of their animals. In 1991, the initial year of intervention, not all sample households in the landless category had cows, buffaloes and goats though all of them reared sheep. However, the average stock of each type of animal had increased sharply by 1997. It was found that each sample landless household had, on an average, 2 cows, 3 buffaloes, 30 sheep and 5 goats. The stocking rate of animals of medium and large farmers also increased steadily during the same year. However, this may not be unusual in view of their landholdings and capacity to rear animals.

The second notable feature of the change in animal stock was that while there was a sharp increase of average stock of all animals in the case of landless households and medium and large farmers, the stock of sheep and goat witnessed a decline in the case of marginal and small farmers. Although there was a decline of all animals during 2002 in all households, the average stock of animals of landless was not as bad as expected in drought conditions. Perhaps the reason for this can be attributed to the non–farm wage employment available in the mines located adjacent to the village.

Table 2.9

Average Number of Animals across Landholdings

Category	*Year*	*Cow*	*Buffalo*	*Sheep*	*Goat*	*Bullock*	*Cattle Unit*
Landless	1991	0.42	0.92	12.50	0.25	—	5.45
	1997	1.75	2.75	29.75	4.92	0.33	16.58
	2002	0.75	1.25	16.08	4.83	0.25	8.98
Marginal/small	1991	0.36	0.24	11.76	5.04	0.08	5.20
	1997	1.48	1.40	7.20	4.84	0.24	7.77
	2002	0.60	0.80	3.20	5.16	0.08	4.45
Medium/large	1991	1.46	0.38	—	3.46	0.77	4.63
	1997	5.31	6.69	4.62	7.23	1.08	23.81
	2002	0.69	2.92	—	1.69	0.54	8.03

Note: Cattle unit is estimated following the criteria of one cow is equal to one cattle, one buffalo is equal to two cattle, one goat is equal to 0.25 cattle and one sheep is equal to 0.25 cattle.

Source: Field survey.

It may be difficult to establish the corresponding impact of the improvement of the degraded pastureland on stocking rates of animals of all varieties based on analysis of data of the three years in Table 2.9. For, the stock of animals at a given point of time depends upon a variety of factors even in normal climatic conditions. As the field survey revealed, animals, especially small ruminants, were sold off at times in order to meet the financial needs and purchased when households were capable of doing so.

However, one must recognise the fact that the stocking rate may be higher in a situation where the pastureland is regularly protected; the carrying capacity of the land may not decline as steeply as one expects in the absence of regular protection in adverse climatic conditions. The effect of adverse conditions on grass production may be outweighed to a certain extent by an increase in land productivity. This is what the community management of Chhota Saradhana village shows. The close correspondence between the carrying capacity and land productivity may confirm sustained development as maintained by the traditional theory of rangeland development (Chopra and Kadekodi, 1999).

The exercise, which is carried out here, does indicate a positive aspect of stocking of animals. Despite persistent drought for the last three consecutive years, even the poor households were able to retain their stock of animals at an acceptable level. Table 2.9 shows that the stocking rate of animals in terms of cattle unit of the poor including landless, small and marginal farmers did not decline to the extent it happened in the case of medium and large farmers. It is equally important to note that the stocking rate was higher during 2002 despite being a drought year as compared to that of 1991 when rainfall was normal. The perception of the households surveyed regarding the role of improved pastureland in improving stocking rate of animals confirms this (Table 2.10).

Table 2.10

Response of the Households Regarding Improvement in the Status of Livestock

Category	*Increase in the Number of Animals*	*Increase in the Production of Milk*
Landless	8 (67.8)	7 (58.0)
Small/marginal	13 (57.0)	17 (68.0)
Medium/large	8 (61.0)	12 (92.0)

Note: Figures in the parentheses are percentages of the sample households of respective category of landholdings.

Source: Field survey.

Table 2.10 shows that 68 per cent of the sample households in the landless category saw a direct benefit of the improvement of the pastureland in the increase of their ability to expand the size of their livestock. In fact, medium and large farmers also benefited in the same way. But they perceived gain more in the form of increase in yield rate of milk as expressed by 92 per cent of the sample households of this category of farmers.

Although the stocking rate of milch animals declined in drought situation, it played a significant role in ensuring the nutritional security of the landless, marginal and small category of farmers. Table 2.11 shows that as much 45 per cent of milk produced in drought year was retained for domestic consumption. The same poor farmers (including the landless, marginal and small farmers) could retain only 31 per cent milk for domestic consumption five years ago when there was no drought. This seems to go against the popular perception that the poor tend to sell more milk in distressed conditions.

Table 2.11

Average Production, Consumption and Sale of Milk Per Household across Different Landholdings Per Year

(in quintal)

Category	*2001-02*			*5 years ago*		
	Production	*Consumption*	*Sale*	*Production*	*Consumption*	*Sale*
Landless	12.76 (100.00)	5.70 (44.67)	7.06 (55.33)	27.14 (100.00)	8.33 (30.69)	18.81 (69.31)
Small/marginal	9.01 (100.00)	4.07 (45.17)	4.94 (54.83)	21.95 (100.00)	6.85 (31.21)	15.10 (68.79)
Medium/large	21.48 (100.00)	7.85 (36.55)	13.63 (63.45)	56.26 (100.00)	13.98 (24.85)	42.28 (75.15)
Average per household	13.15 (100.00)	5.45 (41.42)	7.71 (58.60)	32.12 (100.00)	9.06 (28.21)	23.06 (71.79)

Note: Milk of cows, buffaloes and goats has been taken into consideration.
Source: Field survey.

The retention of milk by almost half of the production during a drought year shows how it could ensure nutritional security of the poor households in the event of bad agriculture production. Besides, retention of milk, as Table 2.11 shows, was higher than that of the medium and large farmer during a drought year. It drives home the point that the pastureland, which was developed, protected and maintained by the village community, had a definite contribution to make in ensuring food security for the poor. Although the sale of milk fetched more income for all households, income from manure and the sale of animals, etc. cannot be ignored (Table 2.12). One must recognise the importance of these sources of income from livestock too, especially during drought years.

Table 2.12

Average Gross Income Per Household from Livestock during 2001-02

(in Rs.)

Category	*Milk*	*Wool*	*Manure*	*Animal Sale*	*Gross Income*
Landless	12,853 (84.80)	250 (1.65)	1,213 (8.00)	842 (5.55)	15,158 (100.00)
Marginal/small	9,211 (75.12)	89 (0.72)	1,318 (10.75)	1,644 (13.41)	12,262 (100.00)
Medium/large	21,070 (73.11)	115 (0.40)	4,058 (14.08)	3,577 (12.41)	28,820 (100.00)
Average per household	13,169 (76.29)	134 (0.78)	2,005 (11.62)	1,954 (11.32)	17,262 (100.00)

Note: Figures in the parenthesis are percentages of gross income from livestock.
Source: Field survey.

And, if the gross value of livestock produce is compared with that from other sources such as agriculture and non-farm, livestock contributed 38 per cent of the gross income of the landless households and 29 per cent of the small and marginal farmers (Table 2.13). It may be noted that the gross value of agricultural produces was estimated for 2001-02, which was one of the drought years. The field survey revealed that both agricultural production and herd size of the livestock witnessed considerable rise in the normal years. Here, agricultural income included both agricultural production

Table 2.13

Average Gross Income Per Household from Various Sources during 2001-02

(in Rs.)

Category	*Agriculture*	*Livestock*	*Non-Farm*	*Total*
Landless	4,940 (12.38)	15,158 (37.99)	19,800 (49.63)	39,898 (100.00)
Marginal/small	6,815 (16.08)	12,262 (28.92)	23,316 (55.00)	42,393 (100.00)
Medium/large	23,228 (28.83)	28,820 (35.77)	28,515 (35.39)	80,563 (100.00)
Average per household	10,633 (20.56)	17,262 (33.38)	23,824 (46.06)	51.719 (100.00)

Note: Figures in the parenthesis are percentages of gross income from various sources
Source: Field survey.

and wage income of those who worked as agricultural labour. Income earned by the sample households through non-farm activities was mainly wages from mining. Earnings from business, manufacturing and the service sectors were insignificant. In any case, Table 2.13 shows how livestock became more important than agriculture as a source of generation of income at the household level during drought years.

II

SINGHARIYA

Singhariya, another village of the same *tehsil* as Chhota Saradhana, presents a few contrasts from Chhota Saradhana in management and development of the village pasture. The conditions that led the villagers of Singhariya to opt for institutional renovation for pastureland development and plantation to meet its fodder and fuelwood crisis were not so dissimilar to that of Chhota Saradhana. However, the social capital, that made a large contribution to democratically building consensus within an institutional framework, differed significantly between two villages in determining the institutional outcome. The study of institutional renovation of Singhariya revolves around these issues and seeks to examine how social capital influences institutional outcome in plantation and grassland development even in a homogeneous social condition. No specific index has been followed to measure the content of social capital in this study. Instead, the assessment is based on the information collected through a household survey, meetings that were held with villagers at different levels, including the members of the executive committee of the village development committee (VDC) and the village *sarpanch*. All these were pointers for comprehending the content of the social capital of the village (Mukherjee, 2002).

In the debate on environment policy, the 'entitlement failure' of the subordinate against the powerful actors is seen as an outcome of their incapacity to make claims (Leach *et al.*, 1997a; 1997b and 1997c). Although it has been suggested institutionalising their claim-making capacity and thus creating space for negotiation (Bradbury *et al.*, 1995), it is not certain how far it is achievable given the pervasive features of social relationships. In such a situation, how does institutional development or renovation at the local level end such an 'entitlement failure'? For, it is apprehended that 'institutions' that appear to be acting for collective good may reproduce same relationships of unequal power and authority as a pervasive feature of social relations (Leach *et al.*, 1997a; 1997b and 1997c). However, this must not reduce the importance of institutions. What is sought is an appropriate development action to get the power process right. In other words, any intervention designed to seek institutional renovation must examine the content of social capital rather than its stock alone (Harriss, 2001). Laying emphasis only on social capital as an instrument to institutionalise collective action without verifying the content of the social capital may obscure power relations (Harriss, 2001). This may lead to result in entitlement failure of the subordinates. One may ask then, do the local institutions, which have been renovated, sustain under such inequitable conditions?

The study of Singhariya is divided into five parts. While the social context in which institutional renovation of the village was sought is briefly examined in part one, the resource base of the village is analysed in part two. Part three focuses on the process through institutional renovation that came about in the village through collective action. Part four examines how institutional outcomes are unequal, while part five is devoted to assess the long-term stability of such a village institution.

The Social Context

The village was uniquely placed in an uncontested social domain with a single caste group called *Rawat*, which was, as claimed, another variant of the *Rajput* community. In order to assess whether such a social condition shapes the content of social capital that generates equitable institutional outcome, one must know its historical roots. Historically, the village saw the early settlement of the Rawat community although the initial inhabitants were a few families from the *Regar* community little more than three centuries ago. However, they could not continue to stay in the face of stiff opposition from the *Rawats*, who settled here when Manohar Singh of Bawra, a neighbouring village, migrated. The community today comprises the descendants of Manohar Singh's two sons Chiman Singh and Nimba Singh, divided into two distinct family lines (or *thok* in local parlance) and dismissive of the conventional norms of collectivity normally practised within a clan. The differences between the two family lines grew such over the years that their social relations became more of that between patron and client.

It began with the strong bias of Manohar Singh, for reasons best known to him, towards his eldest son Chiman Singh, who inherited most of his land. Of the total 810 *bighas*, Chiman Singh was allotted 660 *bighas*. The initial distribution of the agricultural land and the economic power eventually led to a kind of social dichotomy between the two family lines. In the following generations, land was further sub-divided and distributed among their sons and daughters. Until the last generation, Chiman Singh's descendants were economically superior in terms of per capita landholdings. However, in the present generation, owing to the rise in population as revealed by the field survey, the majority of the households of both family lines were reduced to marginal and small landholdings.

Despite this decline, the old dominance continued in the social sphere. The mutual trust, social networking and all that constituted the content of social capital in the village were mainly determined by the historical context deeply influencing the parameters of social relations. Despite this subtle heterogeneity, collective action taken together by the households of both family lines made it possible to restore the degraded pastureland in Singhariya.

As in Chhota Saradhana, the need for restoration occured when villagers found their village commons degraded to a point of no return during the drought of 1987-88. Immediately after Independence, the forest cover of Singhariya village and its surroundings were threatened by rapid industrialisation in and around Beawar town, a few kilometres away. The stone miners quarried away the nearby hillocks; trees were rapidly felled to cater to the increasing demand for timber in the mills thus reducing the scope for regeneration of fodder and fuelwood. This led the 'natural' supply of the latter to fall short of increased demand. The village elders saw this process as a potential threat to livelihood, especially to livestock rearing. However, the immediate casualties were women and animals. Women had to travel long distances for collecting fuelwood now. Paradoxically, the villagers had to now go to Beawar town to purchase fuelwood.

The crisis created a uniform base for all to come together to restore the village pastures through new institutional practices. This was no way different from Wade's relative–need-based explanation (Wade, 1994). Under such circumstances, social capital was not considered a cause for collective action, as rightly argued by Wade, but unity in interest formation and their aggregations were the causes. They enriched the content of social capital and created a condition for change as desired by the majority.

The issue is whether such 'new' social relations that grew in the village around pastureland development ensured long-term stability of the new institutional arrangements. If they succeeded, social capital could then be able to carry forward collective action. In other words, the instrumental role of social capital must be looked upon as a necessary condition for the continuation of collective action. Its intrinsic relation with institutional development had wider implications for the stability of the latter and, therefore, resource conservation and fulfilment of the village needs (Mohapatra, 2001). Of course, how much bearing it had on the nature of functioning of the mediating agency cannot be denied (Krishna, 2003).

However, what was important was whether the existing power relations determined the functioning of the mediating agency. For, there are several evidences that suggest that the former finally tend to be subsumed to the latter (Khan, 1999; Ray, 2003a). Before examining this issue, it may be worthwhile to understand the natural resource base the village is endowed with, and its status in relation to the increasing population pressure.

Resource Base of the Village

That the resource base of the village, including land and livestock and their ownership pattern, played an incredible role in creating the material conditions for collective action is unquestionable. The educational status of the village community, which broadly indicates human resource development, must also be taken into consideration in this context.

Land and Population

The village was (according to Census 2001) inhabited by 199 households of which 197 households belonged to the two family lines, as mentioned earlier. The remaining two belonged to the *Regar* community that seemed to have no voice in the mainstream, as found in the field survey, owing to its peripheral status. Singhariya witnessed a steady rise in the number of households and population in a period of three decades. The number of households almost doubled while the population more than doubled. From 102 in 1971, the number of households increased to 122 in 1981 and then to 199 in 2001 (Table 2.14) accompanied by a steep rise in population. From 672 in 1981, the total population of the village went up to 1047 in 2001.

One may be concerned about the sustainable use of the limited stock of land, held both commonly and privately (Map 4), which produced the grass for animals and plants for fuelwood. With the population growing at the rate of 2.64 per cent per annum during the last three decades or so, the size of landholding declined in this village. This forced the households to opt for non-farm activities and livestock rearing as sources of livelihood generation. The degree of importance being attached to both sources of livelihood generation varied depending upon how the household worked out their relative opportunity costs.

Table 2.14

Stock of Land, Pattern of its Use and Population

(area in hectares)

	1981	2001
Total population of the village	672	1,047
No. of households	122	199
Total reported area	234	234
Irrigated area	49	19
Un-irrigated area	71	99
Permanent pastureland	79	79
Land unavailable for cultivation	35	35

Source: Census of India (Rajasthan) 1981 and 2001.

Strictly speaking, non-farm employment was not substitutable for livestock rearing in drought prone villages like the one under study. It was not only agricultural uncertainty that the villagers were vulnerable to, but also the limited prospect of non-farm wage employment that upset sustainable livelihood in the absence of good agriculture. In such a situation, the only option left to the households was to go in for livestock rearing. However, villagers did not depend much on it since the choice of wage employment in construction activities was available almost throughout the year in Beawar.

Despite this, the villagers treated livestock rearing as crucial since it played a critical role in ascertaining minimum livelihood in the most distressing conditions. In other words, the relative opportunity cost of livestock rearing always scored higher in this village. For, attached to it was the hope that it would at least provide livelihood even if all other legitimate options were closed to the household. In uncertainty brought about by drought, it is this norm that works and encourages livestock population to grow concomitant with the rise in human population. In Singhariya, it grew up to a point where the household looked for an optimal combination of livestock rearing and wage employment. Even if the stock of animals was found to be low at the individual household level as a product of such a combination, the rise of livestock in the village taken together could not be ignored. Its adverse effect was invariably discernible in the steady decline of the carrying capacity of the village commons. Singhariya, endowed with a limited stock of common pastureland of 79 hectares (Table 2.14), was not an exception to it.

However, one has to bear in mind that 24 hectares of this land was forcefully encroached upon by *thok* I for some households who belonged to the same *thok*. Hence, only 55 hectares of the common pastureland, or 0.45 hectares per household in 1981, was available to the villagers for development. This further declined to 0.28 hectares in 2001. Needless to say, the availability of this land per animal declined much more sharply. Of course, 35 hectares of uncultivable land owned by the revenue department was available to the village. The land also had large significance in view of its utility for grazing of animals and fuelwood collection. It could have been developed for meeting their requirements provided permission was sought from the revenue department. Since no effort was made by the villagers to develop it, there was definitely higher pressure on it, resulting in the decline of the carrying capacity of this land too.

As mentioned earlier, as many as 90 per cent of all households (of both *thoks*) are landless, small and marginal farmers while only 10 per cent (20 households) are semi-medium/medium farmers (Table 2.15). However, as shown latter, the medium farmers of *thok* I were dominant and the decision makers of the village on any community-related issue including pastureland development. This was observed even by the village *sarpanch*. He observed that nobody could contest their authority in any decision making at the community level even if it was development work initiated by the local *panchayat*.

Map 4

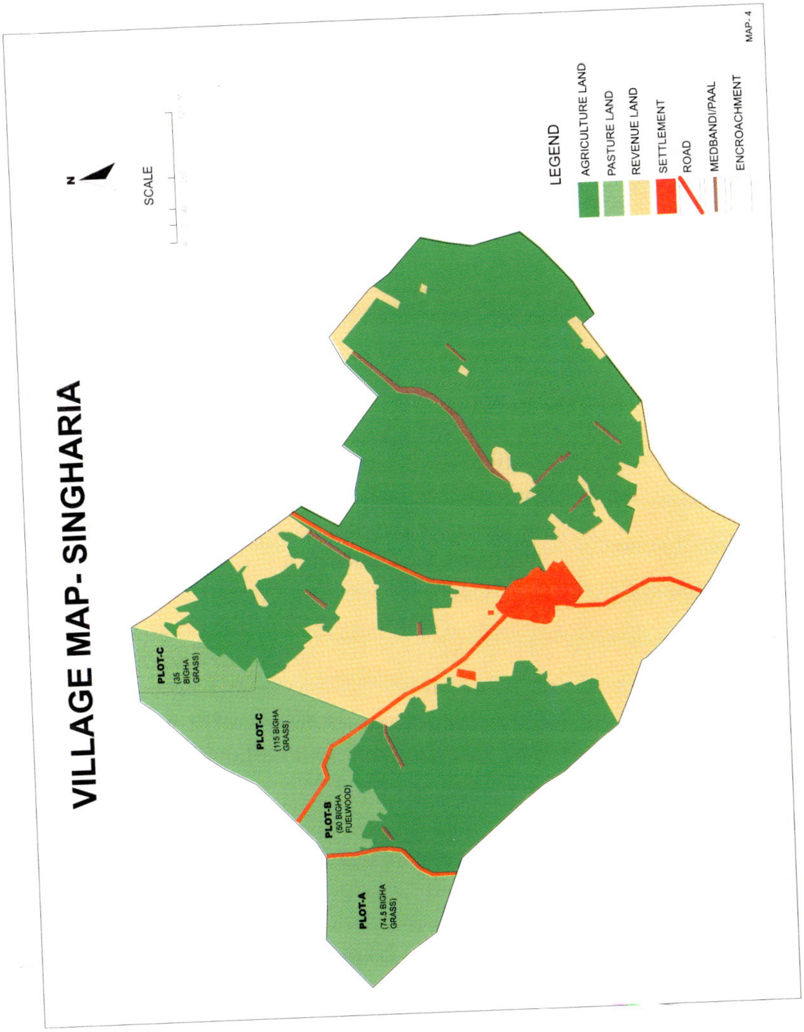
VILLAGE MAP- SINGHARIA
N
SCALE
LEGEND
AGRICULTURE LAND
PASTURE LAND
REVENUE LAND
SETTLEMENT
ROAD
MEDBANDI/PAAL
ENCROACHMENT
MAP-4
PLOT-C (35 BIGHA GRASS)
PLOT-C (115 BIGHA GRASS)
PLOT-B (50 BIGHA FUELWOOD)
PLOT-A (74.5 BIGHA GRASS)

Table 2.17

Educational Background of the Villagers

Age group (in years)	*Illiterate*		*Primary*		*Middle*		*Sr. Sec.*		*Above Sr. Sec.*		*Total*
	M	*F*	*M*	*F*	*M*	*F*	*M*	*F*	*M*	*F*	
Thok I											
5-14	1	7	55	42	8	1	0	0	0	0	114
14-30	7	40	15	21	26	7	26	6	3	0	151
30-60	19	47	18	11	14	0	11	1	0	0	121
Above 60	4	10	7	0	2	0	1	0	0	0	24
Total	31	104	95	74	50	8	38	7	3	0	410
Thok II											
5-14	1	6	8	8	1	0	0	0	0	0	24
14-30	3	24	3	3	8	1	6	3	0	0	51
30-60	6	15	1	1	1	0	2	0	0	0	26
Above 60	2	4	0	0	0	0	0	0	0	0	6
Total	12	49	12	12	10	1	8	3	0	0	107

Note : M = Male; F = Female

Source : Field survey.

may lead one to argue that improvement in educational status may not have much to do with the improvement in vertical social networking. According to Putnam, vertically or hierarchically organised relationships inhibited social capital formation in northern Italy (Putnam *et al.*, 1993). Putnam shows that the stock of social capital expanded through horizontal integration in the rich northern Italy, where improved educational status contributed to heighten civility and, hence, social networking of the people. The Civic Community Index (density of membership in formal organisation, for instance), as a measure of social capital in Putnam's analysis has implicit recognition of the role education plays in increasing the stock of social capital.

The educated youth in the Rajasthan villages were found to be drawing upon the social capital stock of the villagers and took collective action for maximising the benefits from several government programmes that expanded vastly over the past two decades (Krishna, 2003). However, this does not necessarily mean that education contributed to the social capital. For, as Krishna observed, educated youth made use of the existing social capital stock to achieve what the villagers intended. While one must not underrate the contribution that education makes to the increase in the stock as well as its content, excessive reliance on it as a factor for collective action to come about may be misleading. For, there are several instances to show that a rise in the level of social awareness among the illiterate villagers in India brought about a historic democratic movement on several occasions through collective action (Heller, 1996).

Although the educational status of the households of *thok* II was relatively low, they were never hostile to institutional renovation for pastureland development. For they were equally aware of the environmental fallout and potential benefits that they could be deprived of if degeneration was not countered. It was this level of awareness that encouraged them to weave themselves into a new social relationship and join collective action for restoration of the grassland ecosystem.

On the other hand, many members of *thok* I could manage to access government jobs due to their relatively better educational status. This, in fact, reinforced social networking within the same *thok*, but not across *thoks*. More importantly, because contact with the government setup promoted their horizontal integration their hegemony was reinforced. *Thok* I had another reason to exhibit its supremacy over *thok* II. As reported by many households during field survey, the same few relatively affluent households of the same *thok* were engaged in moneylending. The control thus exercised on the informal credit institution in the village buttressed the existing power relations to the disadvantage of the poor, especially of *thok* II.

The poor economic base of *thok* II forced them into debt, taking recourse to credit extended by these powerful farming households. This eventually created a sort of dependency syndrome in that rejection of *thok* I's dominance involved a heavy price that amounted to denial of their entry into the informal credit market. Interestingly, the poor households of *thok* I were also vulnerable to indebtedness. However, in the given undercurrent of dissent, disagreement and rivalry between the two family lines, they did not have to pay as big a price as *thok* II. It was not known whether the self-help groups formed by women of both *thoks* would eventually dilute the hegemony of *thok* I and allow the stock of the social capital to improve. It is in this social and economic context that MMVS made its intervention for sustainable development of village pastureland through institutional development.

Making an Institution

The shift from old to new practices of using village commons under a regulatory regime took place in this social context during late 1980s at the guidance of MMVS. However, leadership for initiating collective action emerged from the same powerful households with their relatively higher educational background from *thok* I. Two school teachers, Radhe Shyam and Mishru Singh, along with Sukhdev Singh (all from *thok* I) sought intervention by MMVS. This goes well, as mentioned earlier, with Krishna's observation (Krishna, 2003). While this may be considered a positive trend, one has to seriously look into whether the leadership is tainted by sectarian attitudes that eventually work against the interest of the community. This was precisely what MMVS, being a mediating agency and external to the village, failed to comprehend during the initial stage of its intervention. MMVS did not care to examine the content of social capital and its correspondence with the existing power relations in the village.

Formation of VDC

Institutional renovation began with the formation of a VDC through consensus building at the general body meeting of the villagers. Sukhdev Singh was selected the president of the committee with households of both *thoks* enthusiastic about it. However, the enthusiasm, especially of *thok* II, began to diminish as they found themselves being marginalised in the decision making process. It was decided in the general body meeting that representation would be made *thok*-wise in both the executive and general body of VDC. (This was revealed in discussions held with the present office bearers and the president of MMVS). However, in practice, the composition of the committees revealed that VDC was under the absolute control of the rich farmers of *thok* I (Tables 2.18 and 2.19).

It may be mentioned that no election of the executive body has taken place for the last five years; the only change that Singhariya witnessed was the replacement of Sukhdev Singh by Mishru Singh as president of the committee. Enquiry revealed that lack of participation of the members, ironically of both *thoks*, was a reason why membership reduced from 25 to 18 as shown in Table 2.18.

This went well with the rules framed by the VDC that any member found absent in three consecutive meetings of VDC would face termination automatically. However, what was irreconcilable was

Table 2.18

Thok-wise Representation of the Members in VDC

Category	*Initial Years*	*At Present*
Thok I	19	14
Thok II	4	2
SC 1	1	
MMVS	1	1
Total	25	18

Source: Field survey.

Table 2.19

Thok-wise Members of the Executive Committee and General Body of VDC

(at present)

S. No	*Name*	*Thok*	*Category*	*Post*
Executive committee				
1	Mishru Singh	I	Medium farmer, retd. teacher	President
2	Sukhdev Singh	I	Medium farmer	Vice-president
3	Mittu Singh	I	Medium farmer	Treasurer
4	Kishan Singh	I	Medium farmer	Secretary
5	Kishan Singh (veterinarian)	I	Medium farmer holding a govt. job	Member
6	Khim Singh	I		Member
7	Gom Singh	I		Member
8	Chittra Singh	I		Member
9	Bhavra Singh	I		Member
10	MMVS Representative	NGO		Member
11	Panchi Devi	I		Member
General Body				
1	Panchu Singh	II		Member
2	Chain Singh	II		Member
3	Mithu Singh	I		Member
4	Sambhu Singh	I		Member
5	Sohan Singh	II		Member
6	Sardar Singh	I		Member
7	Safu Singh	I		Member

Source: Field survey.

the complete rejection of representation of *thok* II in the executive committee of the VDC (Table 2.19). The insignificant representation of *thok* II in the general body did not make any difference to the decision making process of VDC.

The other aspect that needs attention was that women representation in the executive committee was merely symbolic. Discussions with them revealed that they did not even know where did the meeting of VDC took place usually. The members of *thok* II had the same story to tell. Thirdly, the major players in the executive committee, constituted only by the households of *thok* I, were Mishru Singh, Sukhdev Singh, Mithu Singh and Kishan Singh—all from *thok* I and all of them, relatively speaking, affluent farmers.

In fact, it was this minority group that formed the power centre in the village, as the field survey revealed, that made itself crucial to any decisions related to any development programme of the village. Of course, it never had any problem gaining acceptability from its own *thok*. The small and marginal farmers who constituted the largest section of this *thok* did not antagonise it in view of their strong sense of belonging to the same *thok*.

MMVS failed to realise how power relations were interwoven with the social dynamics of this village. It also failed to understand how the same power relations were rooted in the ownership and control of economic resources of a small group of households of *thok* I. However, when it finally understood that the development initiatives were subsumed into the traditional power structure, it had to withdraw from the scene without completing the job. It is significant, however, that their withdrawal did not lead to the collapse of VDC. The latter continued to function independently and enforced its regulatory regime for sustainable development of the village pasture in the village. The issue is whether the regulatory regime or new institutional practices were stable and sustainable as long as their implementation was in the hands of a few powerful rich farmers. Before examining this, it may be worthwhile to take an overview of the steps that were taken by VDC to improve the pastureland.

Steps Taken

Immediately after the formation of VDC, the first step was to abandon the open grazing system being practised for generations and fence the village pasture on all sides, grow grass and plants on it and protect them on a regular basis. The available pastureland was divided into two plots in which the rotational system of open grazing was introduced. The revenue land (3.12 hectares), adjacent to the pastureland and controlled by the revenue department, was earmarked for raising a plantation to meet the fuelwood crisis.

Subsequently, a unique method of plantation was followed accompanied by a ban on tree felling. Women were mobilised for the plantation as well as school children, with 25 *paisa* as an incentive for each plant. Saplings were also distributed as an incentive for private plantation around the agricultural fields. The involvement of school children had a rousing impact on the villagers. It made mobilisation of all villagers easy, irrespective of who belonged to which *thok*, towards development of the village pasture. Although villagers of both *thoks* were equally enthusiastic about grassland development through institutional intervention, the leadership provided by Sukhdev Singh and Mishru Singh was doubted even by some people of the same *thok* while the plantation programme was in progress. They feared that these leaders might finally encroach upon the same land. However, after much persuasion villagers realised that they were to be benefited through such a process. The joint efforts of the villagers with MMVS gained credible support only after this.

Framing of Rules

Subsequent to the fencing of the pastureland, the following rules were framed by VDC to ensure conservation and growth of village pastures.

- No animals would be allowed to graze inside the protected land for five years after plantation was done. (This rule was framed because the species that were planted were expected to grow in five years. Plants could then be saved from small ruminants).
- Within the period of five years, grass must be harvested and distributed equally to all households. The system of stall-feeding of animals would be encouraged.
- Ban on axes or any tree-felling instrument inside the pastureland. Anyone found carrying such instruments had to pay a penalty of Rs. 500.
- Illegal felling of trees was liable for punishment with a fine of Rs. 500.

- No outsider was allowed to trespass on the protected land. In case it was done, VDC would take action against him. Further, anyone found hunting animals inside the protected land must be punished by VDC.
- The meeting of the executive committee of the VDC would be held once in a month. Any member found absent in three consecutive meetings would stand to lose membership.
- If VDC did not have money to pay the appointed guard, *'lakdi system'* would be followed. This was a sort of relay system: two branches of a tree from pastureland were to be handed by the member of the household who guarded it on the previous day over to the next one guarding the pastureland on the next day. If the individual was found to have violated this system of guarding even after accepting the branches, he would be penalised in the event of any loss that might be caused to the plants in his absence.
- If any household was found to have a male member incapable of protecting the pastureland, VDC had to be informed about it on the same day so that it could organise a guard from another household.
- It was also decided that the household must pay a penalty equal to prevailing daily agricultural wages in case its designated member failed to turn up for duty.

This system of protection continued for years, although, initially, two guards were appointed by the VDC at Rs. 600 per month per guard for four months.

While enforcement of these rules was discriminatory, as the subsequent analysis will show, VDC exhibited flexibility in changing its rules as and when it was required to do so. For instance, it was not visualised at the time of framing of rules that pastureland would have to be kept open for grazing. Singhariya witnessed compositional change of livestock with the incorporation of small ruminants. This reinforced the decision of VDC to introduce open grazing system in the protected pastureland at least for five months a year.

Similarly, grazing charges were fixed at Rs. 50 per large animal and Rs. 20 per small ruminant during 1993. However, no charges were fixed for the large animals in the following years as their stock declined sharply. Besides, charges for small ruminant also declined steeply from Rs. 20 to Rs. 2 to facilitate grazing of their rising stock. The response of VDC to three consecutive years of drought again exhibited flexibility in its withdrawal of not only grazing charges on all animals, but also keeping open the protected grazing land for all animals to graze throughout the year except during the rainy season. Auctioning of grass also took place when the yield rate of grass was considerably higher.

Auctioning was adopted even in the case of fuelwood when the initial system of distribution of fuelwood failed to yield desirable results. Initially, each household was allotted fuelwood trees (one or two depending upon the total number of household members). However, this system proved inequitable—plants were not of equal size and density. Those who were allotted bigger and bushier plants benefited more leading to conflicts. Even the households of *thok* I received unfair treatment. It was decided to switch over to auctioning, but the *thok* II households seldom supported even this. For the auction was conducted by the same powerful members of VDC to the disadvantage of the households of *thok* II.

NGO, Panchayat and VDC

The credibility of VDC with the local *panchayats* came under a cloud after the abrupt withdrawal of MMVS from the village. Discussions held with the office-bearers of MMVS revealed that, after working for one-and-a-half year in a collaborative arrangement, VDC began refusing proposals made by MMVS for pastureland development and the plantation programme. MMVS members felt that the committee was geared to serve the interests of their own *thok*, a view supported by the *sarpanch* and the *panchayat*. This negated the objective with which MMVS had intervened, sustained institutional renovation for sustainable development. Faced with no option but to withdraw, MMVS facilitated the framing of rules for the conservation and growth of the village pastures as a last exercise.

The forcible encroachment of the unprotected pastureland by *thok* I households, also continued to be a cause of disagreement between the *panchayat* and VDC. Discussion with the *sarpanch* revealed that VDC was persuaded repeatedly for resettlement of the encroachers outside the village pasture. However, VDC did not pay any heed. On the contrary, Mishru Singh, president of VDC, sought a no-objection certificate (NOC) from the *panchayat* for fencing and developing the land. Apprehending that it would give VDC an opportunity to gain legitimacy for their support for encroachment, the *panchayat* refused to give NOC to VDC.

Instead of fulfilling its role as a village organisation, VDC turned out to be an extended forum for the powerful to exercise their power. Discrimination in enforcement of rules was an inevitable outcome. Its conflict with the local *panchayat* showed that it was incapable of overcoming its own bias that impinged upon the collective character of the village. The issue is whether such a sectarian approach of VDC could ever sustain institutional renovation?

Gains from Institutional Renovation

While there were apprehensions about the stability of institutional renovation, institutional gains that accrued to the villagers must be accounted. One of the significant institutional outcomes was the decline of uncertainty regarding the future prospects of grazing and fuelwood availability. Of course, this would have been substantially higher, had the pastureland been freed from encroachment. Despite the loss, of potential gain, incurred by the villagers as a whole, the intervention made, with their uncritical support, not only produced instrumental value but also existence value. The existence value is derived from simply knowing that the resource exists in a preserved state (Bromley, 1998). However, this was never a sufficient condition for conservation and growth of village pastures. Every individual household did perceive its importance depending upon to what extent the preserved stock of resources could fulfil its requirements in practice. The latter, in turn, depended upon the extent to which access to the improved resource base was given to the households of both *thoks* in this village.

Access to Village Pastures

The extent, to which access was given to the villagers to the improved village pastures, epitomised the enforcement characteristics of the rules. As noted earlier, the village community opted for a rotational grazing system for a few months of the year for both large and small ruminants. It was also noted that within the given institutional norms, fuelwood was allowed to be collected from the pastureland. However, a veiled discrimination was perpetrated by the management of VDC against the households of *thok* II not only in providing access to their animals to the grazing land but also the quantity of fuelwood being allowed to be collected.

Data was collected on the time spent by the households for grazing their animals in different types of lands *viz.*, common protected pastureland, common open land and private land. The protected pastureland was the first choice of all households for grazing their animals as much as they could, since its conditions were comparatively better, especially during drought. However, households of *thok* II appeared to have used this land less because they had less access to it as compared to the access given to the households of *thok* I. This is worked out based on the average number of hours allowed per animal of all households to graze on it during the drought year of 2002 (Table 2.20).

Table 2.20

Average Number of Hours of Grazing Per Animal in the Protected Pastureland vis-à-vis Other Lands

Thok	*CPL*	*COL*	*PL*	*Total*
Thok I	1,040.11 (46.90)	890.32 (40.15)	287.07 (12.95)	2,217.50 (100.00)
Thok II	682.40 (33.57)	1,047.48 (51.53)	302.93 (14.90)	2,032.81 (100.00)
Average per animal (*Thok* I + *Thok* II)	985.94 (44.82)	923.74 (42.00)	289.87 (13.18)	2,199.55 (100.00)

Note: CPL= Common protected land, COL= Common open land, PL= Private land.
Source: Field survey.

Table 2.20 shows that the protected grazing land could meet around 47 per cent of the total hours required by the sample households of *thok* I to graze all their animals. However, it was only 34 per cent in the case of *thok* II. This might be a reason why households of *thok* II had to depend more on the common open land for more than half of their requirements of grazing. The relatively less access given to *thok* II seemed to be a direct consequence of discriminatory enforcement of the rules.

The executive members of VDC, as reported by *thok* II households, were hostile towards them ever since they took over the management of VDC. This discouraged many of *thok* II from grazing their animals on the protected pastureland even when it was free for all. For example, medium farmers of this *thok* could not utilise the protected village pasture for grazing their large animals (not shown in Table 2.20). They did not do so, as the field survey revealed, because they wanted to avoid confrontation with the management of VDC.

The repeatedly discriminatory enforcement of rules benefited mostly a powerful few; as shown latter, it was the medium farmers of *thok* I, who had the privilege of exercising more access to the common protected pastureland as compared to others including even the weaker sections from the same *thok*. With the shift towards rearing small ruminants, grazing gained more importance. Medium farmers had a large number of small ruminants, especially sheep, for which they used the protected village pastures more in their favour.

Table 2.21

Thok-wise Average Collection of Fuelwood Per Household during 2002

(in quintals)

Thok	*Protected Land for Fuelwood and Grazing*	*Common Open Land*	*Private*	*Purchased*	*Total*
Thok I	2.04 (17.35)	2.15 (18.28)	5.04 (42.86)	2.53 (21.51)	11.76 (100.0)
Thok II	1.47 (14.51)	4.75 (46.89)	3.54 (34.94)	0.37 (3.65)	10.13 (100.0)
Average per household (*Thok* I + *Thok* II)	1.91 (16.78)	2.74 (24.08)	4.70 (41.30)	2.03 (17.83)	11.38 (100.0)

Source: Field survey.

Households of *thok* II were no less disadvantaged even in respect of fuelwood collection. Table 2.21 shows that the average collection of fuelwood per household of this *thok* was maximum from the common open land at 47 per cent of their total requirement. However, in the case of *thok* I households, the average collection (43 per cent) was largely from their own private lands. Private land, as it appears from Table 2.21, was the major source of fuelwood for the entire village as compared to common protected and open land.

While taking both *thoks* together, on an average each household collected 4.70 quintals of fuelwood during the year from the same source. The protected land exclusively planted for meeting the fuelwood requirement and the common protected pastureland together could provide only 1.91 quintals of fuelwood per household. This was the lowest of all sources. The main reason was that the size of the land was inadequate to meet the entire requirement. Importantly, the share of *thok* II in this fuelwood was lower than that of *thok* I.

As mentioned earlier, fuelwood from the open pastureland was auctioned while that from the protected land was distributed. Both the systems were biased towards *thok* I. The auction, organised by VDC comprising *thok* I representatives, largely favoured the households of *thok* I. So much so, even the landless household (not shown in the Table 2.21) of the same *thok* could obtain half of its requirement of fuelwood from the protected village pasture, a disproportionately large quantity, because it belonged to *thok* I.

While the benefits derived from the new institutional arrangement met less than one-fifth of the household requirements, households of *thok* I could manage to fulfil their need up to 43 per cent from private land and 22 per cent from the market (Table 2.21). For the poorer households of *thok* II, equal access to either the market or plantation on private land was out of question. The only option was the common land, deteriorating almost to a point of no return in three years of drought.

Table 2.22

Response of the Households on Gains from the Pastureland

Thok	*Grazing Land*				*Fuelwood*			
	A	*B*	*C*	*Total*	*A*	*B*	*C*	*Total*
Thok I	52 (57.78)	30 (33.33)	8 (8.89)	90 (100.0)	46 (51.0)	21 (23.33)	23 (25.56)	90 (100.0)
Thok II	11 (40.74)	13 (48.14)	3 (11.11)	27 (100.0)	14 (51.85)	2 (7.40)	11 (40.74)	27 (100.0)

Note: A= Large benefits, B= Little benefits, C= No benefits. Figures within brackets are percentages of household response of the total households.

Source: Field survey.

Interestingly, analysis of the sample households' responses shows how perceptions about gains from pastureland development differed between *thoks*. Table 2.22 shows that around 58 per cent of the sample households of *thok* I perceived benefits of institutional renovation, while only 41 per cent of households of *thok* II felt the same. It was important to note that the majority of the sample households of *thok* II felt that benefits were little or nil. Similarly, in the case of fuelwood collection, although perception about the benefits was large in respect of both *thoks* (51 per cent of the sample respondents), around 41 per cent of them, belonging to *thok* II, never benefited at all.

The adverse consequence of inequitable access to the improved resource base may be glaring under scarcity conditions. For, as the scarcity conditions worsen, alternative sources such as common grazing land or private land may lose their potential as dependable sources of fuelwood and grass. Private land may have more potential if private investment is affordable for the stakeholders, but that is debatable in scarcity conditions. Naturally, then, protected pastures come under severest pressure. Under such circumstances, unequal access to the village pastures may become a major source of conflict. It may undermine institutional stability.

Status of Livestock

Singhariya witnessed improvement in the livestock over the years in terms of numbers and milk yield through institutional renovation for management of village pastures. The composition of animals also witnessed a shift towards small ruminants. Here too, households of *Thok* I appeared to have gained more from such improvement (Table 2.23).

Table 2.23

Gains through Improvement in the Status of Livestock due to Institutional Renovation

Thok	*Increase in the Size of Livestock*	*Increase in Yield of Milk*	*Increase in the Sale of Small Ruminants*
Thok I	41 (45.56)	52 (57.78)	26 (28.89)
Thok II	6 (22.22)	14 (51.85)	7 (25.92)

Note: Figures within parentheses are the percentages of the response of all sample respondents.
Source: Field Survey

Around 46 per cent of the sample respondents of *thok* I could increase the size of their livestock with the development of the village pastures. However, very few households of *thok* II witnessed such a change. The largest number of households of both *thoks* gained considerably from increase in milk yield, although, again, *thok* I gained more from it. The rearing of more small ruminants also helped *thok* I households raise their earnings relatively more. The rotational system of grazing, contrary to 'cut and carry', might have lent credibility to such a compositional shift. However, it did not mean that the villagers stopped rearing big animals; they continued to do so even when they had to face persistent drought (Table 2.24).

Table 2.24

Average Number of Livestock Per Household of Each Thok during a Normal Year (1997) and Drought Year (2002)

Thok	*1997*				*2002*			
	Cow	*Buffalo*	*Goat*	*Sheep*	*Cow*	*Buffalo*	*Goat*	*Sheep*
Thok I								
Marginal/small	2.02	0.84	6.22	2.77	1.00	0.19	3.99	1.95
Medium	1.78	1.11	4.56	28.89	1.00	1.11	4.67	7.00
Average per household	2.0	0.86	6.06	5.38	1.00	0.28	4.06	2.46
Thok II								
Marginal/small	1.29	0.63	5.33	0.21	1.04	0.25	4.25	2.08
Medium	0.33	1.67	8.00	Nil	0.33	0.33	8.33	Nil
Average per household	1.19	0.74	9.33	0.19	0.96	0.26	4.70	1.85

Source: Field survey.

The stock of animals of both *thoks* declined steeply during the drought years. Strictly speaking, it may be unfair to compare stock position of livestock of drought year with that of the normal year. For, apart from scarcity conditions of green fodder or grass, other factors influence household decisions on both stock and composition during drought years. These include number of household members, the marketing prospect of livestock and its by-products along with their respective prices, etc. Besides, removal of uncertainty about grass availability plays a significant role in influencing household decisions about livestock and its composition. (Ray, 2003b).

Although households of *thok* I were less uncertain about the availability of grass in the protected pastureland, their average stock and composition of animals, as Table 2.24 shows, did not differ much from that of *thok* II during the drought year. Medium farmers of *thok* I could manage to retain the largest number of animals as compared to all other categories of farmers with the largest number of small ruminants. Additionally, they did not witness a decline of average production of milk as much as households of *thok* II did during the drought year (Table 2.25).

Table 2.25

Average Production of Milk Per Household of Each Thok in 1997 and 2002

(in quintals)

Thok	*1997*	*2002*
Thok I	14.50	3.86
Thok II	14.19	2.04
Average per household (*Thok* I + *Thok* II)	14.45	3.58

Source: Field survey.

Table 2.25 shows that the average production of milk per household of both *thoks* was 14-15 quintals during 1997. However, it declined by four times during the drought year. More importantly, the decline was more pronounced for households of *thok* II; compare the decline of 10.64 quintals per *thok* I household to the 12.15 quintals for *thok* II. There may have been many factors such as lesser ability to purchase feed and fodder from the market, inadequate household stock of agricultural by-products to feed animals, etc. responsible for the difference. Unequal access to the households of *thok* II to the village pastures could be another factor.

In 1997, grass was in fair supply since the village received good rainfall. It was available on both private and common open lands. This might be a reason why the average milk production per household of *thok* II was almost at par with that of *thok* I. However, it was altogether a different scenario in 2002 when protected pastures provided better source of grazing. In such a situation it was, basically, the higher entitlement enjoyed by *thok* I households to the village pastures that caused them to achieve relatively higher level of milk production.

Institutional Stability

The issue is how long would institutional practices that failed to ensure equity in the distribution of institutional gains sustain? In other words, how would institutional arrangement and its practices ensure stability? The concept of institutional stability "intends to capture sufficient conditions under which a set of rules generates a constant pattern of behaviour at a low transaction cost over an extended period of time...more precisely, an institution is called stable if compliance is high and rules are rarely changed over an extended period of time" (Chakravarty, 2001).

If one strictly adheres to this concept, Singhariya did not seem to be far away from achieving institutional stability in the management of village pastures. Although rules were changed, it was done so for the convenience of the village community. Moreover, frequency of change of rules was too low in a period of 12 years to make new institutional practices unstable. Despite these achievements, one may have reasons to believe that instability was embedded in the discriminatory enforcement of rules for accessing institutional outcomes. Unequal outcomes carry immeasurable potential for conflict between gainers and losers to a point where losers are left with no choice. This may make the institutional renovation unsustainable (Sengupta, 2002). In this context, as mentioned earlier, households of *thok* II could have the option of grazing their animals and collecting fuelwood from the common open lands. However, this option would cease to exist once livestock population surpassed the carrying capacity of these lands aggravating social tension and conflict.

Table 2.26

Response of the Sample Households on Organisational Matters

Items	*Number of Sample Respondents*
Decisions taken by villagers together	23 (19)
Decisions taken by VDC members	62 (53)
Households that never attended meetings	84 (71)
Households that attended the general meeting	34 (29)
Households that never knew when the meeting took place	47 (40)
Knowledge of rules	119 (100)
In favour of *status quo* in the rules of management	86 (72)

Note: Figures in the bracket are the percentages of the total number of respondents.
Source: Field survey.

Besides, the level of compliance was not as high as one desired against the backdrop of freeriding that the VDC frequented; this was observed by 81 out of 118 sample respondents. The total amount collected as penalty until 2002 was as much as Rs. 15,303. Additionally, the low participation in meetings indicated indifference among a large number of sample households. Even at the risk of over simplification, high level of compliance could actually be misleading where social capital was determined more by hegemonic power relations. Hence, one would probably see a high level of compliance of the *thok* II households more in terms of moral commitment to the cause rather than commitment to the enforcing authority. They were morally committed, but were indifferent to biased enforcement of rules because the cost of voicing their concerns was, as they apprehended, was high.

Indifference to the management can best be assessed in terms of effective participation of the households in the decision making process. Table 2.26 shows that as many as 71 per cent (84 out of 119 sample respondents) never attended the meetings that took place for the development of the village pasture. It is equally important to note that VDC members were all from *thok* I. Despite alienation of majority of the households from the decision making process, 72 per cent (86 out of 118) of the total sample respondents desired no change in the rules of management.

This speaks of their moral commitment to the cause of conservation and growth of grassland of their village. Institutions will remain stable till moral commitment continues to receive overriding importance. This was precisely the reason why no violation of rules of conservation went unnoticed and the grassland ecosystem was not allowed to be upset even in three years of persistent drought (Ray and Upadhyay, 2004). The powerful households of *thok* I would never allow the renovated institution to breakdown since it served their interest more.

While there is no doubt that Singhariya gained through conservation of village pastures by way of renovation of its institutional practices, one may treat it as a sub-optimal achievement of the institution. For, it shows greater fulfilment of the objective of conservation coupled with unequal access to institutional outcome of those who remained outside power relations. The long-term institutional stability, which is a joint product of compliance of higher degree, infrequent changes of rules, consensus-building and equitable institutional outcome may not be agreeable to such sub-optimal solutions that Singhariya attained after having pursued new institutional practices for more than a decade. Conditions under which the same may be achieved are shown in Section IV in this chapter.

III

ANALYSIS OF ECOLOGICAL CHANGES

Two aspects of ecological changes in both villages are examined in this section. While part I of this section analyses the extent and nature of floral diversity achieved in the protected grassland *vis-à-vis* the unprotected one, part II examines whether there has been any improvement in the physico-chemical characteristics of soil in the protected land due to intervention. Part III makes a few comments on the observations on the preceding analysis.

Floral Diversity

While analysing floral diversity, it is imperative to know the biological relationship between species within the framework of inter-species association that exists between them. Species are associated in communities while having definite structure and composition (Singh, 1967). Their inter-species association within the community has a bearing on floral diversity. This section analyses floral diversity that protected pastureland regained due to intervention and compares it with the unprotected one. Standard methods were followed, as explained in Annexure 2. One has to bear in mind that the topography of both pasturelands is similar. Samples were collected during the months of February and August 2003.

It may be noted that the villages under study had passed through severe drought conditions continuously for three years before they received normal rainfall. The floral diversity of grass and plant species was examined based on importance value index and other indices. Importance value index (IVI) was worked out to understand the hierarchical position of species in both types of land based on the frequency, density and dominance of these species. The other indices include similarity index, dominance/diversity index and species

Table 2.27

Importance Value Index of Tree and Shrub Species in the Protected and Unprotected Pasturelands

Importance Value Index	Chhota Saradhana		Singhariya Protected			
	Protected	Unprotected	Plot A	Plot B	Plot C	Unprotected
<10	*Hardwickia, Butea monosperma, Acacia senegal, Azadirachta indica, Rhus mysurensis, Acacia catechu, Salvadora persica, Commiphora wightii, Z. Rotundifolia, Inga Sp., Anogeissus pendula, Tecomella undulata, Ailanthus excelsa, Emblica sp.*		*Dalbergia sissoo, Prosopis cinereria, Anogeissus Pendula, Calotropis procera, Tacomella Undullata, Z. Rotundifolia, Salvadora oleoides, Cordia gharaf, Ailanthus Excelsa, Commiphora Wightii, Bauhinia racemosa*		*Prosopis cinereria, Tacomella undullata,Butea monosperma, Ailanthus excelsa, A. senegal, Cordia gharaf, Commiphora wightii, Salvadora persica*	
10-20	*Ziziphus nummularia, Acacia leucapholea, Prosopis cinereria, Cordia gharaf, Salvadora oleoides, Capparis decidua, Z. mauritiana, Dalbergia sissoo, Euphorbia caducifolia*		*Rhus mysurensis, Azadirachta indica, A. leucaphloea, Z. mauritiana, Adhatoda vasica S. persica, A. catechu A. Senegal*		*Adhatoda vasica, Acacia nilotica, Rhus mysurensis, A. leucaphloea, Calotropis procera, Azadirachta indica, Euphorbia caducifolia*	

contd. ...

contd. ...

Importance Value Index	Chhota Saradhana Protected	Unprotected	Singhariya Protected Plot A	Plot B	Plot C	Unprotected
20-40	*Acacia nilotica*		*Ziziphus nummularia, P.juliflora, Euphorbia caducifolia, Capparis decidua, Acacia nilotica*	*Butea monosperma, Capparis decidua, Euphorbia caducifolia*	*Z. mauritiana, Ziziphus nummularia, Capparis deciduas*	*Euphorbia caducifolia*
40-60	*Prosopis juliflora*	*Ziziphus nummularia*			*Prosopis juliflora*	*Ziziphus nummularia Rhus mysurensis*
60-80		*Rhus mysurensis*				
80-100						
100-150						
Above 150		*Prosopis juliflora*			*Prosopis juliflora*	*Prosopis juliflora*

Source: Field Survey.

Table 2.28

Importance Value Index of Grass and Forb Species of Protected and Unprotected Pasturelands

Importance Value Index	Chhota Saradhana		Singhariya			
	Protected	Unprotected	Protected			Unprotected
			Plot A	Plot B	Plot C	
<10	*Eragrostis ciliaris, Borreria stricta, Aristida funiculata, Salvia aegyptiaca, Panicumm turgidum, Cenchrus ciliaris, Corchorus tridens, Mollugo nudicaulis, Desmostachya bipinnata, Cenchrus setigerus, Commelina benghalensis, Ipomoea pestigrides, Cressa cretica, Heliopropuim benghalensis, Tribulus terrestris, Cyperus kyllinga, Tragus biflorus, Zornia gibbosa, Euphorbia hirta, Mollugo cerviana*	*Euphorbia hirta, Bulbostylis barbata, Dactyloctenium Sindicum, Borreria stricta*	*Indigofera cordifolia, Cenchrus ciliaris, Zornia gibbosa, Dactyloctenium sindicum, Salvia aegyptiaca, Pulicaria angustifolia, Bulbostylis barbata, Boerhavia diffusa, Cynodon dactylon, Alysicarpus monilifer, Eleusine compressa, Aristida funiculata, Eragrostis ciliaris, Lapidagathis trinervis, Heliopropuim benghalensis, Aeluropus lagapoades, Ipomoea pestigrides, Euphorbia hirta, Corchorus tridens, Cyperus kyllinga*	*Desmostachya bipinnata, Chloris virgata, Aeluropus lagapoades, Tragus biflorus, Cleome viscosa*	*Tephrosia apolliniea, Trichodesma indicum, Zornia gibbosa, Cleome viscosa, Aeluropus lagapoades, Aristida funiculata, Eragrostis ciliaris, Chloris virgata, Lasiurus indicus, Alysicarpus monilifer, Mollugo nudicaulis, Eragrostis poaoides, Boerhavia diffusa, Euphorbia hirta, Lapidagathis trinervis, Indigofera cordifolia, Corchorus tridens, Heliopropuim benghalensis*	*Boerhavia diffusa, Lasiurus indicus, Commelina enghalensis, dactyloctenium sindicum, Trichodesma indicum, Eragrostis ciliaris, Tephrosia apolliniea, Eleusine compressa*

contd. ...

contd. ...

Importance Value Index	Chhota Saradhana		Singhariya			
	Protected	Unprotected	Protected			Unprotected
			Plot A	Plot B	Plot C	
10-20	*Tephrosia apolliniea, Cleome viscosa, Cynodon dactylon, Bulbostylis barbata, Dactyloctenium sindicum*	*Cenchrus setigerus, Mollugo nudicaulis, Desmostachaya, Bipinnata, Cleome viscosa*	*Trichodesma indicum, Borreria stricta, Tephrosia apolliniea, Cenchrus setigerus, Cleome viscosa*	*Tephrosia apolliniea, Aristida funiculata, Mollugo nudicaulis, Panicum turgidum, Trichodesma indicum, Borreria stricta*	*Dactyloctenium sindicum, Cenchrus ciliaris*	*Evolvulus alsinoides, Borreria stricta, Aristida funiculate, Bulbostylis barbata, Heliopropuim benghalensis*
20-40	*Indigofera cordifolia, Trichodesma indicum, Eragrostis poaoides, Evolvulus alsinoides*		*Lasiurus indicus, Mollugo nudicaulis*	*Dichanthium annulatum*	*Digera muricata, Cynodon dactylon, Cyperus kyllinga Borreria stricta, Glossocardia bosvelea*	
40-60	*Bothriochloa pertusa*	*Tephrosia apolliniea, Trichodesma indicum, Bothriochloa pertusa, Aristida funiculate, Indigofera cordifolia*	*Dichanthium annulatum, Eragrostis poaoides*			
60-80	*Dichanthium annulatum*				*Dichanthium annulatum*	
80-100						
100-150						
Above 150				*Eragrostis poaoides*		*Eragrostis poaoides*

Source: Field survey.

richness index. The values obtained through IVI analysis were used for working out diversity/dominance index. All these indices are important in their own ways showing different aspects of floral diversity.

Importance Value Index (IVI)

In Chhota Saradhana, 25 trees, shrubs, and 31 grass species were observed. Table 2.27 shows that IVI of *Prosopis juliflora* in protected pastureland was 54.9, which was maximum followed by 25.4 (*Acacia nilotica*) and 16.7 (*Ziziphus nummularia*). The minimum IVI was 3.9 for *Emblica*. The IVI in unprotected pastureland was 167.4 in respect of *Prosopis juliflora,* again the highest.

Although *Prosopis juliflora* had maximum IVI in the protected pastureland, it had much smaller value as compared to that of the unprotected pastureland. The managing committee of VDS of Chhota Saradhana and the NGO were fully aware of the harmful effects of a luxuriant growth of this plant on the growth of other species. Despite this, it was planted on the protected land. For they thought that such plants might help retain the moist condition of the soil (Yadav and Singh, 1970). Of course, it was not allowed to grow more and cover more area in the protected land. However, this was not so in the case of the unprotected land. Further, Table 2.27 shows equal distribution of other plant species in protected pastureland although the IVI of these plants was low. The higher IVI of only three tree/shrub species in the unprotected pastureland of Chhota Saradhana indicates how fragile its ecosystem was.

Among grass and forbs, as Table 2.28 shows, *Dichanthium annulatum* was found to have maximum IVI (71.33) followed by *Bothriochloa pertusa* (49.62) and *Indigofera cordifolia* (26.89) in protected pastureland. In unprotected pastureland IVI was highest for *Tephrosia apolliniea* (57.18) followed by *Bothriochloa pertusa* (49.25), *Trichodesma indicum* (46.29) and *Indigofera cordifolia* (43.49). Here again, one finds a large distribution of grass species with low IVI in the protected land as compared to that of the unprotected one.

In Singhariya, 25 trees, shrubs, and 35 grass species were recorded in its pastureland. As mentioned earlier, this village had three plots that were protected. Plot A and C were used for grazing on rotation while Plot B was exclusively meant for plantation for fuelwood. Table 2.27 shows highest IVI to the tune of 26.0 for *Ziziphus nummularia* in

Plot A followed by 25.5 for *Prosopis juliflora*, 24.3 for *Euphorbia caducifolia*, 20.1 for *Acacia nilotica* and 18.9 for *Rhus mysurensis*. As against this, the *Bauhinia racemosa* (3.55) and *Commiphora wightii* (3.55) had the least IVI. In Plot C, *Prosopis juliflora* (41.2) had the maximum IVI followed by *Ziziphus mauritiana* (36.7), *Z. nummularia* (36.1) and *Capparis decidua* (34.09) with *Salvadora persica* (4.63) recording the minimum. In Plot B, maximum IVI *Prosopis juliflora* was the highest at 222.06; IVI for this plant was expected to be of such high order due to its intensive plantation in this plot for fuelwood.

Prosopis juliflora (149.3) had the highest IVI also in the unprotected pastureland. It is an important plant for the arid climate for a variety of reasons. Apart from retaining soil moisture as mentioned earlier, it can survive under harsh, water distress condition and provide mulch to the soil assuming considerable importance in inhibiting soil erosion. It is proven best for sand dune stabilisation. Besides, villagers find it of utility in fencing and for fuelwood if its luxuriant growth is controlled. In the absence of such control, it may not allow other plants to grow.

Contrary to findings that it increases soil alkalinity, evidence suggests that the soluble salt content declines where *P. juliflora* grows. This, in turn, reduces pH values from 10.8 to 8.2. Besides, the status of organic matter rises from 0.54 to 2.87 per cent. This is accompanied by a rise in the nitrogen content of the soil in the upper 15-cm layer (Yadav and Singh, 1970). This seems to suggest that, if palatable plants and grasses grow along with this tree, its harmful effects can be controlled through inhibiting leaching of salt downward.

The IVI of grasses and forbs in both protected and unprotected pasturelands of Singhariya is shown in Table 2.28. The maximum IVI in Plot A was found in respect of *Dichanthium annulatum* (50.7) and *Eragrostis poaoides* (49.3) followed by *Bothriochloa pertusa* (33.4) and *Mollugo nudicaulis* (20.7). The maximum IVI was recorded for *Eragrostis poaoides* at 171.2 in the unprotected land and 167.6 for Plot B. In Plot C, the maximum IVI was recorded for *Diachanthium annulatum* (77.4) and minimum for *Corchores tridens* and *Heliopropuim benghalensis* (0.67). In respect of both grass and forb species, their pattern of growth did not appear to differ significantly between the protected pasturelands of both villages. As Table 2.28 shows the largest number of these species was found to have IVI, less than 10 per cent in all protected pasturelands. There were, of course, some grass or forb species that were not palatable. However, the incidence of such grass species in the protected pasturelands appeared to be much lower than that in the unprotected pastureland.

Acacia nilotica, Dalbergia sissoo, Salvadora persica and *Prosopis cinereria* are some plant species that help in maintaining moisture and nutrient contents in soil and, therefore, suitable for arid conditions. *Z. nummularia, Rhus mysurensis* and *Capparis decidua* are important shrub species of semi-arid areas, useful even for fruit. The adoption of the xerophytic species of perennial nature, as mentioned above, is relatively more advantageous under stress conditions, while annual species can grow only under suitable conditions with a certain degree of rainfall accompanied by low temperature. It may be mentioned that in the absence of perennials, persistent drought may degenerate seeds. In such a situation, the growth prospects of annual species may be constrained. From this point of view also, perennials act as a source of stability of the dry land ecosystems.

Although the dominance of one or two species with greater importance value shows a type of climax vegetation, a large number of species even with lower IVI bears considerable importance for achieving greater diversity. Variation in the type of species, its morphology, growth pattern and reproductive behaviour, have implications for successional behaviour of the habitat. Complex relations of all species enhance interdependence that encourages survival of plantation of thorny species with reduced transpiration. Villagers appear to have succeeded in their conservation effort as this vegetation survived consecutive three years of drought. Grasses and forbs also showed large variation with IVI.

However, C*enchrus setigerus*, planted at the initial stage of intervention, did not have higher IVI (Table 2.28). A variety of factors might have adversely affected the germination and growth of this species. While one may attribute it to overgrazing during drought years in the recent past, conditions arising out of compaction of soil followed by heavy rainfall during the months of June and July 2003 could be another factor responsible for this. For this might have damaged the mechanical strength of these species to germinate. In such cases, as it appears from field survey, natives were more suitable and adaptable to the microenvironment even if it turned out to be hostile.

Although the unprotected land was not as rich as the protected one, it was richer than many other unprotected lands lying barren. However, what made the difference here was that the density of palatable species was more in the protected pasturelands of both villages as compared to the unprotected ones. The higher stock of non-palatable species in unprotected site is an indication of overgrazing, meaning that when the grassland ecosystem moves towards degradation, the stock of palatable species tends to deplete while that of non-palatable species increases. In this process, a stage is reached where only non-palatable species survive. The intra-species and inter-species relations that exist in such vegetation community cause further degradation of the grassland ecosystem.

The analysis also reveals inter-species relations between several species such as *Dichanthium annulatum* and *Indigofera cordifolia, Borreria stricta* and *Eragrostis poaoides*, and *Evolvulus alsinoides* and *Dactyloctenium sindicum* in the protected pastureland of both the villages. It may be noted that *Dichanthium annulatum* and *Indigofera cordifolia* were dominant but palatable and had positive association with rest of the species mentioned above. This had definitely contributed to the improvement in the ecosystem of both villages' pastureland. The non-palatable species such as *Aristida funiculata*, which is harmful for grasslands, were found largely with greater IVI in the unprotected pasturelands as compared to the protected ones. It inhibits nitrogen-fixing bacteria and thus creates a nitrogen depleted environment. This activity helps the species to avoid competition from those species, which require nitrogen rich habitat (Murthy and Ravindra, 1975). Regular protection being provided to the common pasturelands also made a difference from the unprotected one in this regard.

Similarity Index

The similarity index shows the extent to which both plants and grass species are similar in both protected and unprotected pasturelands. The value of similarity index for grasses was 0.59 in Chhota Saradhana and 0.62 in Plot A and 0.58 in Plot C of Singhariya village (Figure 2.3).

Almost a similar pattern of grasses was found in the pastureland of both villages. Besides, it appeared that the value of similarity index for grasses was more than trees.

Several factors were responsible for this. The area had been facing droughts for three consecutive years. In addition, it may be noted that protected pasturelands were all opened for grazing during drought year, which was why growth of some species was stalled in

Figure 2.3

Similarity Index of Protected and Unprotected Pastureland

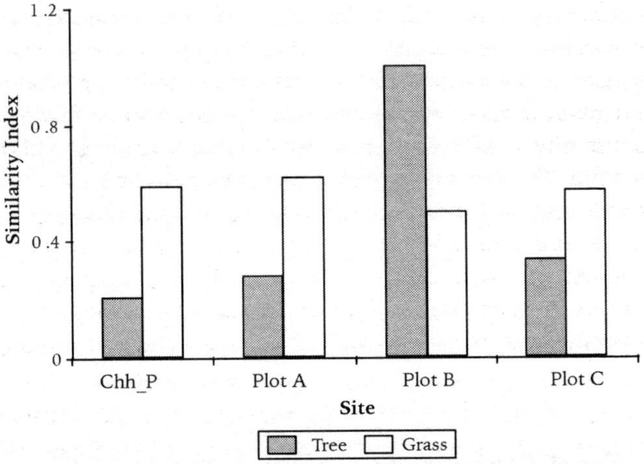

Source: Field survey.

the protected pasturelands. Besides, *Cenchrus setigerus* that was planted initially was almost wiped out, as mentioned earlier, due to drought. This may be a reason why similarity of grasses between protected and unprotected pasturelands was found to be higher. However, the incidence of overgrazing in Singhariya may be an additional factor why similarity of grasses in this village was found to be much higher between protected and unprotected lands as compared to that of Chhota Saradhana.

Regarding plants, a considerable dissimilarity was noticed between protected and unprotected pasturelands of both the villages. Figure 2.3 shows that the value of similarity index was as low as 0.21 in the case of the pastureland of Chhota Saradhana, while it was 0.28 for Plot A, 1.0 for Plot B and 0.34 for Plot C of Singhariya village. The lower the value of similarity index the higher is the qualitative difference between protected and unprotected pasturelands. However, Plot B of Singhariya showed higher value of similarity index. This is desirable because *Prosopis juliflora* that grew in this plot also grew in the unprotected pastureland of this village.

Index of Dominance and Diversity

Floral diversity may be examined in terms of any one of these indices. The higher value of one implies lower value of the other. Plants turn out to be dominant by virtue of their abundance, growth performance and production of biomass. They become conspicuous in a community in that the index of dominance shows importance of each species in relation to the community as a whole. The diversity of species is the ratio of number of species to the importance value. In any plant community, a number of species may be found. However, a community is said to be ecologically more stable and diverse as it grows old. The ecosystem may lose its stability if its natural process of succession is hindered. The case of grassland ecosystem under study is one such case that shows that such a natural process of succession was stalled some years ago. However, after intervention was made to revitalise it, plant communities achieved certain degree of diversity and, therefore, stability, if not a developed climax.

Table 2.29

Index of Dominance and Diversity in the Protected Pastureland vis-à-vis Unprotected One

Index	*Chhota Saradhana*		*Singhariya*			
	Protected	*Unprotected*	*Protected*			*Unprotected*
			A	B	C	
Dominance						
Grass	0.13	0.12	0.07	0.33	0.10	0.34
Trees	0.07	0.40	0.05	0.57	0.06	0.30
Total	0.20	0.52	0.12	0.90	0.16	0.64
Diversity						
Grass	1.2	1.01	1.05	0.68	0.97	0.69
Trees	1.2	0.30	1.22	0.35	1.01	0.52
Total	2.4	1.31	2.27	1.03	1.98	1.21

Source: Field survey.

Table 2.29 shows that the protected lands of both the villages are more diversified as compared to the unprotected ones taking both grass and tree species into consideration. Interestingly, diversity was achieved more in Chhota Saradhana at 2.4 than Singhariya (2.27, 1.03 and 1.98 in Plot A, Plot B and Plot C, respectively). Obviously, the protected pastureland was better managed. Needless to mention, diversity in the unprotected lands was much lower as compared to the protected lands (except Plot B) of both the villages at 1.31 and

1.21 in Chhota Saradhana and Singhariya, respectively. In Singhariya, as it appears from Table 2.29, dominance of plant and grass species together was 0.64 in the unprotected land, which was much higher than the protected Plot A and C. However, it was lower than Plot B. For Plot B was exclusively used for the plantation for *P. juliflora*. Dominance with lower index value indicates presence of variation in the protected lands while higher value indicates lower diversity. Hence, presence of a large number of species with good distribution increases diversity. This has led the protected land to attain higher degree of stability. The species richness index that shows diversity in terms of number of individual species along with their varieties supplements the above findings.

Species Richness Index

The values of species richness index of protected and unprotected pasturelands of both villages can be seen in Figure 2.4 showing an almost uniform pattern for both grass and plant species in both villages. In Chhota Saradhana, species richness index for plants in the protected pastureland was 2.7 while it was 2.2 in the unprotected land. Similarly for Singhariya, it was 2.4 (Plot A), 2.1 (Plot C) and 1.2 for unprotected pastureland. As far as grass species are concerned, the pattern of their presence was almost similar between the protected and unprotected pasturelands in both villages.

In other words, the relatively higher species richness index of grass shows that the richness of grass species was higher in the protected land as compared to the unprotected ones in both the villages. However, it should be noted that the species richness index for trees was much higher than that of grass species in protected areas of both villages. This broadly corroborates with the observations made earlier on similarity index of grass species. As was pointed out earlier, grass species such as *cenchrus* that was planted at the initial stage of intervention was wiped out due to three years of consecutive drought. It made a significant difference to the richness index in that grass species was not much different between the protected and the unprotected lands. Nevertheless, one must not underestimate the contribution made by other grass species to the revival of the protected grassland ecosystem even in the event of persistent drought. The dominance of palatable grass species such as *Dichanthium annulatum* and I*ndigofera cordifolia* seemed to have contributed

Figure 2.4

Species Richness Index of Protected and Unprotected Pastureland

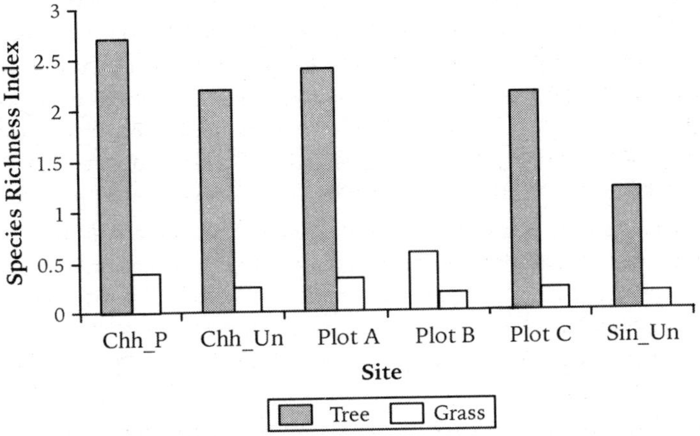

Source: Field survey.

considerably to the enrichment of the vegetation in the protected pastureland of Chhota Saradhana.

Similarly, in Singhariya, where rotational or restricted grazing system was followed, the vegetative reproductive capacity differed depending upon the grazing intensity. Even under rotational grazing system a vegetative reproductive capacity seemed to have been maintained in grass species such as *Dichanthium annulatum* that was accompanied by species such as *Cynodon dactylon* and *Evolvulus alsinoides*. All of them continued to have higher importance value. The vegetative reproductive capacity of *Dichanthium annulatum* increases when seed output and the sexual reproductive capacity decline either in the grazed or transplanted grasslands (Ambasht and Maurya, 1970 a, b). Thus, reproductive vegetative capacity of individual species may have significant influence on changing the vegetation dynamics of the grassland even in the event of persistent drought. This was supplemented by the increase in herbage above the ground in the controlled grazing system such as the one followed in Singhariya (Bhimaya *et al.*, 1967).

Physico-Chemical Characteristics of Soil

Grasses through adventitious roots trap soils and checks erosion.

Grasses and legumes are considered best vegetation means for soil and water conservation. Of course, the growing habit of these species together with their habitat and management imposed on them greatly influence their efficacy in controlling run-off and soil loss. Besides, complementary relations between vegetation and mechanical methods are no less important for soil and water conservation. In the study villages, pasturelands were improved by means of developing vegetation while mechanical methods that included construction of check dams, contour bunds, gully plugging, trenches, etc. were followed to complement the former.

The issue is whether the physico-chemical characteristics of soil underwent changes owing to complementarity between the vegetation and mechanical methods as followed by both village institutions, and, moreover, what difference its absence made to the soil characteristics of the unprotected lands. In order to examine these issues soil samples from both protected and unprotected pasturelands were collected at 25-cm depth in five replicates in April 2003 for laboratory test. The standard methodology of sampling of soil was followed with grid sampling from topographically uniform surface of both protected and unprotected lands. The physico-chemical characteristics such as porosity, water-holding capacity, pH, electrical conductivity, nitrogen, phosphorus, potassium, exchangeable Na, exchangeable K, exchangeable Ca, exchangeable Mg and organic carbon were analysed by following standard methods (Jackson, 1958; Black, 1965; Chapman and Pratt, 1961).

Porosity and Water-Holding Capacity

Pore spaces are responsible for better growth of plants for they contain air moisture. Porosity raises water-holding capacity. As the porosity of soil increases, the moisture content and humus create suitable microenvironment for the survival of microorganisms, which are integral parts of the biogeochemical cycle in that they make nutrients available for plants. They also degrade the biodegradable substances and help in weathering and soil formation. The porosity of soil declines as the soil particles become much smaller due to the reduction of pore spaces.

In Chhota Saradhana, porosity was recorded at 44.92±0.47 and 38.2±0.88 per cent in protected and unprotected pasturelands, respectively (Figure 2.5). A relatively lower percentage of porosity in the unprotected pastureland indicates higher availability of sodium content in the soil. This happens because of deflocculating of clay particles; the high sodium content adversely affects the physical properties of soil, reduces permeability and pore space (Black, 1968). The sodium content of soil in the unprotected pastureland was found to be 0.42±0.01 me/100 gm as compared to that of the protected pastureland (0.39 ±0.008 me/100 gm). The analysis of variance shows that the difference in this respect between two lands was significant (p<. 000).

Figure 2.5

Porosity and Water Holding Capacity of Protected and Unprotected Pastureland (in percentage)

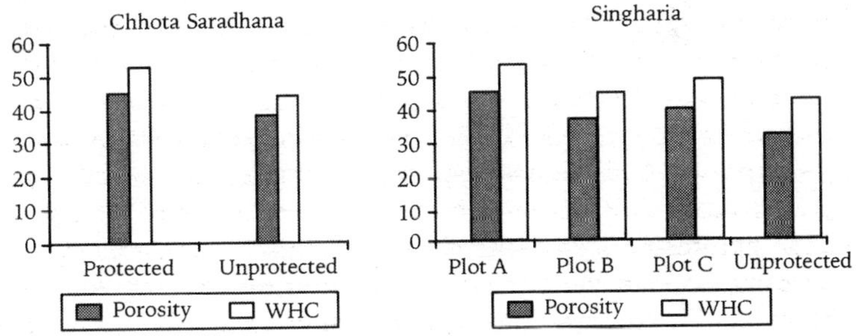

Source: Field survey.

In Singhariya, the results of porosity of the protected land were similar to that of Chhota Saradhana (Figure 2.5) with maximum at 44.14±0.85 per cent in Plot A, followed by 38.5±0.85 per cent in Plot C and 35.85±2.33 per cent in Plot B. The value of porosity in the unprotected land was as low as 31.1±1.27 per cent. The analysis of variance revealed significant difference between Plot A and unprotected land (p<. 001), Plot C and unprotected (p<. 021). However, the variance between Plot B and unprotected land was insignificant p<. 127.

Water-holding capacity is directly related to the porosity of the soil. In Chhota Saradhana, the water-holding capacity was recorded 53.02±0.86 and 44.3±1.59 per cent for protected and unprotected pastureland, respectively (Figure 2.5). In Singhariya it was recorded at 52.04±1.59 per cent, 43.65±0.35 per cent and 47.85±0.21 per cent in Plot A, B and C, respectively. In the unprotected land, it was as low as 41.35±0.5 per cent. The difference of values of the water-holding capacity between both types of lands in both the villages was found to be statistically significant at p<. 000. As shown above, the sodium content of the soil of the unprotected land was higher. This might have changed the physical properties of the soil and adversely affected its water-holding capacity, as in this situation soil particles are reduced into hard prismatic surface leaving very little space thus reducing the water-holding capacity (Hesse, 1970).

pH and Electrical Conductivity

pH indicates status of acidity and alkalinity while electrical conductivity describes the presence of ions in soil through which salinity is determined. Soil pH does not change with the application of chemical input up to a certain level because of its buffering capacity. The rise in pH is found to be associated with increasing concentration of carbonate and sodium in the soil. pH was recorded 7.02±0.34 for protected and 6.89±0.71 for unprotected pastureland in Chhota Saradhana. Although pH did not differ significantly between the two sites, it was found to be lower in the unprotected site, perhaps, due to drier climatic conditions for three drought years. The upper layer of the soil contains higher sodium content during dry months. In these months, rapid upward movement of soil moisture and evaporation may bring sodium to the top layer of the soil. In any case, to whatever extent difference was found to exist between protected and unprotected pasturelands in terms of soil pH, it was not statistically significant.

Figure 2.6

pH and Electrical Conductivity (dsm⁻1/cm) of Protected and Unprotected Pastureland

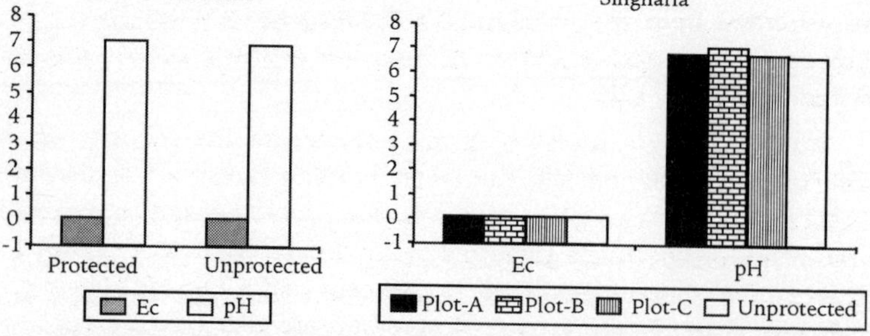

Source: Field survey.

The scenario in Singhariya was not different in this regard. Soil pH was recorded more or less same on both sites. While, it was 6.64±0.4 in respect of unprotected land, it was 6.85±0.88, 7.08±0.16, 6.78±0.35 in respect of Plot A, Plot B and Plot C, respectively. Normally, pH in soil is observed within a range of 6.5 to 8.5. Any deviation from this range either along the lower or upper range causes serious concern for plantation or/and reclamation due to the chances of it being vulnerable either to acidity or alkalinity. Nutrients become non-available at very high or low pH. For example, potassium, calcium and magnesium become deficient at low pH. Since no significant difference was found between pH status of the soil of the two sites, both types of land may be said to be less prone to acidity or alkalinity. Nevertheless, one must not ignore relatively improved pH of soil in the protected area even if it was inconsequential.

It is in this context that one may distinguish between soil of silvipastoral land and agriculture land. Agricultural soils are particularly prone to gradual acidification for caution such as Ca^{2+}, Mg^{2+} and K^{+}, which counterbalance excessive acidity. Due to harvesting, these cautions are transported away and, hence, they do not return to the soil naturally. In the absence of these cautions, soil becomes increasingly acidic. However, in the case of silivipastoral land chances of acidification of soil is relatively less because of regular litter fall that instantaneously nitrify the land. This is more so with the protected pastureland.

No significant difference between the two sites was found to exist even in terms of electrical conductivity. As can be seen from Figure 2.6, electrical conductivity was recorded at 0.12± 0.03 and 0.10±0.01 dsm¯1/cm for unprotected and protected pastureland, respectively. Ec was recorded at 0.11±0.02, 0.1±0.01 and 0.11±0.007 dsm¯1/cm in Plot A, B and C, respectively. Hence, both protected and unprotected sites were free from the problem of salinity since Ec here had values <0.75 dsm¯1/cm which indicates normal and salt free soils (Ashman and Puri, 2002).

N, P, K and Organic Carbon

The richness of soil in terms of fertility can be indicated also in terms of macronutrients including nitrogen, phosphorus, potassium and organic carbon. Table 2.30 shows there had been remarkable accumulation of organic carbon in all the protected pasturelands as

Table 2.30

Composition of N, P, K and Organic Carbon in Protected and Unprotected Pasturelands

Parameter	Chhota Saradhana			Singhariya						
	Unprotected	Protected	% Increase in the Protected One	Unprotected	Plot A	% Increase in the Protected One	Plot B	% Increase in the Protected One	Plot C	% Increase in the Protected One
N Kg/Ha	118.6±5.03	142.6±4.09	20.2	98.5±2.12	139.4±2.07	41.5	105.5±0.71	7.1	134.0±0	36.0
P Kg/Ha	9.84±0.24	11.94±0.15	21.4	6.9±0.28	11.9±0.08	72.4	7.8±0.28	13.0	11.8±0.14	71.0
K Kg/Ha	108.58±1.08	160.2±1.48	47.5	106.5±2.12	151.0±2.12	41.7	116.5±2.12	9.3	147.0±1.41	38.0
Organic carbon	0.19±0.02	0.42±0.02	121.0	0.11±0.007	0.40±0.01	263.6	0.18±0.007	63.6	0.40±0.014	263.6

Source: Field survey.

compared to the unprotected ones. For example, in Chhota Saradhana its accumulation was to the extent of 121 per cent in the protected land over the unprotected one, while it was 263.6 per cent in both Plot A and C of Singhariya village. The accumulation of the organic carbon in Plot B was 63.6 per cent, much lower than other protected pasturelands for obvious reasons. As noted earlier, *Prosopis juliflora* was the only plant grown on this plot. The absence of inter-species competition in this plot appeared to be one of the factors for over-growth of this species. However, one should bear in mind that this plant also increases organic as well as nitrogen contents of the soil (Yadav and Singh, 1970). Their uses facilitate luxuriant growth of this plant. This may be the reason why increase in other nutrients under consideration was comparatively less than what was found in other protected pasturelands. Even in respect of nitrogen, phosphorus and potassium there had been perceptible increase in all these nutrients in soil of the protected pastureland *vis-à-vis* unprotected ones (Table 2.30).

Another important observation that one makes is that the nutrient status of soil in the protected pastureland of Chhota Saradhana was higher than that of the protected lands of Singhariya. For example, Table 2.30 shows that nitrogen content of the soil in the pastureland of Chhota Saradhana was 142.6±4.09 kg/ha while it was 139.4±2.07 kg/ha in Plot A followed by Plot C (134.0±0) of village Singhariya. Similarly, in respect of potassium, it was 160.2±1.48 kg/ha in Chhota Saradhana while it was 151.0±2.12 in Plot A, 116.5±2.12 in Plot B and 147.0±1.41 kg/ha in Plot C. The comparatively higher degree of floral diversity achieved through better planning and management in Chhota Saradhana as shown earlier has more to explain why rate of accumulation of nutrients in the pastureland here was higher. Analysis of Variance (ANOVA) for all nutrients in respect of both the villages were found to be significant ($p<.000$).

Exchangeable Na, K, Ca and Mg

The preceding analysis on pH and Ec in soil of both protected and unprotected lands of the concerned villages under study has already shown that the incidence of alkanity and salinity was within the permissible limit. It fairly indicates stability of pH and Ec in soil, which is desirable from the point of view of improvement in the status of soil. However, one should bear in mind that stability is conditioned by exchangeable ions. Therefore, it is necessary for one to understand the status of soil in terms of exchangeable Na, K, Ca and Mg also. It appears from Table 2.31 that the exchangeable calcium content in soil was almost same (me/100 gm) for all protected and unprotected lands. Limestone was the parent material available in both protected and unprotected lands of the villages concerned. While this mineral was utilised by flora in the protected land, it was continuously deposited in the unprotected land due to relatively lower vegetation cover and, therefore, it largely remained unutilised.

However, deposition of this mineral in the protected land was no less. In fact, it must have been higher than that in the unprotected one. But some part of it was consumed by flora and brought down its concentration to the level of unprotected land. This, along with other minerals, tends to exhibit higher deposit in lands where floral density is higher. For in such cases physical, chemical and biological processes of weathering for soil formation are higher. However, it is not known whether higher concentration of such ions in the unprotected site would increase alkalinity in future for the same has poor organic content.

Results obtained on exchangeable Na and K seemed to have excellent conformity with what was observed in the preceding analysis on Na and K. In Chhota Saradhana, exchangeable K was 4.34 ± 0.29 and 3.42 ± 0.42 me/100 gm in the protected and unprotected sites, respectively indicating a difference to the tune of almost 1 me/100 gm. Similarly, in Singhariya, as Table 2.31 shows exchangeable K was more in Plot A and Plot C to the tune of 1.43 me/100 gm and 1 me/100 gm, respectively. Plot B, however, showed a reduction of exchangeable K to the tune of .03 me/100 gm for obvious reasons. Although exchangeable Mg was found higher in all protected plots of Singhariya *vis-à-vis* the unprotected land, ANOVA shows that the difference was not significant at five per cent level of significance. However, the unprotected land of Chhota Saradhana showed higher concentration of exchangeable Mg as compared to that of its protected land.

In a dry climatic condition where evaporation exceeds precipitation, minerals such as exchangeable Ca and Mg have a tendency to move upward in the soil in any type of land. However, presence of rich vegetation that maintains soil-water balance may be a reason why concentration of exchangeable Mg in the protected land of Chhota Saradhana was lower. Although concentration of exchangeable Mg was more in the protected land of Singhariya as compared to the unprotected one, the process through which accumulation of the same took place was similar. However, its lower concentration in the unprotected pastureland might be the result of leaching.

Table 2.31

Composition of Exchangeable Ions in Protected and Unprotected Pasturelands

Parameter	*Chhota Saradhana*		*Singhariya*			
	Protected	*Unprotected*	*Protected*			*Unprotected*
			Plot A	*Plot B*	*Plot C*	
Ex Na me/100 gm	0.39±0.008	0.42±0.01	0.36±0.03	0.42±0.01	0.37±.04	0.44±0.04
Ex K me/100 gm	4.34±0.29	3.42±0.42	4.48±0.32	3.02±0.02	4.05±0.21	3.05±0.07
Ex Ca me/100 gm	16.18±0.13	16.78±0.28	16.4±0.56	16.7±0.0	16.8±0.21	16.7±0.14
Ex Mg me/100 gm	2.08±0.11	2.60±0.27	2.04±0.21	1.95±0.21	2.05±0.07	1.70±0.14

Source: Field survey.

Table 2.31 shows that although concentration of sodium was higher in the unprotected lands calcium content did not differ from that of the protected land. It was around 16 me/100gm in both protected and unprotected lands of both the villages. Therefore, soils in both types of lands of both villages did not appear to be on the wrong side of sodium content, as calcium deficiency was not observed. Hence, it is not always true that higher concentration of sodium necessarily leads to calcium deficiency in soil as shown by Threshow (1970). It depends on the degree of concentration of sodium and availability of other cautions. Agarwal *et al.* has found similar relationship between sodium with potassium and calcium (Agarwal *et al.*, 1982).

Commentary

The ecological analysis made above shows how a degraded pastureland ecosystem was restored through adoption of silvipastoral mode of grassland development. It also shows how improved status of the ecosystem in terms of floral diversity and soil characteristics was maintained despite persistent drought that the state witnessed for three years or so. A much higher availability of nitrogen, phosphorus and potassium in the protected sites as compared to the unprotected ones accompanied by tree species such as *Acacia, Prosopis, Salvadora, Dalbergia, Inga, Tecomella* and grasses such as *Cenchrus, Lasiurus, Dichanthium,* etc. sufficiently indicated how effective and appropriate this mode of development was.

These species are important not only for forage production but also for improving soil fertility and changing soil properties favourably.

Apart from the natural benefit that accrued to soil due to litter fall from trees, leaves of leguminous trees that were naturally used by the pastureland, partially substituted fertiliser nitrogen (Murthy *et al.*, 1990). The higher nitrogen content and organic carbon in soil as found in the protected lands can be cited as some examples in this regard.

One of the noticeable changes that took place in the pastureland was that the vegetative cover in it created friendly environment conducive for growth of plant species. This was possible to achieve due to the increased moisture content in soil and humid substances through litter fall, as mentioned above, and, of course, restricted cutting of grass approximately four inches above the ground. Survival of vegetation in the same area and grass production despite having no rain for three years explains the enhanced resilience that the grassland ecosystem achieved due to institutional intervention (Holling, 1973; Holling and Meffe, 1999). However, this was absolutely missing in the unprotected grazing land.

Along with the regulatory pattern of use of grassland, one may take note of the conservation steps that did not allow run-off to flow down and soil to erode. On the unprotected land, no such step was taken, which was why neither was the soil conserved nor moisture retained. The soil was gravelly and stony, while it was yellowish-red and sandy loam on the protected land. All this made considerable difference in the nutritional content of the soil that increased consequent to the presence of biomass content in the soil. A symbiotic relationship between soil quality and vegetation cover was established. For grass provided adequate canopy against the action of rain drops. It obstructed easy flow of water through adventitious roots and bounds soils (Hazra, 1995). As a result, neither was there any indication of decline in soil nutrients and humus nor was loss of soil caused due to wind and water run-off. Such a relationship assumed great importance in the management of dry land ecosystem (Dasgupta, 2001b).

Another important aspect of improvement in the ecosystem was the growth of native plant and grass species. Some of the grass species such as *Cyanodon barbasi, C. dactylon, Cenchrus ciliaris, C. setigerus* and *Lasirus sindicus* were grown on the pastureland. They were observed in patches in association with perennial tree species. Common plant species grown on the protected land were *Acacia senega, A. nilotica, A.*

leucaphaloe, Prosopis cineraria, Commiphora wightii, Dalbergia sissoo, Salvadora persica, S.oleoides, Tecomella undulata, Ailanthus excelsa, Anogerissus pendula, Cordia gharaf, Euphorbia caducifolia, Calotropis procera, Zizyphus mauritiana, and *Z. nummularia*. Floral diversity to such an extent enhanced the capacity of the ecosystem to deepen its resilience.

In the unprotected land, plant such as *Aristida adscensionsis* was observed predominantly. It was an indicator of overgrazing and erosion. As trees surround protected pastureland, adverse effects of the particulate matter from mining are minimised. However, this was not so with the unprotected land due to the absence of vegetative cover.

The improvement in the ecosystem can further be attributed to the silvipastoral development. For silvipastoral development improves natural variation through combining good grass, legumes and variety of tree species. This enhances the value of the protected *vis-à-vis* unprotected grassland (Hazra and Singh, 1992; Hazra *et al.*, 1994; Yadav and Rajora, 1995). Trees supported growth of small plant species (herbs, grass, etc.) by way of providing shades to the latter that helped in reducing evapo-transpiration. The water that was saved was captured by small plants and naturally utilised for their growth. Although the ecosystem was revived in the protected land, it was not lush green. However, it exhibited accumulation of strength for survival of trees and grasses even in the event of drought for three years consecutively. It happened due to circulation of natural cycles in the protected land that witnessed microbial reactions (biogeochemical cycles such as N, P, and S cycles, etc.) It is needless to mention that biogeochemical cycles perform better owing to the presence of organic content and moisture in soil.

The improved status of grassland ecosystem of this village was maintained at a low level of regenerative capacity. However, the regenerative capacity was sure to witness definite improvement under favourable climatic conditions due to the presence of biogeochemical cycles. Rejuvenation of regenerative capacity of the ecosystem of unprotected land might be difficult to achieve for such cycles and vegetative covers were absent there.

The role of protection of grassland through fencing in enhancing floral diversity must not be ignored. It helps in combining good (palatable) grass, legumes and a variety of tree species. Several studies show that such a unique combination enhances the quality of protected pastureland *vis-à-vis* the unprotected one (Hazra and Singh, 1992; Hazra *et al.*, 1994; Yadav and Rajora, 1995). However, these studies were conducted on experimental sites. The present study goes further to show that the quality of the same can improve greatly even if local communities were to manage these lands under new institutional practices. Needless to mention, the village community, including the poor, no longer faced fodder crisis and was able to maintain livestock as a sustainable source of livelihood even in perpetual drought conditions. The critical need for revival of the pastureland ecosystem in a state like Rajasthan is inescapable. While the state cannot dispense with its development implications, appropriate measures must be taken to restore vast stock of degraded pastureland in the light of the lessons that one learns from the present study.

IV

INSTITUTIONAL PERFORMANCE: A SCENARIO ANALYSIS

The methodology followed to conduct the scenario analysis has already been indicated in Chapter I and explained in Annexure 3. As many as 14 independent variables X_i (i =1,2,3...n) that are incorporated into the present analysis are shown in Table 2.32. All these variables are expected to influence the institutional performance directly or indirectly. Many of them are chosen in conformity with the design principles as constructed by Ostrom for collective action to sustain (Ostrom, 2000). However, those, which are missing in the design principles for example equity (that defies inequitable enforcement of rules perpetrated by the powerful), landholding pattern (an indicator of disadvantaged and disadvantaged members of the community), etc. are also incorporated. The landholding pattern has a bearing on the power relations in rural India and more particularly in Rajasthan. A relatively higher concentration of land and other resources in a few hands gives more power to them to perpetuate their domination over the landless and the poor who have less command over land and other resources. Despite such heterogeneous economic base of the households in the village and concentration of economic power in a few hands, the issue is whether institutional outcome could be equitable and participatory through collective action or not. It is needless to mention that it is the equity that finally ensures institutional sustainability. If it is inequitable, resentment of the deprived ones is bound to accumulate. This may, at some stage, lead to result in breakdown of the institution. These give enough justification for the independent variables to enter in the model.

Table 2.32 shows that only eight variables were statistically significant in influencing institutional outcome either positively or negatively in respect of both villages. While the sample size of Chhota Saradhana was 50 households, it was 120 in respect of Singhariya village (60 per cent of the total households of each village). These variables were chosen on the assumption that they can sufficiently explain failure or success of the institutional outcomes (Agrawal, 2001; Damodaran, 1991). Responses elicited from the sample households on the variables as shown in Table 2.32 have been taken

Table 2.32

Results of Significance Test of the Parameters

Parameters	Type of response
Regular attendance in the meeting *	Yes=1, No= 0
Effective participation in decision making process*	Yes=1, No= 0
Awareness about intervention*	Yes=1, No= 0
Women's participation in local institution##	Yes=1, No= 0
Equity in enforcement of rules*	Yes =1, No= 0
Freeriding*	Yes =1, No= 0
Satisfied with the distribution pattern##	Yes= 1, No= 0
Desiring change in the existing rules*	Yes= 1, No= 0
Conflict between villagers and the members of the VDC ##	Yes= 1, No= 0
Improvement of living standard of the household through pastureland management ##	Yes =1, No= 0
Fulfillment of the household requirement through Pastureland development ##	Yes = 1, No= 0
Landholding pattern##	Landless=1, Marginal/Small =2, Medium/Large =3
Membership in the committee*	Yes=1, No=0
New decision against the old ones*	Yes=1, No=0

Notes: * Significant ## Not Significant

Goodness-of-Fit of the model as explained in equation (iii) is tested by following Hosmer and Lemeshow Goodness-of-Fit Test.

into consideration for analysis. It is needless to mention that the only significant variables are incorporated into the backward (LR) logistic regression model. To determine their validity and relevance, this model is used on the independent variables that are responsible for probability (Y=1) to result in.

Analysis of Probable Scenarios

Before estimating the results of the probability (Y=1), the values of the independent variables (predictors) that are found significant in Table 2.32 are estimated and shown in Table 2.33.

Table 2.33

Estimated Value of the Parameters

Parameters	*Estimates*	
	Chhota Saradhana	*Singhariya*
Regular attendance in the meeting	8.3445	0.353
Membership in the Committee	1.5150	0.204
Awareness about the intervention	0.0061	0.795
Effective participation in decision making process	-2.3545	0.200
Equity in enforcement of rules	0.0168	0.065
Freeriding (no compliance)	0.2273	-0.492
Desiring change in the existing rules of management about pastureland development	1.1856	-0.794
New decision about pastureland development against the old one	0.3211	-0.017

As explained earlier, objective is to indicate likely scenarios regarding institutional outcome that each village under study may witness under different conditions. Different conditions are represented by different combination of independent variables as shown in Table 2.33. It may be worthwhile to know whether the same combination of independent variables generates different scenario of institutional success and failure in both villages during the same time. Such an exercise helps one to identify the best possible combination of independent variables (predictors) that yield highest achievable institutional success. Table 2.34 shows the goodness-of-fit of the parameters, which are estimated in Table 2.33. It shows that the parameters that are finally included in the model are statistically significant, implying thereby that the predictors are chosen correctly for the estimation of the probability of institutional success.

Table 2.34

Results of Goodness-of-fit Test of the Model

Chhota Saradhana	*Singhariya*
Estimation terminated at iteration number 9 because log-likelihood decreased by less than 0.01 per cent.	Estimation terminated at iteration number 3 because log-likelihood decreased by less than .01 per cent.
Cox & Snell –R^2 = 0.341	Cox & Snell -R^2 = 0.080
Nagelkerke –$R^{\wedge 2}$ = 0.523	Nagelkerke -$R^{\wedge 2}$ = 0.118
Goodness-of-Fit = 29.686	Goodness-of-Fit = 115.936
$x_{(8)}{}^2$= 20.833 (p >0.05)	$x_{(8)}{}^2$ = 9.819 (p=0..3336)
Measure of Association η=0.582 (η^2 =0.339)	Measure of Association η=0.229 (η^2 = 0.53)

Note: R^2: Coefficient of determination.

Scenarios of institutional success in terms of its outcomes were worked out within the probability of 0 to 1 and shown in Table 2.35 for Chhota Saradhana and Table 2.36 for Singhariya.

Table 2.35

Probable Scenarios of Institutional Outcome under Different Conditions in Chhota Saradhana

Variables	*Scenarios*						
	A	*B*	*C*	*D*	*E*	*F*	*G*
Regular attendance in meeting	1	1	1	1	1	1	1
Member of the VDC	1	1	1	1	1	1	0
Awareness about intervention	1	1	0	1	1	1	1
Effective participation in decision-making process	1	1	1	0	1	0	0
Equity in enforcement of rules	1	0	1	0	1	0	0
Freeriding (no compliance)	0	0	0	1	1	1	1
Desiring Change in the existing-rules related to management-of pastureland	0	0	0	0	1	1	0
New decision about pastureland-development against the old one	0	0	1	0	1	0	0
Probability of Success	**0.850**	**0.810**	**0.768**	**0.349**	**0.190**	**0.141**	**0.106**

Table 2.36

Probable Scenarios of Institutional Outcome under Different Conditions in Singhariya

Variables	*Scenarios*						
	A	*B*	*C*	*D*	*E*	*F*	*G*
Regular attendance in meeting	1	1	1	1	1	1	1
Member of the VDC	1	1	1	1	1	1	0
Awareness about intervention	1	1	0	1	1	1	1
Effective participation in decision-making process	1	1	1	0	1	0	0
Equity in enforcement of rules	1	0	1	0	1	0	0
Freeriding (no compliance)	0	0	0	1	1	1	1
Desiring change in the existing rules-related to management of pastureland	0	0	0	0	1	1	0
New decision about pastureland-development against the old one	0	0	1	0	1	0	0
Probability of Success	**0.789**	**0.778**	**0.623**	**0.636**	**0.503**	**0.441**	**0.588**

Both Tables 2.35 and 2.36 show seven scenarios indicating different levels of probable achievement of institutional success for Chhota Saradhana as well as Singhariya under different combinations of independent variables. In other words, seven combinations of independent variables represent seven scenarios. Each scenario represents different levels of institutional outcome.

Table 2.35 shows scenario A as the best possible one that yields maximum attainable institutional success for Chhota Saradhana in term of its outcome under the given combination of predictors. Chhota Saradhana was able to achieve this under the present management regime. The probability of success of institutional outcome in this case is 0.850 under the condition when the existing rules are not allowed to change and no new decision is taken against the earlier decision taken by the community. Of course, a high level of compliance or absence of freeriding is an important factor responsible for appropriate combination to emerge and generate scenario A. It definitely plays significant role in achieving highest level of institutional success.

In any case, the most significant factor that generates scenario A and needs to be underlined is equitable enforcement of rules of management of conservation of village pastures (Baland and Platteau,

1997). This is what we observed earlier while analysing the extent to which community members across all landholding groups in Chhota Saradhana village gained access to the improved resource base in terms of actual collection of grass and fuelwood. It is significant because probability of institutional sustainability will go down if the village confronts inequity as shown by scenario B. In the absence of equity, the probability of institutional success goes down as is indicated by scenario B in Table 2.35. In scenario B, value of equity assumes 0 instead of 1 given the values of other predictors same. The probability of institutional success resultantly declines to 0.81. This shows how significant is the role of equity in bringing higher level of institutional success. Its importance can be shown in another way also. For example, if other predictors assume the value of 0 and equity assumes value 1 in scenario B, the probability of success will not be as low as indicated by B. In other words, one cannot dispense with equity in enforcement of rules (equitable distribution of environmental entitlements) if the highest level of institutional success is to be achieved and ensure institutional stability in the long run. Although in terms of magnitude, decline in the probability of institutional success is low, it is desirable for one to concentrate on the direction of the change.

Institutional successful performance may decline from scenario A to scenario C if equity is not accompanied by awareness about intervention and the absence of unpalatable decision even within the existing rules. Scenario C shows a lower probability of institutional success with a value of 0.768 when awareness is 0 and new decision is taken against the institutional practices after renovation.

The regulatory behavioural pattern about the use of grassland of the community at large was well established and practised for last 12 years or so. Now, if any new decision is taken that goes against the established behavioural norms of using the pastureland, outcome may effect lower institutional success. It means that the established norms are useful and, hence, acceptable to the villagers at large. Similarly, correct understanding of the nature of intervention is an important prerequisite for higher level of institutional success that, in turn, ensures institutional stability. Any combination of the predictors below scenario C yields institutional success below average. Hence, such combinations of factors may be said to be weak enough to ensure institutional success and stability for resource conservation in a village like Chhota Saradhana.

In the case of Singhariya, what is intriguing to observe is that the same combination of factors yields different level of institutional success from that of Chhota Saradhana. Here again, there are seven probable scenarios, each of which correspond to different level of institutional success (Table 2.36). Table 2.36 shows that the highest probability of institutional success under the same combination of predictors as in Chhota Saradhana is 0.789.

Singhariya is one of those villages where social networking and relations are largely determined, as mentioned earlier, by hegemonic power relations. While institutional renovation and its implementation for more than a decade have made it possible to improve the conditions of village pasture, environmental entitlements was inequitable. In such a situation, the probability of institutional success as achieved by Singhariya at present corresponds to scenario D with a value of 0.636 in Table 2.36. It means that the present level of institutional success in Singhariya as indicated by scenario D is much lower than what can be achieved as indicated by scenario A. In other words, if Singhariya where to achieve scenario A that indicates highest probability of institution success which is already achieved by Chhota Saradhana, it has to ensure equity in addition to higher level of compliance. Besides, effective participation in decision making process needs to be ensured even if village community is completely aware of the nature of intervention being made for conservation and growth of village pasture. As the field survey revealed, decisions related to the management of village pastures were never taken through consensus. It was mostly a few affluent farmers who acted as change agents of the village and took decision.

Inequity which was built in the process of institutional renovation in Singhariya is so significant that even if equity is presumed to exist as shown in Table 2.36 and generates scenario A, institutional success is much lower as compared to that of scenario A of Chhota Saradhana. It is so because the present scenario in Singhariya as shown by D is too weak to raise the probability of institutional success to the level achieved by Chhota Saradhana as represented by scenario A of this village. It can rise only up to the level indicated by scenario A of this village as shown in Table 2.36.

The interesting aspect is that the probability of institutional success below average (0.50) under the same combination of factors is much less promising for Chhota Saradhana as compared to that of

Figure 2.7

Probable Scenarios of Institutional Performance

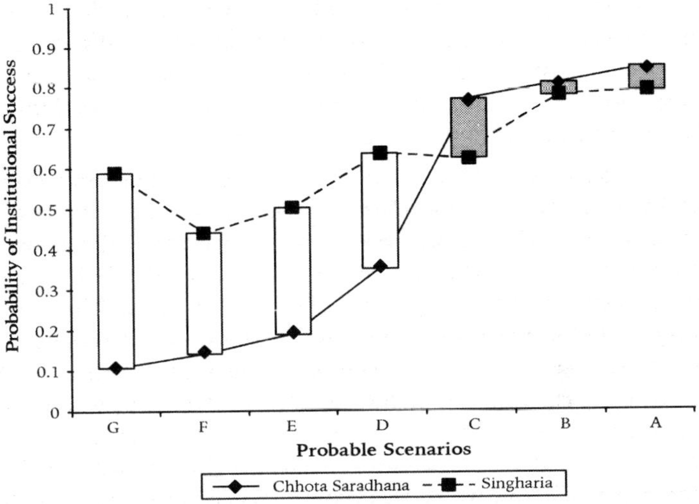

Singhariya. This is shown clearly in Figure 2.7 where the estimated values of the probability of institutional success are plotted. As can be seen from Figure 2.7, difference between the probable scenarios of the two villages under the same combination of factors goes wider after scenario B. Moreover, the probability of institutional success with respect to Chhota Saradhana successively becomes much lower after scenario C. The successive scenarios after C (E, F and G) do not represent institutional performance of Singhariya as poorly as that of Chhota Saradhana. However, combinations of the factors as shown by scenario A, B and C indicate more institutional success for Chhota Saradhana as compared to that of Singhariya.

Of course, scenarios such as E, F and G represent low institutional performance in respect of both the villages as compared to the scenarios such as A, B and C. But, what is important to observe is that institutional success, as Figure 2.7 shows, in respect of Chhota Saradhana falls steeply below average under any combination of factors while this does not happen in the case of Singhariya. This may lead to open a fresh debate on the choice of intervention in the field of natural resource management in two different social contexts. (However, one must bear in mind that this debate is different from that of private *versus* community management). While a few affluent farmers were the ones who held the key to power relations in Singhariya, discrimination was perpetrated against the disadvantaged ones in gaining more access to the improved resource base. However, no such discrimination was found to exist in Chhota Saradhana and decision making process was never authoritarian like that of Singhariya.

The question is, therefore, whether one should opt for a management regime that is run by a few and imposed on others or a regime that has evolved through consensus building. In Singhariya, the authoritarian management of a few definitely contributed to the improvement of the pastureland ecosystem. However, such an improvement was desired by VDC because it largely served the interest of the powerful households. Institutional outcomes in such a situation, one may argue, are sub-optimal. Despite this being so, VDC never allowed its institutional performance to fall below average. If it does so, it is the powerful ones whose interest would be impaired mostly. Comparatively very low value of institutional success (below average) in Chhota Saradhana may be explained by the fact that once democratic norms are violated with low compliance as represented by scenario D, E, F, G in Figure 2.7, institutional performance is likely to be much lower than average. However, the significant aspect of intervention through consensus building in this village is that it ensures institutional success of a very high order that authoritarian management, as that of Singhariya, can never achieve.

Authoritarian approach to institutional development is inherently constrained to achieve as much institutional success as it is possible to achieve through consensus approach. Singhariya that follows authoritarian approach for development of village pastures can move upward from scenario D (present level of institutional success) to A (potentially higher level of institutional success) in Figure 2.7 by means of ensuring equity of institutional outcomes, effective participation of all, etc. However, still, the level of its potential institutional success is lower than that of Chhota Saradhana. It will continue to remain so, as long as the village VDC determines the management norms within the ambit of hegemonic power relations. Under such circumstances, even if equity of institutional outcomes and effective participation, etc. are ensured, they will be mere lip service. This may make transition from D to A for Singhariya. However, it may not make the same from D to A of Chhota Saradhana unless VDC follows the underlying principles of consensus building for conservation and development of village pastures.

The institution may sustain if the transaction cost is too low for any collective action taken. Lip service to equity and other democratic practices by the authoritarian approach can never eliminate the discriminatory enforcement characteristics of the rules of management of pastureland permanently. It may yield social tension that may be seemingly absent in the short run. In such a situation, the transaction cost of bringing unity among the villagers and collective action may be untenable. The sustainability of the new institutional practices may then be doubted.

V

Conclusion

Local institutions have the potential to offer a development approach to the management of common property resources even in heterogeneous social and economic conditions if they are renovated. In such a situation, the institutional outcomes are less likely to be concentrated in the rich. The case of Chhota Saradhana reveals that the poor can have equal access to the improved resource endowment and share the institutional outcome equally. This happens if disadvantaged sections of people across all social groups polarised even when not all including rich and poor are equally vulnerable to natural calamity such as drought. Of course, one has to recognise the role of external intervention; it is imperative that it does not develop and perpetuate a dependency syndrome. Chhota Saradhana is no longer in such a syndrome. Furthermore, flexibility that allows space for combining traditional and modern scientific knowledge for revival of the regenerative capacity of the grassland ecosystem should be considered as an institutional achievement. The division of pastureland into different plots allocated to different groups of households who are inclined to follow different norms of management of the land may eventually reduce the significance of institutional approach. While asserting against the possible backlash on the collective ideology of the community due to such division, a woman from the scheduled caste community says, "We are all benefiting from our protected pastureland for the last 10 years. This land has given incredible strength to the growing unity of the village community. We will never allow it to disintegrate." It shows how overriding importance is given to the institutional choice.

The local leadership had an immense contribution to it. It stabilised the institutional practices in a democratic manner in Chhota Saradhana and helped the villagers to form SHGs that eventually gave considerable strength to the poor to assert their right at each stage of their activities from participation in the VDC to accessing the village pastureland. However, the case of Singhariya provides a different lesson about pastureland management through institutional development. Here, the relatively affluent households provided local leadership. No doubt, their leadership had a definite contribution to the stability of the institutional practices; but it was much against the interest of the disadvantaged. Against this backdrop, institutional sustainability of Singhariya, as indicated by the scenario analysis was much less than that of Chhota Saradhana.

3

Rainwater Harvesting

Introduction

Rainwater harvesting, a traditional practice of trapping monsoon rainwater has gained serious attention in the development discourse in the recent past. The increasing importance of its revival is now echoed in almost all hydro-poor parts of the state of Rajasthan. In the absence of this practice when rainfall is more erratic than ever, groundwater aquifers are not sufficiently recharged to meet increasing demand. In such a scenario, the accompanying deforestation causes loss of topsoil and moisture content, and decline in agriculture production. Very often, these factors combine to push families to migrate.

These conditions dominated the rural economy of some water-deficient parts of the state, like Bhanwata-Koliyala and Samra villages of the Thanagazi block of Alwar district, during one of the severest drought years in the mid-80s. It was against this backdrop that the significance of the entry of Tarun Bharat Sangh (TBS), an NGO, in these areas could be underlined. Since then, TBS has entered as many as 700 villages in Ajmer, Alwar, Bharatpur, Bikaner, Dausa, Jaipur, Jaisalmer, Jodhpur, Sawai Madhopur, Tonk and Udaipur districts of Rajasthan (Pangare *et al.*, 2002). The rationale was to reverse the prevailing law of backwardness by reviving the traditional practice of harvesting rainwater and rehabilitating the intrinsic link between forest, water and agriculture.

This approach to rainwater harvesting was central to the intervention made in the Gangzhu province of China, where the biggest achievement was the removal of food insecurity that the people of this province suffered from almost perennially (Qiang, 2003). However, the Chinese intervention was not based on community management of common resources; rather the government collaborated with private household management. Several such examples can be cited the world over, including community management of rainwater harvesting with different technologies, with several states realising the importance of rainwater harvesting during the last two decades (Padre, 2002). While the efficacy of this age-old system is unquestionable in terms of material gains that accrue to the community, several conflicting issues that have serious development implications are surfacing. Prominent among them is how participatory and, therefore, sustainable is community management and how allocation of water is economically efficient and socially fair. As Kelman observed in the Brazilian context, these objectives are frequently in conflict (Kelman and Kelman, 2002).

It is in this context that this study examines the question of riparian rights of farmers downstream *versus* upstream. Should structures be built upstream without a proper understanding of their effect downstream and on the larger drainage system (Pangare *et al.*, 2002)? Should there be a system of priority of access based on some rationality (Kelman and Kelman, 2002)? And then, it is not only the question of equitable allocation of water between upstream and downstream, but, also, who benefits from the structures. This issue assumes considerable importance in view of whether there exists integrative and comprehensive land water management of the complete catchment, not upstream or downstream alone. For, in addition to the issue of inequitable access to the water, partial improvement of the ecosystem in the entire catchment never fits into the development paradigm that any interventionist seeks to justify.

The second set of issues is related more to community *versus* private initiative or combination of initiatives of both. It is in the context of community initiative that one must examine whether the intervention is mainly geared to promote participation of the users or the community as a whole (Pangare *et al.*, 2002). If it is for the first to the neglect of the second, conflict is bound to result and turn the intervention unsustainable. If the intervention combines the initiatives of both, it may be of great concern how gains accruing from the intervention are distributed to the stakeholders. The relevant question, therefore, is how the poor and the disadvantaged, including women, benefit from such interventions. The third set of issues pertains to the relational framework of the state and the community in that how does the state figure its role in the management of rainwater harvesting when water rights refer to both ownership rights and right to use water (Singh and Jain, 2002). There are several such issues to be probed and lessons learnt from, which may help in reversing the law of backwardness perpetrated in many parts of the state.

It is in this context that the present case study was undertaken to examine how revival of traditional practices of rainwater harvesting through community participation restored the agro-ecology and economy of Bhanwata-Koliyala and Samra villages. The study is organised in four parts with part I briefly describing the context in which intervention was made by TBS and examining the socio-economic and demographic profiles of the villages under study, part II focusing on the steps taken by the village community, part III examining the changes witnessed after the interventions and part IV concluding.

I

SOCIOECONOMIC AND DEMOGRAPHIC PROFILE

The Study Area: A Brief Outline

Bhanwata-Koliyala and Samra had a history of community initiatives in rainwater conservation through construction of small dams and *johads*. However, the sense of communal responsibility towards natural resources eroded over the years as community reliance gradually shifted to government for maintenance of these structures. The study areas had to witness a devastating drought in 1944 that took a heavy toll on human and natural resources (Shresth and Devidas, 2001). The depletion of forest resources reportedly began during that time when livestock resources became the major source of livelihood and there were no restrictions in place to regulate the grazing of animals. Overgrazing became the normal practice of the village community continuing until the TBS initiative in the mid-80s for reviving traditional rainwater harvesting practices.

Bhanwata and Koliyala were earlier hamlets of Bhuriawas village. In 1994-95, however, Bhanwata attained the separate status of a revenue village, though Koliyala continued to be a hamlet of Bhuriawas village.[1]

---

1. Koliyala was a hamlet adjacent to Bhanwata. Although it belonged to another revenue village, some of its agricultural lands were located in the same area as that of Bhanwata was located. This was the reason why geographical areas of both were taken together for interventions.

Villages of Thanagazi block are covered by flat-topped Aravali hills that run north to south ranging in height from 300-600 metres between which lie fertile valleys. The reason for seeking joint management of Bhanwata and Koliyala for rainwater harvesting was that the agricultural lands of most farmers of Koliyala were located adjacent to Bhanwata. Secondly, the grazing routes of animals, on which the villagers depended, were the same for both villages. Bhanwata-Koliyala and Samra were located upstream and downstream in the Arwari river basin. Apart from topographical differences, the villages had dissimilar formation of rock structures with slate in Bhanwata and quartzite in Samra (Map 5 for location).

Undulating terrain and limited availability of cultivable land restricted the scope for agricultural development here. For example, only 21 per cent of total geographical area of Bhanwata-Koliyala and 9 per cent of Samra were under actual cultivation during the late 1980s (before intervention).

Table 3.1 shows that as much as 87 per cent of the total area was unavailable for cultivation in Samra during 1989, while it was 68 per cent in the upstream villages. This type of land included forest, mountain, barren and uncultivable land, land put to non-agricultural uses and pasturelands (Map 6).

Table 3.1

Area under Actual Cultivation of Bhanwata-Koliyala and Samra in 1988-89

(land in hectares)

	Bhanwata-Koliyala			*Samra*
	Bhanwata	*Koliyala*	*Total*	
Total area	221 (100.00)	118.19 (100.00)	339.19 (100.00)	2057.00 (100.00)
Land not available for cultivation	151 (68.33)	80.08 (67.76)	231.08 (68.13)	1798.00 (87.41)
Land available for cultivation, but not cultivated	28 (12.67)	9.26 (7.83)	37.26 (10.98)	80.00 (3.89)
Land actually cultivated	42 (19.00)	28.85 (24.41)	70.85 (20.89)	179.00 (8.70)

Note: Area that falls under Koliyala has been taken out from Bhuriyawas revenue village and added with the lands of Bhanwata revenue village. This is done in consultation with the local *patwari*.

Source: *Tehsil* office, Thanagazi (Alwar).

Map 5

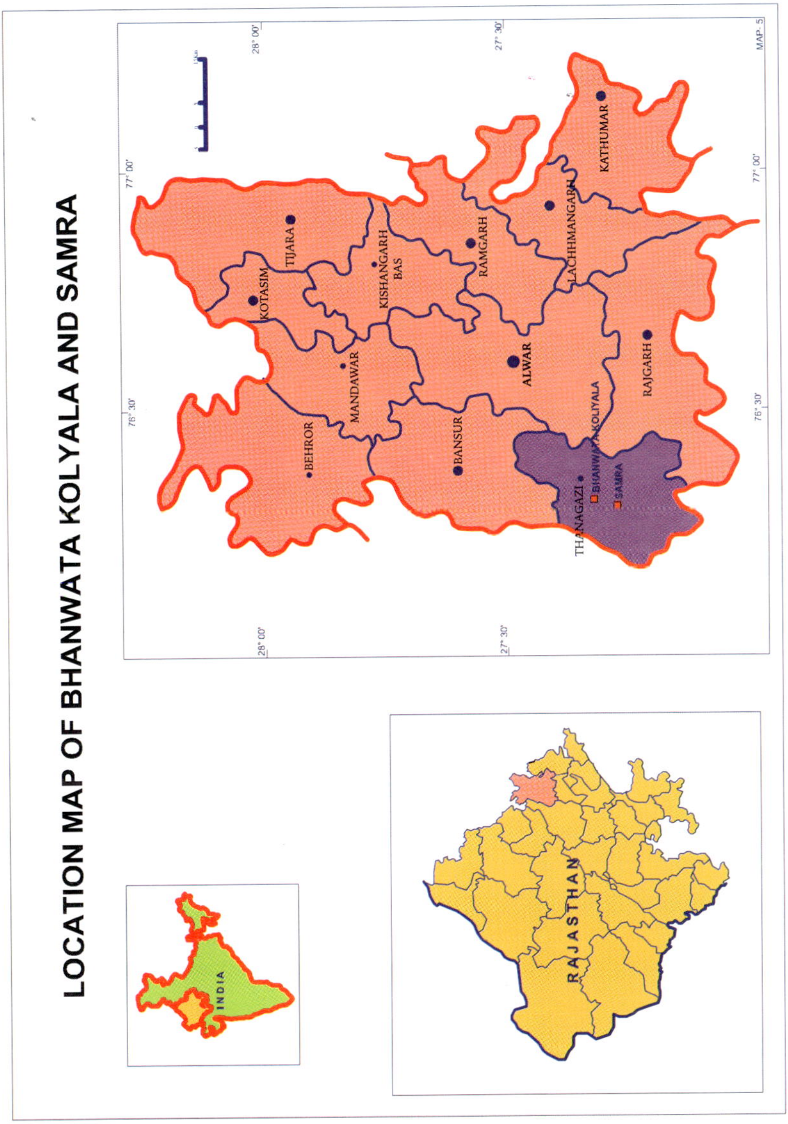
LOCATION MAP OF BHANWATA KOLYALA AND SAMRA
28° 00'
27° 30'
77° 00'
76° 30'
MAP-5
KOTASIM
TIJARA
KISHANGARH BAS
RAMGARH
LACHHMANGARH
KATHUMAR
MANDAWAR
ALWAR
RAJGARH
BEHROR
BANSUR
THANAGAZI
BHANWATA KOLYALA
SAMRA
INDIA
RAJASTHAN

Map 6

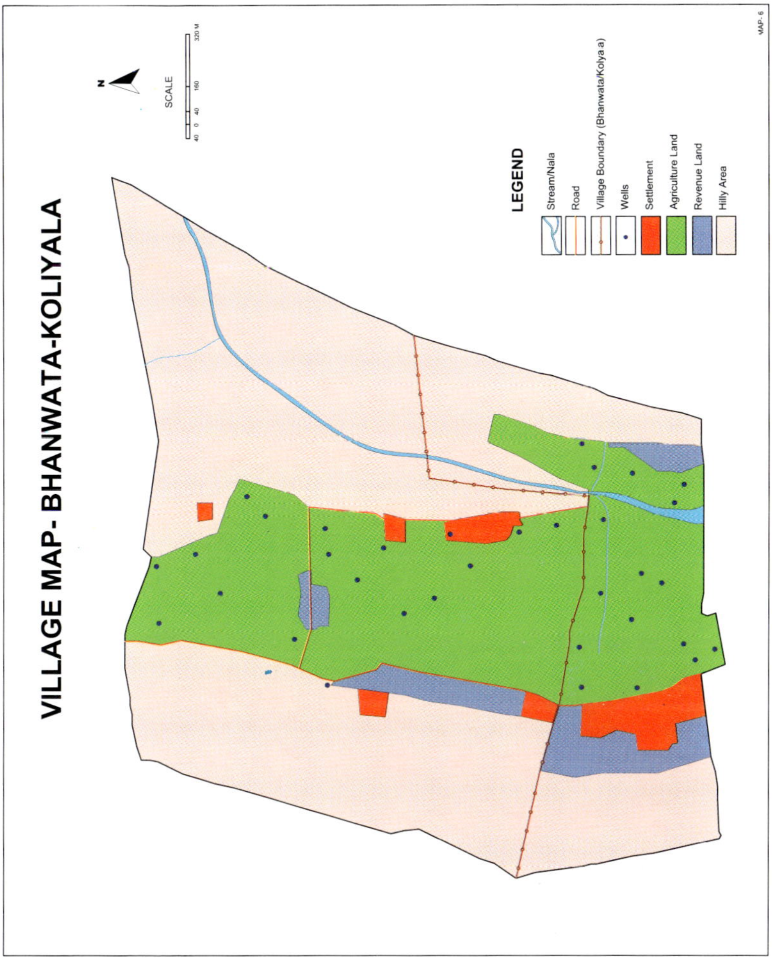
VILLAGE MAP- BHANWATA-KOLIYALA
N
SCALE
40 0 40 160 320 M
LEGEND
Stream/Nala
Road
Village Boundary (Bhanwata/Kolya a)
Wells
Settlement
Agriculture Land
Revenue Land
Hilly Area
MAP- 6

The limited availability of cultivable land with a growing population made adequate and sustainable livelihood difficult for the villagers. The stock of human population rose by 2.91 per cent annually in the upstream village (only Bhanwata), while it did so by 1.91 per cent in the downstream village during 1981-1991 (Table 3.2). The all-pervading effect of drought ensured that migration became the only way to survive for majority of the villagers. Apart from putting agricultural lands under pressure, the drought degraded forest and pasturelands and threatened the sustainability of the livestock economy. All these together reinforced the livelihood crisis of the villagers who realised that restoration of local natural resource base was the only route through which such a crisis could be mitigated. They turned back to the traditional practice of rainwater harvesting through collective action.

Table 3.2

Population Size of Bhanwata and Samra

Year	*Bhanwata*	*Samra*
1981	216	1316
1991	288	1590
2001	417	2195

Source: Population Census of India, Rajasthan.

Social Composition and Landholding Pattern

That social capital, particularly its content, plays a crucial role for collective action to succeed is widely recognised in development literature (Harriss, 2001). However, there may be a lack of insight into the constituents of social capital such as mutual respect, social networking, reciprocity, etc. among the social groups outside of the power framework. It is in this context that power relations must be addressed; the backdrop of social and economic forces to which the dominant social group(s), concentration of land resources and higher access to irrigation lend credible support. If we look at Bhanwata-Koliyala, amongst three social groups the *Gurjar* were the largest numerically and with the biggest landholdings by virtue of their numbers (Table 3.3).

Table 3.3

Social Composition and Landholding Pattern of Bhanwata-Koliyala

Caste	*Landless*	*Marginal*	*Small*	*Medium*	*Total*
Rajput	—	—	3	6	9
Gurjar	—	21	31	9	61
Balai	—	19	—	—	19
Total	—	40	34	15	89

Source: Field survey.

Significantly, the numerical strength rather than landholdings made considerable difference to the *Gurjars* in influencing the decision making process for the development of this village. For, though, at 11.3 *bighas* per household, the socially superior *Rajputs* had the largest landholdings, the *Gurjars*, at 5.20 *bighas* per household, were more assertive and the major players in any development activity. As the field survey revealed, they were also more active, as compared to others, in the rainwater harvesting initiated by TBS.

The *Rajputs*, who were major beneficiaries of the *jagir* system, were the first ones to settle down in this area. However, this was contested by the *Gurjars*, who claimed to have been the early settlers of this village. (The *Balai* community, the poorest one, was brought by the *Rajputs*.) In any case, the *Rajput* supremacy was diluted over the years due to continuous fragmentation of land within the family and lesser returns from agriculture with declining yields. During the field survey, it was argued by some that the Rajputs might regain their power in the years to come with the improved agricultural scenario. With the highest per capita landholding, they appeared to be the largest beneficiaries of the revival of traditional practices of rainwater harvesting (Shresth and Devidas, 2001).

The social composition in Samra, the downstream village, was somewhat different in that *Gurjars* was the second largest of all 11 social groups who settled here at different points of time. The early settlers, *Meenas*, were followed by *Gurjars* and the rest. With the rise in population and new social groups settling in, the *Meenas* and *Gurjars* shifted their dwelling place nearer to their agricultural fields. Other communities followed suit subsequently, which was why the geographical area of the village expanded in due course. The area of the village finally ended up with the formation of 17 hamlets (*dhanis*) spread within a radius of 20.57 sq km (Map 8).

Map 7

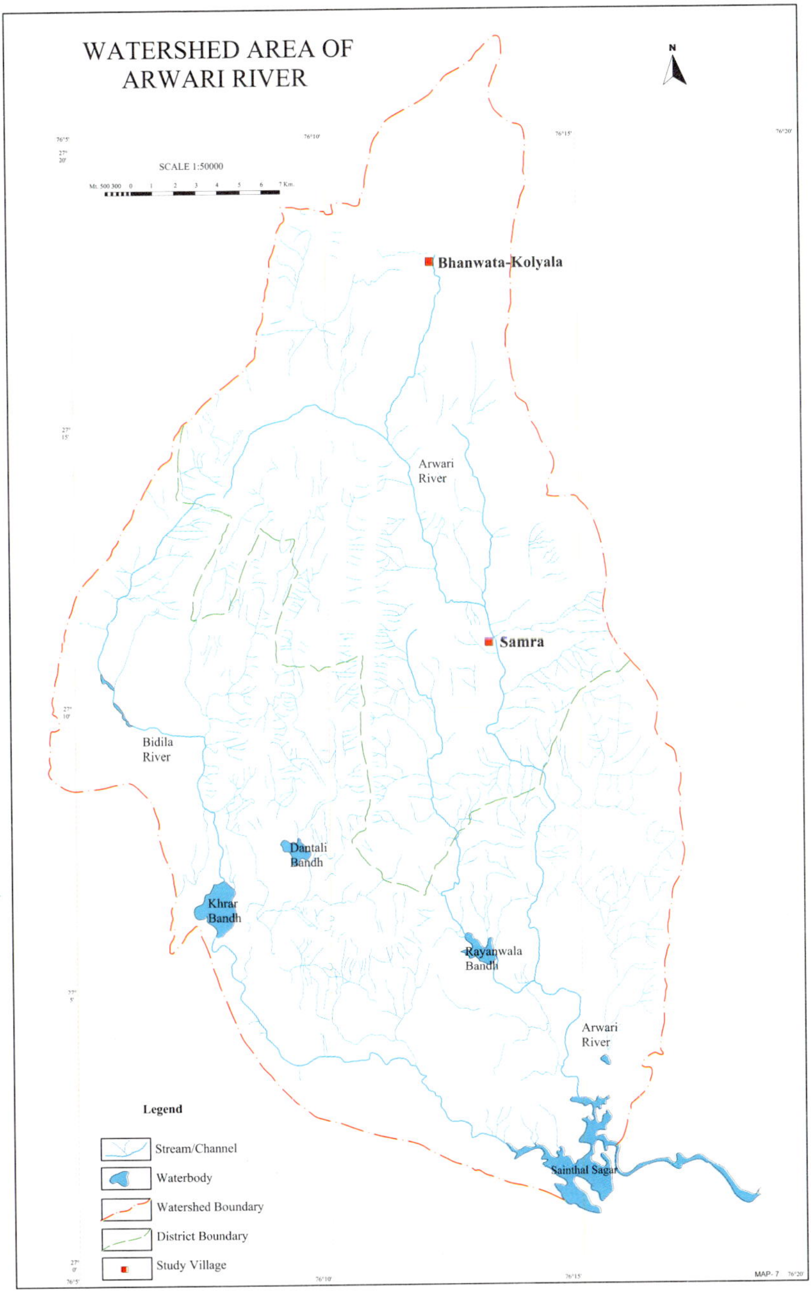
WATERSHED AREA OF
ARWARI RIVER
N
SCALE 1:50000
Bhanwata-Kolyala
Arwari
River
Samra
Bidila
River
Dantali
Bandh
Khrar
Bandh
Rayanwala
Bandh
Arwari
River
Sainthal Sagar
Legend
Stream/Channel
Waterbody
Watershed Boundary
District Boundary
Study Village
MAP- 7

Table 3.4

Social Composition and Landholding Pattern of Samra

Category	*SC*	*ST*	*OBC*	*Others*	*Total*
Landless	13	1	—	1	15
Marginal	23	48	61	1	133
Small	Nil	37	33	1	71
Medium	Nil	57	8	1	66
Large	Nil	2	Nil	Nil	2
Total	36	145	102	4	287

Note: SC-*Balai* & *Koli*; ST- *Meena* (*Zamindar* and *Chowkidar* together); OBC- *Gurjar, Swami, Jogi, Kumhar* and Others; General- *Brahmin* & *Bania*.

Source: Field survey.

Table 3.4 shows that the majority of the *Meenas* were medium farmers with two large farmers. Next came the OBCs (other backward castes), mainly constituted by the *Gurjar* community. It may be noted that these two communities were not competitive. They were rather complementary to each other, building a sort of power nexus within the village community. It was this combination of social forces that determined power relations in the village. No decision related to the welfare of the village community was ever taken without the approval of this power nexus. The members of the Scheduled Caste (SC) community complained that even when government initiated any development works such as the drought relief work, wage labourers were hardly drawn from their community. Rather, the village *patels* of the *Gurjar* community and the leaders of the *Meena* community were contacted, who normally preferred to provide wage employment to members of their own castes. The same practice was repeated when rainwater-harvesting structures were constructed.

One may tend to oversimplify power relations if they are examined without looking at the ownership pattern of agricultural land. In other words, one has to recognise the fact that households that were poor, including the landless, marginal and small farmers, might polarise cutting across their respective social backgrounds. This happens when rich farmers, irrespective of their caste background, gain at the cost of others including those who come from their own social milieu. The sustainability of collective action that the villagers here embarked upon more than a decade ago had much to do in the polarisation of all social forces, irrespective of the ownership pattern of land resources.

One may, therefore, ask why there was such polarisation that defied traditional power relations. For if the content of social capital fails to defy traditional power relations, institutional outcome brought about by intervention may turn out to be unequal (Ray and Sharma, 2004). One may then have reasons to doubt the sustainability of collective action. Before analysing these issues in the present context, it may be worthwhile to examine the steps that were taken by the villagers at the initiative of TBS to revive the harvesting practices of rainwater.

II

THE INTERVENTION

Entry of TBS

The Tarun Bharat Sangh entered the scene when the livelihood crisis was deepest and chose to undertake the formidable task of revamping social space beyond altruistic individualism. It found that the livelihood crisis of the rural community was rooted in the steady depletion of groundwater. Conservation of rainwater along with soil and forest was the way through which groundwater stock could be augmented. The message that went straight to the residents of Gopalpura, the village, where it launched its movement, was essentially revival of the link between human beings and nature. This was what was required to ensure generation of sustainable livelihood.

To demonstrate the impact of conservation of rainwater on the groundwater levels, a *johad* was constructed during 1986 and the result was quick (Shresth and Devidas, 2001). It worked. The simple activity had a spiralling effect on the process of sensitisation, which gained an overriding importance as a tool to create social space in the village community for water and forest conservation. Mutual trust and solidarity that had been pushed to the periphery of any social action rejuvenated during 1988 when the construction of the *johad* began in this village. Continuous dialogue with the villagers, promotion of participatory collective action, fuller use of collective wisdom, traditional knowledge and skill were the guiding principles that were followed by TBS while carrying out construction activities. It was decided to seek outside help only to catalyse (Patel, 1997).

Johads and Other Structures

Johad, a remnant of the rich heritage of traditional rainwater harvesting in Rajasthan, particularly around the Aravali hills, fell into disuse due to lack of maintenance. Most of the *johads* in the Aravali were built along the contours of mountain slopes for arresting and storing the flow of rainwater. *Johads* are earthen concave dams having high embankment on three sides. The fourth side is left open for the entry of rainwater. They are usually constructed on common lands at the foothills and are much smaller than anicuts. However, they were constructed on private lands also in the study areas. *Johads* improve moisture level at subsoil level in the fields, particularly downstream, and recharge groundwater. The recharged groundwater serves as the major source of irrigation in the region through wells (Samantray, 1998).

There were other structures that were constructed in the study area such as anicuts, *medbandhi, bandh* (small earthen dam), *khet talai* (farm pond), etc. Each one has a distinctly different purpose. For example, anicut, which is a stone dam, is constructed on riverbeds. It has the largest water holding capacity among all types of structures. The purpose of constructing *paal* and *medhbandi* is to conserve soil moisture. Built on edge of the field to prevent water from flowing out, these structures are constructed on private lands. Apart from constructing new structures, TBS took up the task of repairing old, dilapidated structures and transformed them into effective means of water conservation.

Private versus Community Structures

TBS encouraged construction of both private and community structures simultaneously. Table 3.5 shows that as many as 19 new structures were constructed to meet private needs while only 4 structures were constructed at the community level in Bhanwata-Koliyala. Similarly, 14 new private structures were constructed in Samra, while the number of community structures was 7. Also, the community mostly owned old structures that were constructed earlier, but repaired at the initiative of TBS in the recent past. This leads one to ask what did the renovated village institution have to do with the construction of private structures while promoting collective action for community welfare?

Table 3.5

Rainwater Harvesting Structures in Bhanwata-Koliyala and Samra

Structures	*Bhanwata-Koliyala*				*Samra*			
	Community		*Private*		*Community*		*Private*	
	New	*Old*	*New*	*Old*	*New*	*Old*	*New*	*Old*
Johad	2	4	—	1	5	19	1	2
Bandh	1	1	1	—	—	—	4	1
Anicut	1	0	2	1	2	1	3	—
Medhbandi	—	—	15	—	—	—	6	2
Baori	—	—	1	—	—	—	—	—
Total	4	5	19	2	7	20	14	5

Source: Field survey.

In view of the improvement of soil moisture and reduction of soil erosion, one may not see any reason why *medhbandi* should not be constructed on private agricultural land. However, while constructing them one has to bear in mind the topography of the area, limit up to which such construction works may be undertaken and their respective heights (more particularly, the slope of the land). These are certain crucial decisions that must be taken at the community level. They must not be left to the individual decision of the farmers who calculate their private gains without paying heed to the negative externalities of their actions.

The issue, therefore, is: does private construction of any type narrow down social space for collective action to sustain at the initiative of the renovated community institution. In a choice theoretic framework, one may look upon this problem in terms of social choice *versus* individual choice. If individual choice is a subset of the social choice in such a situation, one might see resolution of the conflicts even in the immediate terms. For, the individual even while pursuing individual goals would follow common rules of behaviour. However, interventions that allow individual choice to override community choice might finally lead to result in erosion of a natural resource base such as groundwater. The non-cooperative strategy followed by each individual will end up in a situation inferior to following the cooperative strategy (Sen, 1987).

One might, therefore, have reasons to doubt the sustainability of the village institution that consolidated social capital for restoration of natural resource base. In the event of large investment being made by TBS to raise as many private structures as possible, how does one expect community institution to become strong enough to maintain community structures? The study conducted by Indian Network in Participatory Irrigation Management (India, NIPM) observed that the large percentage of funds spent for constructing private structures could never contribute towards strengthening village institutions. (Pangare *et al.*, 2002). Besides, the cost-sharing method as followed by TBS did have an in-built tendency to reduce the scope of the poor to access institutional gain equally. For those private individuals who could afford to invest money to share the cost of construction would obviously go in for construction of water harvesting structures. Table 3.6 shows that dominant social groups in both villages under study owned the maximum number of structures—the *Gurjars* in Bhanwata-Koliyala, and *Meenas* in Samra.

Table 3.6

Private Ownership of Structures by Different Social Groups

Structures	*Bhanwata-Koliyala*			*Samra*		
	Rajput	*Gurjar*	*Balai*	*Meena*	*Gurjar*	*Others*
Johad	—	1	Nil	1	2	—
Anicut	1	2	Nil	1	2	—
Bandh	1	—	Nil	4	—	1
Medhbandi	1	14	Nil	6	1	1
Baori	—	1	Nil	—	—	—
Total	3	18	Nil	12	5	2

Source: Field survey.

As the field survey revealed besides, that, having largest number of private structures, it was the powerful *Gurjar* community that emerged as the main decision maker in Bhanwata-Koliyala for land and water management. The following will illustrate how intra-village conflicts related to conservation and management of groundwater pushed community interests to the back and reduced community institutions to ineffectual bodies.

- The agricultural land of the *Balais* was located upstream of the river channel towards the neighbouring village of Agar. The

Balais wanted to shift from Bhanwata and settle near their agricultural plots towards the Koliyala side. However, the *Gurjars* of both Bhanwata and Koliyala constructed *johads* in the area. The area adjacent was taken for plantation, therefore, prohibiting such settlement. Although the *Balais* made a case at the *panchayat* and *tehsil* office, they were thwarted by the *Gurjars*. When they again attempted to settle near Chariwala *johad*, the *Gurjars* lodged a complaint against them to the forest ranger (Pangare *et al.*, 2002). This problem that remained unresolved shows how the *gram sabha* was an ineffective and redundant body (Pangare *et al.*, 2002).

- When the President of India donated Rs. 1 lakh to the *gram kosh* on his visit to the village in 2000, it turned out to be a source of conflict between the residents of Bhanwata and Koliyala. While the former wanted to use this money for repairing Sankrawala *bandh*, the latter desired that the money should be utilised to repair Bhenruji *ka bandh*. It was basically a conflict of interest between the villagers—the lands of Koliyala were located downstream of Bhenruji *ka bandh* and the villagers of Bhanwata felt Sankrawala *bandh* could recharge their wells—although both these dams were common. The conflict was finally won by the *Gurjars* of Koliyala.

Gram Sabha

At the initiative of TBS, a village council, or *gram sabha*, was formed in Bhanwata-Koliyala. However, it practically became a forum to maintain the links between TBS and the villagers and to control the construction and maintenance of the rainwater harvesting structures. Although, the ideals with which the *gram sabha* was formed were laudable, it saw a decline over the years because it was co-opted. The *gram sabha* turned out to be more an affair of the socially dominant *Gurjar* community rather than that of the village community as a whole. With *Gurjars* having maximum representation in the council, the role of other social groups was reduced to insignificance (Table 3.7).

Even women members, who hardly participated in any meeting, were mostly from the *Gurjar* community. It appeared from the field survey that the *gram sabha* had nothing much to do with planning for

Table 3.7

Composition of Gram Sabha in Bhanwata-Koliyala

Caste group	Members* (Initially)	Members* (2001)
Gurjar	9	8
Rajput	0	1
Balai	0	0
Women		
Gurjar	2	4
Others	0	3
Total	11	16

Note: * Members include president, vice president and secretary also.
Source: Field survey.

construction of the structures in terms of location of the sites, height, etc. As mentioned above, it provided a forum to facilitate the entry of TBS to construct community structures based on the terms of agreements provided by TBS. While TBS entered in such contracts for private construction, it also did the same for the community construction. No concrete rules were framed by the *gram sabha* for water and soil conservation, which would eventually lead the village community to achieve self-reliance. Of course, villagers were made aware of the adverse consequence if the harvested water was utilised for irrigation; it was to be used only for drinking purposes. Some rules were framed for forest conservation, however:

1. Tree felling in the forest area along the slopes and catchment area of the river was prohibited.
2. If caught felling, a penalty of Rs. 21 was to be imposed on the offender, going up to Rs. 51 and Rs. 101 for second and third-time offences, respectively.
3. The use of axe inside the protected area was banned; only dry wood was to be collected.
4. An individual could collect wood for house construction or for fuel for wedding ceremonies with the permission of the *gram sabha*.
5. The *gram sabha* was to meet every month on *Amavasya* day.
6. Any other issue relevant to the village society could be discussed.

While the village community of Samra framed almost the same rules, they did not go down well with the dominant *Meenas* and *Gurjars*. Samra and its hamlets together had a traditional village institution represented mainly by a group of elders characteristically identified as village *patels* (major decision makers). Such an institution was active in settling intra-village disputes of any kind. TBS extended its cooperation to strengthen the institution only for forest conservation.

Table 3.8

Composition of Gram Sabha in Samra (2001)

Caste group	*Members*
Meena	12
Gurjar	5
Kumhar	2
Balai	1
Jogi	1
Swami (Brahmin)	3
Baniya	1
Total	25

Source: Field survey.

Although the initial years of conservation yielded positive results, Samra gradually witnessed collapse of conservation efforts in the subsequent years leading to depletion of forest cover perhaps because the rules framed were never followed by the villagers:

1. A penalty of Rs. 1100 to be paid for felling of trees in the protected area.
2. Selling of wood and wooden products made from felled trees to nearby mining areas was liable for serious penalty.
3. Anybody going to the protected area for animal grazing with axe was liable to pay a fine of Rs. 101.
4. Grazing of goats from the neighbouring villages was stopped.
5. Protection of forest must continue from the threat of tree felling in the neighbouring villages.

While one may not dispute, *per se*, the attempts made by TBS to build on existing village institutions for conservation of water, soil and forest, one may have to critically examine why they failed to take off. It is true that initially, for a few years after inception, the village community was greatly enthusiastic about going in for what TBS sought to achieve. However, such enthusiasm did not help social capital to deepen for collective action to sustain.

One of the serious limitations was that no cognisance was taken of the existing power relations or content of the social capital of the villages while the intervention took place and the *gram sabha* was formed. As discussed earlier, the *Meena* and *Gurjar* communities had the tradition of taking leadership in the villages. The villages also had a tradition of resolving conflicts, including land disputes, by these village leaders. This was especially true in the case of Samra. When construction of rainwater harvesting structures was initiated by TBS, these powerful social groups readily came forward and took the leadership. All renovations that were sought in the traditional institutions were co-opted by them only to see that *status quo* of the existing power relations remained undisturbed.

For example, most of the free riders in Samra came from *Meena* community. All of them were members of the *gram sabha*. However, they never paid penalty for their offences while the minority social groups had to pay up. Discriminatory enforcement threatened mutual trust in this village and resultantly other communities also stopped paying penalty. Its ramification on the collective action was so adverse that the *gram sabha* lost its essence in practice. The other aspect was that no concrete step was taken for construction/repairing of rainwater harvesting structures on community basis in the village. The irony was, as the documents on the proceedings of the meetings showed, proposals were made for construction, anicuts, etc. and individual household contribution discussed. However, the decisions taken were never implemented. The old practice continued of the leaders from the *Meena* and *Gurjar* communities taking all the decisions with the rest of the villagers being asked to follow.

In Bhanwata, the *gram sabha* was relatively more active, in the initial stages at least. People would collect every *Amavasya* in each month and hold discussions pertaining to forest conservation. However, the field survey revealed that this kind of collective forum hardly existed now in Bhanwata-Koliyala. This was admitted frankly by Arjun Gurjar, chairman of the Arwari Sansad and resident of Koliyala, "The *gram sabha* used to hold meetings regularly earlier. However, it has stopped now functioning regularly because it has failed to resolve several conflicts. It has become virtually non-existent." Almost similar observations were echoed by the members of the *Balai* community. According to them, the *gram sabha* did exist, but it had lost its relevance and no longer remained functional in the real sense. Only a few members were found to be serious about it. Not surprisingly, the incidence of freeriding had risen and no penalty reported to have been imposed.

The *gram sabha* also had nothing much to do with the village *panchayat*. A similar observation was made by the in-depth NIPM study on rainwater harvesting in several villages where TBS intervened. The study observed that *gram sabhas* that were found functional in some villages were more like water users' associations. However, they had no formal status for interacting with other government agencies (Pangare *et al.*, 2002). Besides, as the field survey revealed, there was a wide gap between what was proposed and implemented. According to TBS norms, *gram sabhas* had to constitute various sub-committees, including resource committee, construction committee, forest committee, *gram kosh* committee, grazing committee, water committee and women's associations. However, in practice, no such committee, except *gram kosh*, was found to be in existence in Bhanwata-Koliyala. *Gram kosh* also collapsed after a few years of operation.

Gram Kosh

The *gram kosh* (village fund) that was under the direct control and supervision of the *gram sabha* was envisaged as seed bank of the village in that the earnings of *gram sabha* would be kept under the supervision of *gram kosh* and utilised for the community. Villagers could borrow food grain from it and return with excess quantity at the rate of 5 kg per 40 kg of food grain borrowed. It is a form of interest, locally known as *Bandi* system. Besides this, each household had to contribute 5-10 kg of food grain to the *gram kosh* every year. Table 3.9 shows the total collection of food grain in the *gram kosh* received from the households of *Gurjar* community alone from 1991-1999.

Gram kosh ceased to exist after 1999. The entire stock that was collected was finally used for repairing the Bhenruji ka bandh as the *Gurjars* desired. As mentioned earlier, the rift that was created between the members of the executive committee on the use of prize money of Rs. 2 lakh (Rs. 1 lakh from the President of India and Rs. 1 lakh from the Ford Foundation for the initiative taken by the villagers for water conservation) was the root cause of this. Hereafter, the *Gurjars* lost interest in contributing to *gram kosh* anymore.

Table 3.9

Collection of Food Grains for Gram Kosh

Year	*Caste*	*No. of Households*	*Total Collection*
1991	*Gurjar*	44	220 kg
1992	*Gurjar*	44	220 kg
1993	*Gurjar*	36	180 kg
1994	*Gurjar*	40	200 kg
1997	*Gurjar*	28	220 kg
1999	*Gurjar*	30	300 kg

Source: Field survey.

However, the attitudinal change of the villagers towards the use of water was an important achievement of the intervention that must not be discounted. The household survey shows that majority of the sample respondents wanted to use the water collected in *johads*, anicuts and *bandhs* for drinking needs rather than irrigation (Table 3.10).

Table 3.10

Awareness about the Use of Water for Irrigation

	Bhanwata-Koliyala	*Samra*
Want to use	16	14
Do not want to use	37	59
Total	53	73

Source: Field survey.

However, it was dismal on the question of institutional performance (as long as the *gram sabha* remained active initially) against the backdrop of participation of the villagers in the decision-making process in the *gram sabha* (Table 3.11).

Table 3.11

Participation of the Households in the Decision making Process

	Bhanwata-Koliyala	*Samra*
Participated in meetings	15	15
Never participated in the meetings	38	58
Total	53	73

Source: Field survey.

Hence, participation in collective action in both villages was short-lived, temporary in nature and mostly guided by the immediate requirements for construction of water harvesting structures at the community level. One may argue that the importance of *gram sabha* no longer remained once the construction of structures was over. *Gram sabha* should have then been replaced by water users' association that did not require involvement of the entire village community. Even in this respect, none of the villages showed encouraging results.

In other words, the communal involvement of the villagers in any real sense at any level was now hard to find in these villages as far as conservation of soil, water and forest was concerned. While the rising trend of freeriding reduced forest conservation into insignificance, the management of community structures by social groups that developed vested interest in them, as it happened in the case of Bhenruji ka *bandh* in Bhanwata-Koliyala, was increasing. All these together entailed social costs in the form of declining social cohesiveness and environmental costs in the form of degeneration of the resource base. The equal importance given to private management of soil and rainwater harvesting weakened collective action at the community level. While institutional development, as a result, has lost its meaning one is yet to be certain when the degeneration of the resource base would reach the point of no return.

III

THE CHANGE

Development Contrasts: Upstream *Versus* Downstream

While traditional practices of rainwater harvesting were revived in both upstream and downstream villages, the issue is how did groundwater scenarios of both locations undergo a change and what was their development implication for: (1) irrigation and agricultural development; (2) off-farm activities including livestock and others; and, finally (3) biodiversity (floral).

Changing Groundwater Scenario

The rationale that one seeks to justify revival of traditional practices of rainwater harvesting in a drought prone area, needless to mention, may override other development initiatives. The practices are simple, ecologically sensitive, people friendly and proven through centuries old tradition. However, the colonial legacy that had a remarkable impact on designing development intervention pushed this tradition to the periphery (Shresth and Devidas, 2001). What underlies this powerful tradition is that it is based on the fundamental law of hydrology. It views hydrological cycle in its entirety in that no superficial distinction is made between ground and surface water (Chopra and Kadekodi, 1999). It recharges underground aquifers and makes groundwater reserves possible depending upon the nature of the former. Therefore, it is a 'dynamic resource' strictly from the viewpoint of a hydrologist (Chopra and Kadekodi, 1999). One may argue that the approach to development and management of water, as followed after Independence, failed to appreciate this basic principle of hydrology. In its absence, forces of economic change in the drought prone areas turned retrogressive as a sequel to repetitive ecological backlash. The synergy that existed earlier between ecology and economy broke down because water was a significant component of the former.

While one acclaims revival of traditional practices of rainwater harvesting as a step in the right direction, it is imperative to examine whether such a practice leads to uneven recharge of groundwater between upstream and downstream. In other words, does recharge of groundwater justify equitable allocation of groundwater between upstream and downstream when rainwater is harvested upstream of a river? Or, is it the 'law of the water jungle' that says, 'he who is upstream is allowed anything; whosoever is downstream better get used to it' (Kelman and Kelman, 2002).

It is in this context that the present study is undertaken to examine what happens to the status of groundwater availability through recharge in the downstream *vis-à-vis* upstream in the same river basin when rainwater is harvested simultaneously at both locations. Equally important is to examine its development implications especially for agriculture through intensification of irrigation at both locations. At the initiative of TBS, village communities of a large number of villages located upstream and downstream in the Arwari river basin in Thanagazi block of Alwar district brought back their traditional practices of rainwater harvesting. As many as 700 villages in 12 districts of Rajasthan were covered for conservation and management of water resources, which it was claimed, resulted in the regeneration of 6,500 sq km of land

and an increase in forest cover (Pangare *et al.*, 2002). Bhanwata (upstream) and the other one, Samra (downstream) were two such villages of Alwar district where village communities constructed a large number of rainwater harvesting structures besides repairing old ones during last one and half decades. The present exercise concentrates on these two villages and examines how revival of traditional practices of rainwater harvesting affected groundwater recharge and brought about change in agriculture.

Physiography and Drainage System

One of the most influential factors that determine groundwater recharge is the hydro-geological formation (rock structure) given the rainfall of the area. Villages of Thanagazi block, including Bhanwata-Koliyala and Samra, are covered by Aravali hills that run north to south and range from 300-600 m in height. The region has more or less flat-topped hills between which lie fertile valleys. Apart from topographical differences, the villages also had dissimilar formation of rock structures slate in Bhanwata and quartzite in Samra.

No significant difference is discernible between the specific yields of these rock structures. However, differences might occur in case faults develop in the rock leading to groundwater recharge increases. The upstream village of Bhanwata has small faults in its rock structure and its dips are more inclined towards the village. There is no fault in the rock structure of Samra. In other words, the capacity to recharge groundwater is 'naturally' under favourable condition upstream as compared to downstream.

The soil in both upstream and downstream villages is primarily brown light loam that varies from reddish brown to dark reddish brown. Bhanwata, a tiny village with a total number of 55 households, is located nearer one of the sources from where Arwari River originates. Samra, which is downstream, has 291 households.

One of the major sources of the river, as the drainage system shows, originates near Bhanwata and flows from north to south where it joins Santhal sagar (Map 7). The other source of the river originates near the village Agar. There are many tributaries joining Arwari river in Samra, of which two are main tributaries—one flows from the Jhiri side (north of Samra) and the other one flows from the Piplai Jagannathpura side on the east of Samra. The main stream of the river stretches for 45 km covering a catchment area of 503 sq km. The river is not perennial one, it flows only during the rainy season. About 70 villages, 35 villages each of Thanagazi and Jamwa Ramgarh blocks, are situated in the catchment area of this river for last several centuries.

Besides repairing the old ones, different types of structures, including *johads*, bunds, anicuts, *paal/medhbandi*, *talai*, *talab*, etc. were constructed at different points of time in both villages. Bhanwata-Koliyala had 30 such structures within an area of 3.39 sq km, while Samra had 46 structures in an area of 20.57 sq km. Individuals and the village community owned both these structures.

Estimation of Groundwater Recharge and Availability

Guidelines recommended by the Groundwater Estimation Committee appointed by the Ministry of Water Resources, Government of India in 1997 were followed in order to estimate groundwater recharge in both villages.[2] The 1997 guidelines were a revised version of the Ministry's recommendations in 1984. The revised version is illustrated briefly in the following model (Annexure 1).

Model

Several factors were taken into consideration for the estimation of groundwater recharge in each year from 1988 to 2001. These included the geographical area of the village, fluctuations in the groundwater level, crop yield, drafting of groundwater (agriculture and domestic), and rainfall during monsoon and non-monsoon period etc.[3] Data on specific yield of groundwater with the given underground rock structure of both villages was collected from groundwater department, Government of Rajasthan—it was 0.03 mcm (million cubic metre) per sq km of the area in both villages. Besides, new variables were generated from the basic data and incorporated into the model.

One must bear in mind that recharge of groundwater of any year is assessed on the data available of each factor in the previous five years. This is done to normalise the recharge of that particular year as per the Groundwater Estimation Committee recommendations. Hence, the groundwater recharge in both villages after 1989-90 might be treated more as an outcome of the efforts made for harvesting

2. Guidelines were collected from the Groundwater Department, Government of Rajasthan. It was followed by a series of discussion with the hydrologists and geo-hydrologist of this department and Central Groundwater Board, Jaipur.

3. In the absence of data on rainfall in the villages under study, the same was collected for the block to which these villages belonged.

rainwater, given the same rainfall. The method for estimating total annual groundwater recharge for any assessment year is as follows:

TGWR = TRM + TRNM (i)

(TGWR=Total normal groundwater recharge, TRM=Total recharge during monsoon season, TRNM=Total recharge in non-monsoon season)

While,

TRM = ANMR + TRS (ii)

(ANMR= Accepted value of normal monsoon rainfall recharge, TRS=Total recharge from other sources)

Accepted value of normal monsoon rainfall recharge is obtained by rainfall infiltration factor (RIF), which is constant.

MRIF = PA * NMRA * R I Factor (iii)

(MRIF=Monsoon recharge by RI Factor, PA=Potential Area, NMRA=Normal monsoon rainfall of the assessment year, RIF=Rainfall infiltration factor)

The rainfall infiltration factor is 0.08, which is constant for Thanagazi block (Dixit *et al.*, 2001). The following linear regression equation is used in order to estimate normal recharge of groundwater during monsoon

Normal recharge = A* Normal monsoon rainfall + B (iv)

(Where A and B are the coefficient in the regression analysis)

While estimating the accepted value of normal monsoon rainfall recharge, RIF, as mentioned earlier, assumes an important role. As per the Estimation Committee, normally there is a deviation between normal recharge and recharge by RI Factor. However, if the percentage deviation of recharge by RI Factor from normal recharge lies between ± 20, recharge by RIF factor may be accepted. And, if percentage deviation is less than –20, then 80 per cent of recharge by RI Factor is acceptable. However, if it is more than +20, then 120 per cent of recharge by RI Factor may be taken into consideration.

Similarly, for groundwater recharge during non-monsoon period,

TRNM (Pre monsoon + post monsoon) = NRRNM + RFOS (*Rgw* + *Rt* + *Rsw*)

{NRRNM=Normal recharge from rainfall during non-monsoon season, RFOS=Recharge from other sources, Rgw=Monsoon recharge from groundwater irrigation, *Rt*=Monsoon recharge from tank and pond, *Rsw*=Monsoon recharge from surface water irrigation}

As per the Estimation Committee, if normal non-monsoon

Map 8

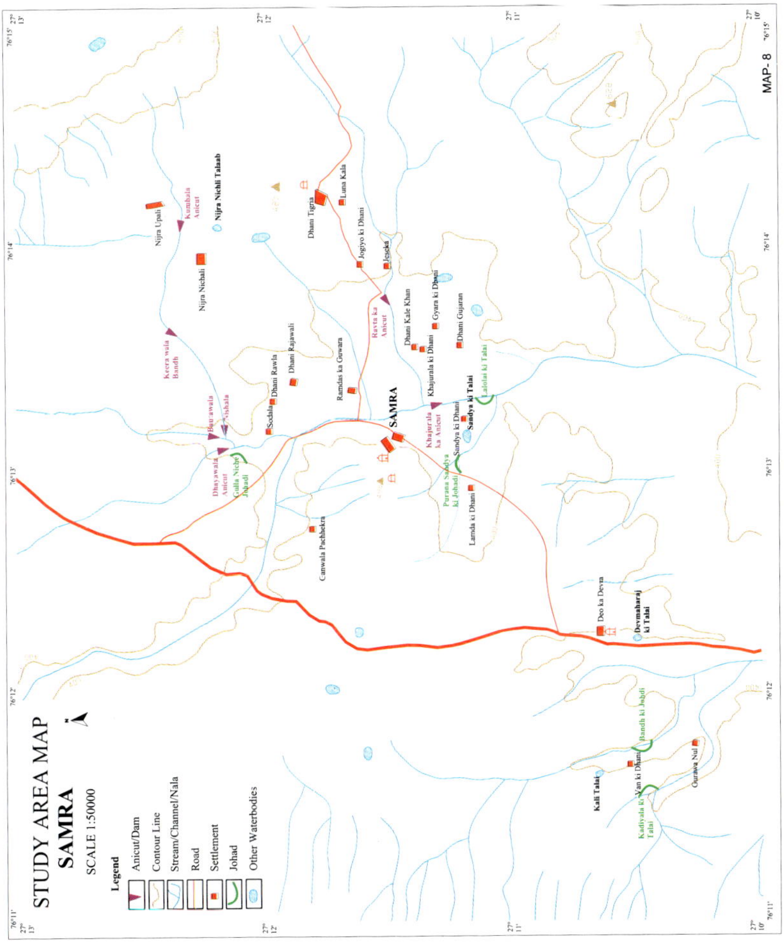
STUDY AREA MAP
SAMRA
SCALE 1:50000
Legend
Anicut/Dam
Contour Line
Stream/Channel/Nala
Road
Settlement
Johad
Other Waterbodies
MAP- 8
SAMRA
Nijra Upali
Nijra Nichali
Nijra Nichli Talaab
Dhani Tigria
Luna Kala
Jogiyo ki Dhani
Dhani Rawla
Dhani Rajawali
Ramdas ka Guwara
Dhani Kale Khan
Khajurala ki Dhani
Gyara ki Dhani
Dhani Gujaran
Canwala Pachhekra
Lamda ki Dhani
Deo ka Devra
Guráwa Nul

rainfall is equal to or more than 10 per cent of normal annual rainfall of the assessment year, then

TRNM = RIF * PA * NNMR. (v)

(NNMR=Normal non-monsoon rainfall)

If normal non-monsoon rainfall is less than 10 per cent of normal annual rainfall of the assessment year, it may be assumed that there is no recharge during non-monsoon season.

NGWA = TGWR–NDNM (vi)

SGWD = NAV–TD. (vii)

(NGWA=Net groundwater availability, NDNM=Natural discharge during non-monsoon season, SGWD=Stage of groundwater development, NAV= Net availability, TD=Total draft for all uses.)

Based on the methodology above, the total annual groundwater recharge per sq km, including monsoon and non-monsoon recharge, is estimated for both upstream and downstream villages and shown in Table 3.12. Interventions made by TBS in these villages took place during 1987-88. Hence, the status of annual groundwater recharges roughly little before 1990 and after can fairly explain the improvement in groundwater recharge. The estimated results are shown in respect of each year from 1988 to 2001 for both villages in Table 3.12.

Table 3.12 shows that groundwater recharge during monsoon, non-monsoon and total annual recharge was almost same for both villages before 1990. The total annual recharge of groundwater was 0.0733 mcm per sq km in Bhanwata during 1988-1989, while it was 0.0708 mcm in Samra. However, after 1989, the volume of groundwater recharge varied widely between both the villages in that it declined in Samra much below that of Bhanwata. One can make out its trend clearly from Figures 3.1a, 3.1b and 3.1c that depict the estimated results of monsoon, non-monsoon and annual groundwater recharge, respectively.

Figure 3.1c shows that the gap in the volume of groundwater recharge per sq km began increasing between upstream and downstream after 1989 when water-harvesting structures made an impact. It widened further in the following years as revival of rainwater harvesting intensified upstream. It may be noted that the new impounding structures were also constructed downstream after 1988. Despite this, the scale of recharge of groundwater was much lower here than upstream. Interestingly, the gap narrowed down during 2000-01 (Figure 3.1c). This seemed to have happened firstly, because these villages received more than the average rainfall during

Table 3.12

Annual Groundwater Recharge Including Monsoon, Non-monsoon in Upstream and Downstream

(1988-2001)

Year	Upstream (Bhanwata)			Downstream (Samra)		
	Recharge in Monsoon Per sq km (Mcm)	Recharge in Non-monsoon Per sq km (Mcm)	Total Recharge Per sq km (Mcm)	Recharge Monsoon Per sq km (Mcm)	Recharge in Non-monsoon Per sq km (Mcm)	Total Recharge Per sq km (Mcm)
1988	0.0691	0.0042	0.0733	0.0685	0.0023	0.0708
1989	0.0691	0.0042	0.0733	0.0685	0.0023	0.0708
1990	0.0685	0.0153	0.0838	0.0457	0.0146	0.0604
1991	0.0691	0.0042	0.0733	0.0457	0.0018	0.0475
1992	0.0691	0.0155	0.0846	0.0457	0.0131	0.0589
1993	0.0703	0.0077	0.0780	0.0460	0.0026	0.0486
1994	0.0702	0.0165	0.0868	0.0454	0.0098	0.0552
1995	0.0697	0.0130	0.0828	0.0461	0.0097	0.0558
1996	0.0697	0.0127	0.0825	0.0461	0.0093	0.0554
1997	0.0702	0.0324	0.1025	0.0511	0.0278	0.0789
1998	0.0706	0.0164	0.0870	0.0571	0.0104	0.0676
1999	0.0714	0.0111	0.0826	0.0572	0.0029	0.0600
2000	0.0711	0.0101	0.0812	0.0686	0.0027	0.0713
2001	0.0707	0.0287	0.0994	0.0686	0.0222	0.0908

Source: Estimated based on the data collected from groundwater and irrigation departments, Government of Rajasthan.

the previous years. Secondly, there was no question of arresting run-off water upstream during the drought years of 1999-2000, which could have increased its groundwater recharge more than downstream. The all-pervading drought conditions did not allow the gap to aggravate, especially after 2000. This seems to indicate that the status of recharge of groundwater remained unaffected in both upstream and downstream villages when both were affected equally by drought. Almost a similar condition existed before 1990, which was why, as Figure 3.1c shows, recharge of groundwater during this period was almost same in both locations.

Downstream was likely to encounter three scenarios after 2001. The first one was the decline in the recharge of groundwater in case drought continued to pervade these villages after 2001. In such a situation, the question of rainwater run-off flowing towards downstream from upstream would never arise. The condition of recharge upstream would be same as downstream. The second,

Figure 3.1a

Groundwater Recharge Per Sq Km in Two Villages during Monsoon

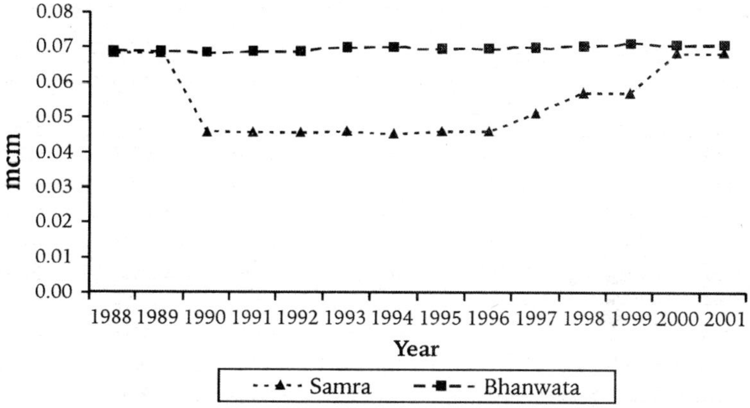

Source: Same as for Table 3.12

Figure 3.1b

Groundwater Recharge Per Sq Km in Two Villages during Non-monsoon

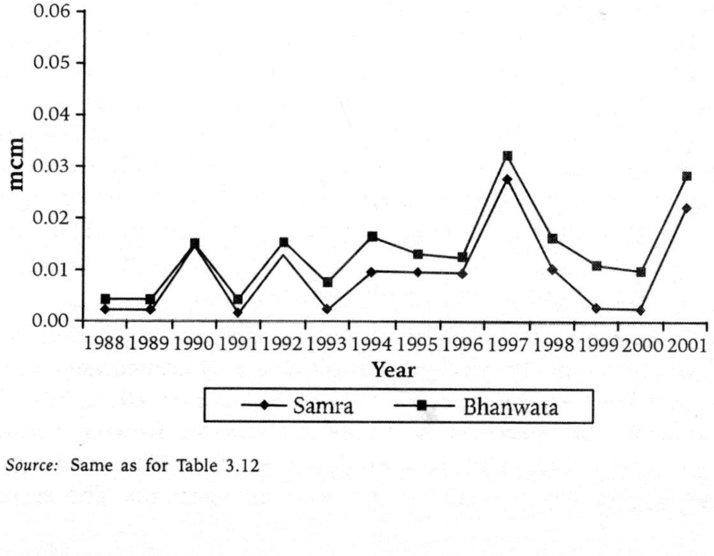

Source: Same as for Table 3.12

Figure 3.1c

Annual Groundwater Recharge Per Sq Km in Two Villages

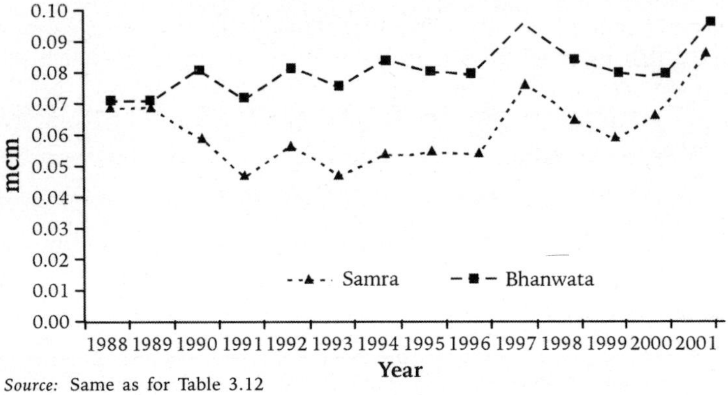

Source: Same as for Table 3.12

however, would be different from the first if rainfall was normal (not exceeding average rainfall) and inadequate quantity of run-off was allowed to flow downstream. In such a situation, recharge of groundwater upstream might go up while it might go down downstream widening the gap between the two further. The third scenario was the one that might be an outcome of excessive rainfall (more than average) in the area. This was the most favourable condition for increasing the scale of recharge of groundwater downstream. For, it was only under such condition that run-off could flow downstream from upstream.

Hence, it is not only a question of how many water-harvesting structures are constructed in the downstream—their importance can hardly be overstated, and, therefore, they do fulfil necessary conditions for recharging groundwater. However, the sufficient condition for groundwater recharge in the same river basin with the same physiography is how much run-off is allowed to flow downstream from upstream. Figure 3.1a shows that groundwater recharge per sq km during monsoon was same in both upstream and downstream villages before 1990. However, after 1990, upstream recharge of groundwater per sq km during monsoon was much higher than that of downstream at 0.07 mcm in Bhanwata and 0.04-0.05 mcm in Samra although both villages received the same amount of rainfall. The status continued to remain almost same in both locations until 1999.

It was only after 1999 that downstream could witness a rise in groundwater along with upstream due to excessive rainfall, as mentioned above, in the previous years. Bhanwata, however, did not exhibit any noticeable change in the status of groundwater recharge during monsoon (although its level was higher than that of Samra). The normal rainfall is 705.4 mm in Thanagazi block. However, the actual rainfall received during 1993-1998 varied from 797.0 mm, to 1,194.0 mm (Table 3.13). The rise of recharge of groundwater in the downstream village may be explained more in terms of cumulative effect of excessive rainfall (more than average) during these years.

Table 3.13

Rainfall (mm) in Thanagazi Block, 1984–2001

Year	*Pre-Monsoon*	*Monsoon*	*Post-Monsoon*
1984	0.0	703.0	0.0
1985	0.0	650.0	76.0
1986	37.0	178.0	11.0
1987	143.0	331.0	17.0
1988	46.0	510.0	5.0
1989	6.0	418.0	0.0
1990	124.0	587.6	37.0
1991	5.0	279.0	38.0
1992	115.2	621.8	26.0
1993	73.0	845.0	0.0
1994	112.0	797.0	0.0
1995	86.0	1,016.0	0.0
1996	75.0	1,194.0	7.0
1997	27.0	810.0	286.0
1998	63.0	711.0	33.0
1999	49.0	629.0	0.0
2000	66.0	499.0	0.0
2001	220.0	334.0	24.0

Note: Normal rainfall for Thanagazi block is 705.4 mm.

Source: Irrigation Department (hydrology), Rajasthan.

The downstream village of Samra had reasons to benefit from excessive rainfall. For excess run-off is naturally let off towards downstream once the maximum capacity of the existing structures to harvest rainwater is filled up upstream. This explains why groundwater recharge in Bhanwata was almost the same throughout 1990 and early 2000. This also explains why groundwater recharge in Samra rose steeply in the late 1990s. One may note that average rainfall before 1993 was lower than normal rainfall (Table 3.13). It means that there was no excessive run-off flowing down during this period and, therefore, no positive impact was made on groundwater recharge in Samra. However, it did not make any serious difference to Bhanwata. For it could harvest rainwater to whatever extent it could, depending upon the number and type of structures, no matter what the amount of rainfall was.

Figure 3.1b shows no significant difference in groundwater recharge during non-monsoon period between upstream and downstream villages. However, even at a lower level, the recharge downstream was consistently below upstream. The comparatively higher volume of recharge in both villages in 1997 and 2001 was because non-monsoon rainfall received during these years was much higher than 10 per cent of the normal rainfall (as explained in the model).

In order to gain more insight into the changing status of groundwater availability, it may be equally important to examine the fluctuations in the water level corresponding to recharge of groundwater in the same period. For water level and its fluctuations have a significant bearing on agro-ecological change.

Groundwater Level and Its Fluctuations

Groundwater level is assessed on the depth from the surface it is available at. It provides an appropriate condition for recharge of underground aquifers. In other words, the deeper the level of groundwater, the higher is the capacity of the aquifers to recharge through greater infiltration of rainwater if favourable hydrological conditions exist. The higher recharge, in turn, contributes to the enhancement of groundwater level. Apart from recharge, other hydrological factors contribute to the enhancement of the groundwater level. In order to examine water level fluctuations both upstream and downstream, and its status before and after water- harvesting structures were constructed, data on water level was collected from hydrographic stations located at the upstream Agar ki Dhani, half-a-km away from Bhanwata, and the downstream Samra. Data on water level was collected in the pre- and post-monsoon season.

The seasonal (pre- and post-monsoon) fluctuations of groundwater level for almost two decades are shown in Figure 3.2a and 3.2b. It appears that the groundwater level of downstream Samra in both

Figure 3.2a

Groundwater Level (Pre-monsoon)

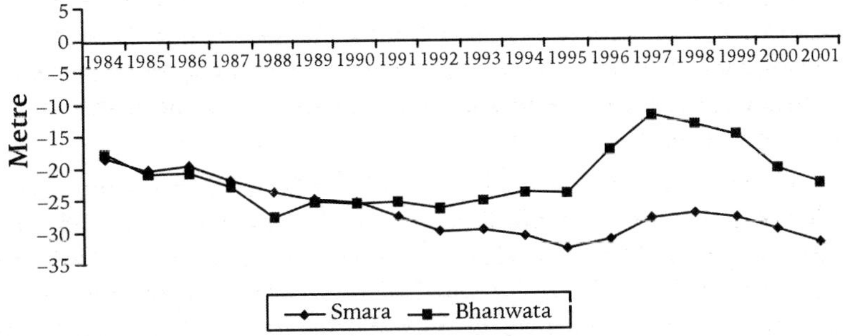

Source: Groundwater Department, Government of Rajasthan.

Figure 3.2b

Groundwater Level (Post-monsoon)

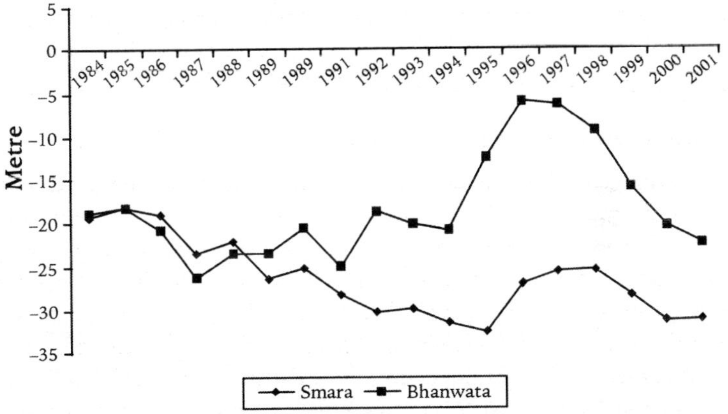

Source: Groundwater Department, Government of Rajasthan.

seasons was higher than that of upstream Bhanwata until 1988-89 when water-harvesting structures were constructed and began to prevent run-off rainwater from flowing downstream. However, after 1990, its groundwater level was much lower than that of Bhanwata and continuously declined until 1994. As seen earlier, this part of the state received more than average rainfall consecutively for five years from 1994-1998. This might be the reason why groundwater level exhibited a rise in both villages, although the level at which it increased downstream was lower than that of upstream. A long drought may be the reason for the decline of groundwater level in the following years.

In other words, had the flow of water in the Arwari river basin remained the same (in the absence of human intervention upstream), the groundwater level of the downstream village would have been 'naturally' above that of the upstream village. However, the reversal of the trend of groundwater level of both villages, as Figures 3.2a and 3.2b show, seemed to have pushed the downstream village's groundwater lower than before. The statistical results, as shown in Table 3.14, indicate that the fluctuation were insignificant before the intervention. After the water-harvesting structures were constructed, it was significant at 1 per cent level of confidence.

Table 3.14

Fluctuation of Groundwater before and after Intervention

	C.I.#		*Significance (2-tailed)*
	Lower	*Upper*	
Before intervention	-3.57	0.97	0.211
After intervention	-6.47	-1.37	0.007*

Note: # 95% Confidence interval * Significant at 1%.
Source: Calculated based on the data collected from the Ground Water Department, Governement of Rajasthan.

Such a fluctuating trend indicated better availability of groundwater upstream. For increasing availability of groundwater definitely contributed to agriculture in terms of expansion of irrigation, agricultural production and livestock development. However, the rise in agricultural activity, it seemed, did not complement the quantum of recharge of groundwater under the given hydrological conditions. This might lead one to doubt whether such a positive change is sustainable.

As per the Groundwater Estimate Committee, if the rate of utilisation of groundwater is above 70 per cent of the available groundwater, regeneration (recharge) of groundwater becomes unsustainable. In view of this, it is possible that the present trend of availability of groundwater upstream may be reversed in the future. In such a situation, one is not certain how agricultural development

Figure 3.3

Net Annual Recharge and Draft of Groundwater in Upstream Village

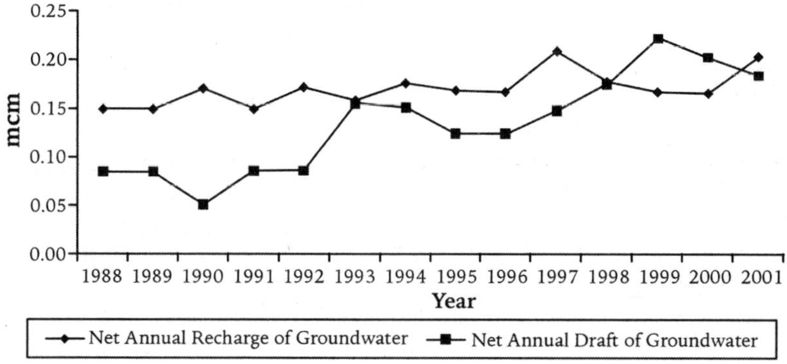

Source: Data collected from groundwater and irrigation departments, Government of Rajasthan.

will be adversely affected in Bhanwata. Figure 3.3 shows that the gap between annual recharge and draft was quite comfortable until 1993 when the annual draft was much lower than the recharge. However, after 1993, the gap narrowed down implying that the draft of groundwater increased steadily, so much so that it exceeded recharge after 1998.

It may be noted that groundwater draft in the upstream village went beyond 70 per cent of the recharge in almost all years after 1993. In other words, upstream village became vulnerable to shortage of groundwater after this year and reached a critical stage in 1998 when draft was more than recharge. The trend analysis shows that groundwater was drafted upstream annually at the rate of 0.011 mcm against the annual recharge of 0.0027 mcm during 1988-2001 (Table 3.15).

Samra, however, presents a different scenario altogether. Here, consumption of groundwater never exceeded 70 per cent of its availability. It was mainly because groundwater level went down by 30 metres after 1991 (Figures 3.2a and 3.2b). In Bhanwata it was available at a depth of 25 metres during the same year and much less than 25 metres in the subsequent years. This might be the reason why almost 90 per cent of the sample respondents of Bhanwata observed that the groundwater table had risen while only 15 per cent said this in Samra. It must be noted, however, that residents of Bhanwata-

Table 3.15

Analysis of Annual Recharge and Draft of Groundwater in Upstream and Downstream Villages

Upstream	Downstream
Annual Recharge = 0.0027* (Year) + 0.1495	Annual Recharge = 0.0217* (Year) + 1.0428
Annual Draft = 0.0108 * (Year) + 0.0527	Annual Draft = 0.013* (Year) + 0.3649

Source: Data collected from groundwater and irrigation departments, Government of Rajasthan.

Koliyala could gain access to that level of groundwater by deepening more than 50 per cent of the existing wells here. For water level in this village, as shown in Figure 3.2a and 3.2b, declined continuously after 1996. So long as the deepening of wells remains cost-effective, farmers will continue to make efforts to reach this level of groundwater. If more cropped area is brought under irrigation than at present, it is quite likely that the farmers of the upstream village may be pushed into the situation of the downstream farmers.

In the case of downstream Samra, drafting of groundwater was less cost-effective for two reasons, the lower water level and recharge. In fact, the draft was so low that it hardly exceeded 0.50 mcm (Figure 3.4). This seemed to be the reason why annual recharge of groundwater remained above its annual draft; the volume of recharge remained almost constant ranging between 1.0-1.5 mcm throughout the period under study.

The inequitable status of groundwater availability in two villages is also discernible in the availability of rainwater harvesting structure. Table 3.16 shows that one structure was available in each 11.3 hectares in Bhanwata-Koliyala, while the same was available in 45 hectares of land in Samra. It also shows that the density of wells was much less in Samra than Bhanwata-Koliyala. While one well was constructed for every 10 hectares of land in Bhanwata-Koliyala, the same was done for every 18 hectares of land in Samra. Bhanwata-Koliyala had 35 wells spread over 339 hectares of land, while Samra had 116 wells in 2,057 hectares of land.

Changing Agricultural Scenario

The inequitable availability of groundwater after water-harvesting structures were constructed appeared to have led to the emergence of

Figure 3.4

Net Annual Recharge and Draft of Groundwater in Downstream Village

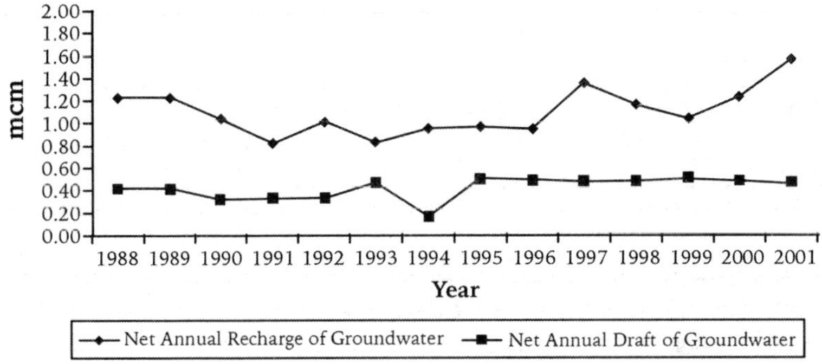

Source: Data collected from groundwater and irrigation departments, Government of Rajasthan.

Table 3.16

Availability of Rainwater Harvesting Structure and Wells in Upstream and Downstream Villages

Village	*Bhanwata-Koliyala*	*Samra*
Hectares of land per structure	11.3	44.72
Hectares of land per well	9.69	17.73

Source: Field survey.

disparity in agricultural prospects between upstream and downstream villages. If examined from the point of view of bringing more land under irrigation, Bhanwata, the upstream village, may be considered to have benefited by it immensely. This is analysed based on the data collected from secondary sources and through field survey.

Figure 3.5 shows that the share of irrigated area of total cropped area of Samra was little more than 60 per cent before 1990-91, roughly the year before intervention was scaled up. However, in subsequent years, the irrigated area declined and continuously remained lower than upstream Bhanwata until 2002-03. About 75-80 per cent of the total cropped area was under irrigation in Bhanwata during three years of drought from 2000 to 2003, while it was around 40-50 per cent in Samra during the same period (Figure 3.5). Even when the water table witnessed a declining trend during these years (as shown earlier in

Figure 3.5

Irrigated Area as a Percentage of Total Cropped Area in Bhanwata and Samra

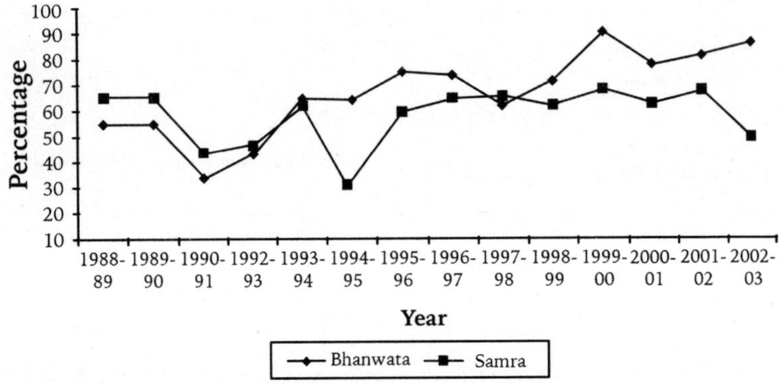

Source: Tehsil Office, Revenue department, Government of Rajasthan.

Figures 3.2a and 3.2b) such a large percentage of area was brought under irrigation in Bhanwata. The household analysis substantially supports these observations. Each sample household in Bhanwata-Koliyala increased its area of irrigated land by around 44 per cent during 2002-03, a drought year, as compared to the initial years of intervention (roughly around 1988-89). However, it steeply declined by 27 per cent in Samra during the same period (Table 3.17).

Table 3.17

Irrigated Land Per Sample Household in Both Upstream and Downstream Villages During Rabi Season

(land in bighas)

	Bhanwata-Koliyala	*Samra*
Initial years of intervention (1988-89)	2.52	6.00
During field survey (2002-03)	3.62	4.37
Increase (per cent)	43.65	-27.17

Source: Field survey.

In order to examine it further, data was collected from secondary sources randomly from a sample of 14 other villages from upstream and downstream of Arwari River. The sample upstream villages from Thanagazi block were Bhanwata, Bhuriyawas, Jagannathpura,

Chausala, Jhiri, Dumoli, Khardata, Sanwatsar and Chandpura. The downstream sample villages were Nimla, Hingawala, Kaljpuri, Shri Ramgopalpura and Ramyawala, all from Jamwa Ramgarh block.

These villages were located in an area of five km around the Arwari River. The analysis of growth of irrigated area shows that the percentage of irrigated area to the cropped area was higher at 50 per cent in downstream villages than in upstream villages (less than 50 per cent) before the practice of rainwater harvesting was revived in the late 1980s (Figure 3.6). However, in the following years, the irrigated area in all downstream villages declined (below 50 per cent) and that of the upstream villages expanded. This continued until 1997-98. It increased in the following years, though at a slower rate than upstream, which witnessed a sharp rise—the irrigated area here was more than 70 per cent of the cropped area.

It was possible to increase the irrigated area to such an extent even in drought conditions because, as mentioned earlier, the groundwater recharge increased with excessive rainfall in the previous years. The decline of the irrigated area during 2002-03 in both upstream and downstream villages can be said to be an outcome of decline in groundwater table thanks to low recharge; quite likely during a long drawn drought. It was possible that the rise of irrigated area in upstream village like Bhanwata, under such conditions, could be unsustainable. In any case, the area under irrigation differed significantly between upstream and downstream villages even during drought years. The sample upstream villages had 46 per cent of the cropped area under irrigation, while it was 35 per cent of the same in the downstream villages during 2002-03 (Figure 3.6).

The trend analysis of the growth of area under irrigation also shows that the latter had grown annually by 0.65 per cent in the case of the downstream village, while it increased by 1.99 per cent in the upstream villages. This was three times more than that of the downstream village.

Cropping Pattern and Crop Yield

The preceding analysis shows that the downstream Samra witnessed steady decline of area under irrigation. Such a decline seemed to have largely contributed to the low generation of income from agriculture as compared to that of Bhanwata. The analyses of cropping pattern and crop yield of both upstream and downstream villages present extremely varying status of agricultural development.

Figure 3.6

Percentage Share of Irrigated Area to Total Cropped Area

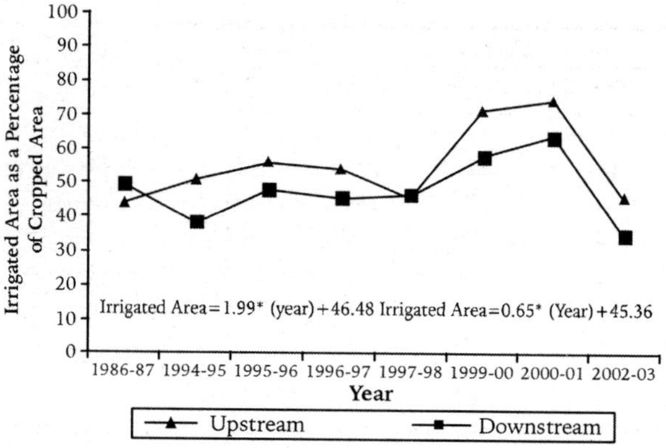

Source: Tehsil Office, Revenue department, Government of Rajasthan.

In order to examine whether the area under each crop changed over the years after rainwater harvesting began in both locations, data was collected for 1988-89, the initial years of intervention, 1995-96 when the rainfall was much above normal and 2002-03 which had very low rainfall. The results of the analysis are shown in Table 3.18.[4]

4. A similar analysis that was conducted based on the data collected through household survey shows almost same results. This is shown in the following table.

Area under different crops in upstream and downstream village (*in bighas*)

Crop	*Bhanwata-Koliyala*			*Samra*		
	Initial Years of Intervention (1988-89)	*(1996-97)*	*During Field Survey (2002-03)*	*Initial Years of Intervention (1988-89)*	*(1996-97)*	*During Field Survey (2002-03)*
Maize	99.00	113.60	110.90	218.95	220.00	174.67
Bajra	47.03	54.83	69.38	171.39	181.89	156.90
Wheat	84.05	96.40	91.90	209.00	226.30	115.50
Barley	33.50	40.25	42.25	129.95	128.75	83.00
Mustard	23.90	33.90	45.50	56.33	56.33	44.88

Note: The significant observation that one makes is that the area under almost each crop in the upstream village increased consistently during the years under consideration. In sharp contrast, the area under each crop in the downstream village declined during the same period. This was true in respect of both *rabi* and *kharif* crops. Sample respondents of the upstream village could afford to expand the area even under perpetual drought conditions by way of, as observed in the preceding section, exploiting groundwater over and above the volume of its recharge. However, this was not possible for the sample respondents of the downstream village.

Source: Field survey

Table 3.18

Area under Different Crops in Upstream and Downstream Villages

(in hectares)

	Bhanwata			Samra		
Crops	1988-89	1995-96	2002-03	1988-89	1995-96	2002-03
Kharif						
Maize	18	22	22	83	78	110
Bazra	8	7	11	36	42	27
Gawar	0	0	0	6	0	0
Jawar	9	2	1	1	0	0
Till	2	0	0	10	8	0
Jawar Chari	0	0	0	4	8	0
Kalijiri	0	0	0	7	4	3
Vegetables	0	0	2	1	0	0
Groundnut	0	0	0	0	2	0
San	0	0	0	0	3	0
Tobacco	0	0	0	0	1	0
Total *Kharif*	37	31	36	148	146	140
Rabi						
Wheat	6	14	13	38	62	28
Barley	6	10	9	33	30	27
Gram	4	4	0	6	71	0
Mustard	6	12	9	31	39	7
Methi	0	0	0	2	3	0
Taramira	0	0	0	1	10	0
Dhaniya	0	0	0	5	0	0
Jeera	0	0	0	1	0	0
Chari	1	0	0	0	0	0
Total *Rabi*	23	40	31	117	215	62

Source: Tehsil office, Thanagazi, Alwar district.

Significantly, the area under almost all crops, except maize, declined in the downstream village during 2002-03. Maize is a rain-fed crop that expanded in area over the years. However, the noticeable feature of the cropping pattern was that the area under all *rabi* crops, including wheat, barley and mustard that are normally cultivated through irrigation, declined in the downstream village in this poor rainfall year while upstream village expanded the area under cultivation of these crops. In other words, the area under *rabi* crops in the downstream village declined by 45.61 per cent, while the same increased by 40.90 per cent in the upstream village. Table 3.18 further shows that during 1995-96 when rainfall was much above average, the area under cultivation of almost all these crops in the downstream village increased like that of the upstream village. The only exception was the area under barley, which declined marginally.

However, this is not the issue. The issue is whether agriculture could do well even during the poor rainfall year with construction of rainwater-harvesting structures. It was in this context that one finds, as it appears from Table 3.18, the downstream village failed to perform better in agriculture in terms of expansion of the area under cultivation, especially under *rabi* crops. The total area under *rabi* crops in the downstream village declined sharply from 215 hectares in 1995-1996 to 62 hectares in 2002-03. In the upstream village, however, the decline was marginal, falling from 40 hectares to 31 hectares in the same period. But it could resist this decline, even when facing drought, by exploiting groundwater over and above the volume of its recharge. Downstream, the water level was beyond the reach of an average farmer.

The scenario of agricultural yield was equally disappointing for the downstream village. The yield of five major crops as reported by the sample households located upstream and downstream was taken into consideration. The responses pertained to yields of crops, including *bajra*, maize (*kharif*), wheat, barley and mustard (*rabi*) in the years 1988-89 and 2002-03. In the absence of data from secondary sources on yield rate of these crops in these villages, interviews were held with the members of the sample households. Questions were asked about crop yield per *bigha* during the initial years of intervention and 2002-03 (a year before the field survey was conducted).

The yield rate in terms of crop output per *bigha* of the sample households is estimated and shown in Table 3.19. The results are plotted in Figure 3.7; the yield rate of all *rabi* crops witnessed a decline in the downstream village, while the same increased in the upstream village. It is important to note that the decline of yield rate downstream was the most in respect of mustard at -4.68 per cent. However, upstream witnessed a maximum rise of the same crop by 16.45 per cent. Of all crops, mustard turned out to be a major source of gain in agriculture in terms of increased yield after rainwater harvesting began in Bhanwata. Table 3.19 shows that, although the yield rate of rainfed crops such as *bajra* and maize increased by 1.13 per cent and 2.49 per cent, respectively in Samra, the rise was much more in respect of the same crop in the upstream village at 13.82 per cent and 4.65 per cent for *bajra* and maize, respectively.

Figure 3.7

Percentage Increase of Yield Per Bigha in Both Upstream and Downstream

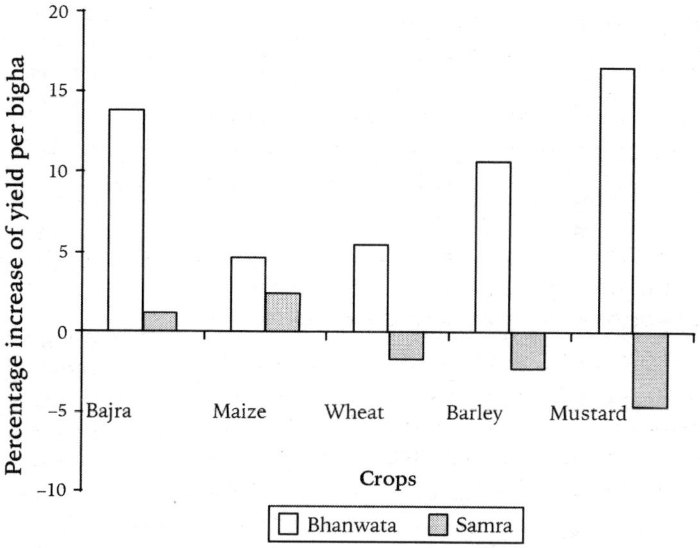

Source: Field survey.

Table 3.19

Yield Rate of Main Agricultural Crops of Sample Households of Bhanwata-Koliyala and Samra

(per bigha)

Crop	*Bhanwata-Koliyala*			*Samra*		
	Initial Years of Intervention (1989-90)	*During Field Survey (2002-03)*	*Percentage Increase*	*Initial of Years Intervention (1989-90)*	*At the Time of Field Survey (2002-03)*	*Percentage Increase*
Bajra	2.46	2.80	13.82	2.65	2.68	1.13
Maize	4.30	4.50	4.65	3.61	3.70	2.49
Wheat	7.20	7.59	5.42	5.39	5.30	-1.67
Barley	5.20	5.75	10.58	4.41	4.31	-2.27
Mustard	3.10	3.61	16.45	1.71	1.63	-4.68

Source: Field survey.

While other factors, such as use of fertiliser, had a role to play in enhancing crop yield, availability of water for irrigation was the primary consideration. The field survey revealed that the sample households of the downstream village used less than half the quantity of fertiliser that was used in the upstream village during 2002-03. On an average, upstream farmers used 9.05 kg fertiliser (DAP+urea) per *bigha* for all crops, while downstream farmers used only 4.01 kg. In other words, the agricultural scenario, even in respect of improvement in crop yield, seemed to be more encouraging for upstream. The relative disparity in agricultural performance associated with intensive irrigation and improvement in crop yield between upstream and downstream in the same river basin fairly indicates a development dichotomy in agriculture in that one grew at the expense of the other.

If development gains were to be assessed based on valuing agriculture crops and by-products in market prices, downstream Samra did not prosper as Bhanwata-Koliyala did.

Table 3.20 shows that the average annual gross value generated by agriculture per household in Samra was around one-third of that of Bhanwata-Koliyala during 2002-03 at Rs. 8,952.56 for the latter's Rs. 23,778.96. Besides, more importantly, the contribution of agricultural by-products, such as barley husk, wheat husk, barley straw, wheat straw, etc. to the gross agricultural income was as much as 40 per cent in the case of Samra and around 30 per cent for Bhanwata-Koliyala. It appears that there was no significant difference between different categories of farmers in terms of production of agricultural by-products either upstream or downstream. However, what differed was the volume of their production—a comparatively large production and, therefore, its higher share in agricultural income in Samra also implied lower productivity of its agriculture.

Livestock Development

Assured supply of water acts as an incentive to the farmers to improve their livestock, especially large animals. Data on livestock holding was collected for both drought and normal years in order to obtain a comparative picture. Table 3.21 shows that livestock holding of Samra was better in respect of all animals during normal years (five years ago). The average holding of livestock in terms of cattle unit per household was 21.43. However, this declined to 9.63 cattle

Table 3.20

Gross Value of Agricultural Crops and By-products

Village	*Category*	*Agriculture Crop Value (in Rs.)*	*Agricultural By-products Value (in Rs.)*	*Agriculture Gross Value (in Rs.)*
Bhanwata-Koliyala	Marginal	8,068.33 (69.13)	3,602.08 (30.87)	11,670.42 (100.00)
	Small	16,967.75 (70.66)	7,045.00 (29.34)	24,012.75 (100.00)
	Medium	40,382.22 (72.70)	15,166.67 (27.30)	55,548.89 (100.00)
	Average per household	16,913.87 (71.13)	6,865.09 (28.87)	23,778.96 (100.00)
Samra	Marginal	3,242.35 (60.56)	2,111.25 (39.44)	5,353.60 (100.00)
	Small	4,090.26 (58.19)	2,939.47 (41.81)	7,029.74 (100.00)
	Medium	8,225.81 (61.76)	5,093.55 (38.24)	13,319.35 (100.00)
	Average per household	5,446.05 (60.83)	3,506.51 (39.17)	8,952.56 (100.00)

Note: Figures in parentheses show percentage of gross value of agricultural crops and by-products.

Source: Field survey.

Table 3.21

Average Livestock Holding Per Household

Animal	*Bhanwata-Koliyala*				*Samra*			
	At present (2002-03)	*Cattle Unit*	*5 Years Ago*	*Cattle Unit*	*At Present (2002-03)*	*Cattle Unit*	*5 Years Ago*	*Cattle Unit*
Cow	0.92	0.92	1.75	1.75	1.37	1.37	4.88	4.88
Buffalo	2.75	5.50	3.45	6.90	1.78	3.56	4.05	8.10
Sheep	0.38	0.10	1.06	0.27	0.89	0.22	1.1	0.28
Goat	8.87	2.22	18.53	4.63	9.12	2.28	20.52	5.13
Bullock	0.64	1.28	1.25	2.50	1.1	2.20	1.52	3.04
Average per Household	13.56	10.01	26.04	16.05	14.26	9.63	32.07	21.43

Note: Cattle Unit: 1 Cow = 1 Cattle, 1 Buffalo = 2 Cattle, 1 Sheep = 0.25 Cattle, 1 Goat = 0.25 Cattle and 1 Bullock = 2 Cattle.

Source: Field survey.

units in the drought year. The decline appeared to be more pronounced in Samra than Bhanwata-Koliyala. The stock of animals declined by 6.04 in Bhanwata-Koliyala, in Samra it was 11.80. Significantly, each sample household of the upstream village had one buffalo more than the average stock of holding of buffalo per household in Samra. The upstream village could afford it due to better performance of agriculture even during the drought year.

It is somewhat difficult to gain an accurate understanding of the changing status of livestock at the household level unless one keeps track of the cycle of the changing stock in terms of birth and death of animal, purchase, sale, etc. over a period. However, an analysis of the livestock in both places based on the data collected through field survey at a particular point of time only shows how it varied consequent to the availability of groundwater. In order to gain more analytical insights into the differences that existed between upstream and downstream villages with regard to the livestock status, sample households were asked about how they evaluated the changes in the status of their own livestock over the years since the intervention took place. The responses elicited are analysed and shown in Table 3.22.

Table 3.22

Response of the Samples Households about the Status of Livestock after the Intervention

Response	*Bhanwata-Koliyala*		*Samra*	
	Status of Livestock	*Milk Productivity*	*Status of Livestock*	*Milk Productivity*
Declined	13 (24.53)	6 (11.32)	50 (68.50)	36 (49.32)
Increased	18 (33.96)	28 (52.83)	4 (5.48)	2 (2.74)
Same as before	21 (39.62)	15 (28.30)	4 (5.48)	12 (16.44)
No response	1 (1.89)	4 (7.55)	15 (20.55)	23 (31.51)
Total	53 (100.0)	53 (100.0)	73 (100.0)	73 (100.0)

Note: The figures in parentheses are percentages.
Source: Field survey.

It is clear that the status of livestock in Samra was never encouraging after the practise of rainwater harvesting was revived.

Nearly 70 per cent of sample households in Samra observed that the livestock declined in the village over the years. The upstream village appeared to have gained substantially from the intervention as compared to the downstream one.

Table 3.22 shows that the majority of the respondents of Bhanwata-Koliyala gained through increased milk yield. Few of them did so in Samra. One may attribute it largely to better agriculture production conditions created through greater availability of groundwater and green fodder in the forest. Even small farmers could comparatively gain more from milk production upstream than downstream.

Table 3.23

Average Requirements Per Household of Cattle Feed and Fodder for Livestock

(*in Rs.*)

	Bhanwata-Koliyala			*Samra*		
	Own Source	*Purchased from the Village*	*Total*	*Own Source*	*Purchased from the Village*	*Total*
Grass	4,560	0	4,560 (20.75)	3,581	0	3,581 (21.36)
Green fodder	1,913	10	1,923 (8.75)	171	264	435 (2.59)
Dry fodder	8,504	1,407	9,911 (45.11)	4,358	2,932	7,290 (43.48)
Food grain	375	116	491 (2.23)	352	656	1,008 (6.01)
Concentrate	0	3,846	3,846 (17.50)	0	2,404	2,404 (14.34)
Tree lopping	1,222	18	1,240 (5.64)	1,937	110	2,047 (12.21)
Total	16,574 (75.43)	5,397 (24.56)	21,971 (100.0)	10,399 (61.30)	6,366 (38.70)	16,765 (100.0)

Note: Price of each item was estimated on the basis of its prevailing rate in the village market at the time of field survey. Green fodder, such as 'rajka', 'jawar chari', (local terms), was grown in the field. Dry fodder included agriculture by-products, such as paddy and wheat husk, etc.

Source: Field survey.

The relatively higher availability of green fodder upstream is discernible from Table 3.23. Around 9.0 per cent of the average requirement of feed were met by green fodder in the same village as against just 2.56 per cent in Samra. In terms of quantity, while one farmer fed 10 quintals of green fodder to the animals in Bhanwata-Koliyala, in Samra it was only 2.54 quintals (not shown in the Table). Similarly, the average consumption of grass of livestock per household in Bhanwata-Koliyala was 45.6 quintals and 35.81 quintals in Samra (not shown in the Table). Recharge of groundwater helped improve moisture content of the soil and create better conditions for growth of grass production in Bhanwata-Koliyala. Not so in Samra. Nonetheless, dependence on the market for feed and fodder in both the villages was comparatively much less; it was around 25 per cent in Bhanwata-Koliyala and 39 per cent in Samra (Table 3.23). The quantity of feed and fodder is expressed here in monetary terms in order to gain insights into the comparative picture between the items consumed by livestock, on the one hand, and consumption pattern of all items together, on the other. Unit prices were collected from the villagers concerned.

Table 3.24

Production, Consumption and Sale of Milk Across Different Landholdings in 2002

(in quintal)

Category	*Bhanwata-Koliyala*			*Samra*		
	Production	*Consumption*	*Sale*	*Production*	*Consumption*	*Sale*
Landless	—	—	—	1.22	1.22	0.00
Marginal	13.00	13.00	0.00	4.32	4.32	0.00
Small	17.51	17.51	0.00	6.46	6.46	0.00
Medium	35.37	33.23	2.13	15.35	14.92	0.41
Average per Household	18.49	18.14	0.36	9.43	9.25	0.17

Source: Field survey.

Table 3.24 shows that the average milk production per household of the medium farmers of Samra at 15.35 quintals per year was lower than even that of small farmers of Bhanwata-Koliyala at 17.51 quintals. Figure 3.8 clearly shows how both upstream and downstream villages witnessed two different levels of performance of livestock during the drought year.

Figure 3.8

Difference in Livestock Performance Between Upstream and Downstream in Terms of Production, Consumption and Sale of Milk (in quintal)

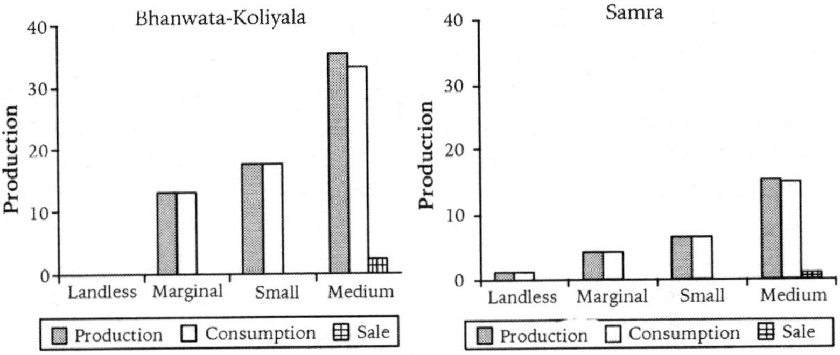

Source: Field survey.

Since milk productivity was lower downstream, the gross value of milk produced here was half of the value of milk produced upstream, which, in 2002, was estimated to be Rs. 18,485.66, while it was Rs. 9,433.15 downstream. Figure 3.9 shows that livestock contributed around 34 per cent of the gross value generated from the three main sources of income at the household level in Bhanwata and 32.57 per cent in Samra. Interestingly, its relative contribution was higher than agriculture in both villages. However, the largest contribution was made by the non-farm sector at 36.32 per cent in Bhanwata-Koliyala and 46.86 per cent in Samra (Figure 3.9).

Figure 3.9

Relative Contribution of Agriculture, Non-farm and Livestock to the Household Economy

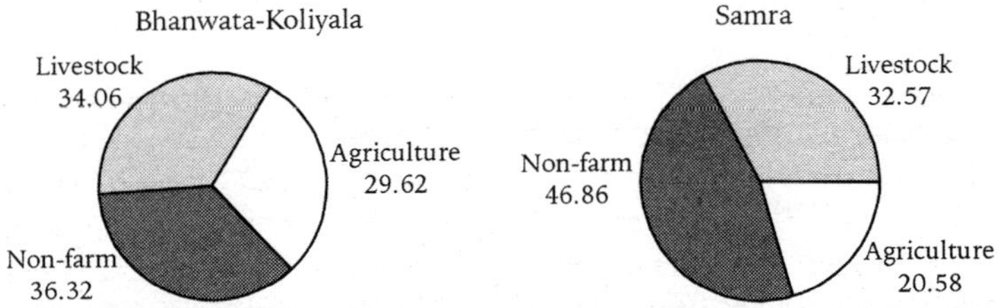

Source: Field survey.

Table 3.25

Relative Contribution of Livestock vis-à-vis Other Sources to Annual Gross Income of Different Category of Farming Households

(in Rs.)

Village	*Category*	*Agriculture*	*Non-farm*	*Livestock*	*Total*
	Marginal	11,670.42	31,845.83	21,295.96	64,812.21
Bhanwata	Small	24,012.75	23,277.50	25,946.00	73,236.25
	Medium	55,548.89	35,122.22	46,637.78	1,37,308.89
	Average per household	23,778.96 (29.61)	29,168.87 (36.32)	27,354.02 (34.06)	80,301.85 (100.0)
	Landless	0.00	29,000.00	2,620.00	31,620.00
	Marginal	5,353.60	21,011.75	6,527.60	32,892.95
Samra	Small	7,029.74	25,244.74	11,295.37	43,569.85
	Medium	13,319.35	16,168.90	21,977.23	51,465.48
	Average per household	8,952.56 (20.58)	20,385.22 (46.86)	14,168.74 (32.57)	43,506.52 (100.0

Note: Figures in parentheses are percentages.
Source: Field survey.

The difference of income between non-farm and the rest of the sources was not significantly higher in Bhanwata-Koliyala. What it implied was that the farming households in Samra saw non-farm sector as a more dependable source of livelihood generation than the other two sub-sectors. For neither agriculture nor livestock in this village was as prosperous as they were in Bhanwata-Koliyala. This seemed to have forced the villagers to widen the range of activities in the non-farm sector, mainly to the mines located in the neighbouring area. Besides, quite a few members of the village community regularly migrated to several places including Jaipur, Andhi, Gujarat, Madhya Pradesh and others as wage earners. In the case of Bhanwata-Koliyala, sources of wage earnings were different. Apart from being engaged in handicraft production and petty trades, the residents also migrated as wage labour in Delhi.

Again, incidence of migration, for employment and otherwise, after intervention was relatively more pronounced downstream than upstream jumping from 38.4 per cent (of the sample respondents) to 65.8 per cent after the intervention. In Bhanwata-Koliyala, it increased from 60.4 per cent to 71.7 per cent during the same period (Table 3.26).

Table 3.26

Incidence of Migration before and after Intervention

Options	*Before Intervention*		*After Intervention*	
	Bhanwata-Koliyala	*Samra*	*Bhanwata-Koliyala*	*Samra*
Number of households-migrating for-animal grazing	10	2	12	7
Number of households-migrating for-employment	26	27	35	46
Total number of-sample households	53	73	53	73
Incidence of-migration*	32 (60.4)	28 (38.4)	38 (71.7)	48 (65.8)

**Note:* Some households of each village migrated for both seeking employment and grazing animals.

Figures in parentheses are percentages.

Source: Field survey.

Floral Diversity

Several water-harvesting structures were constructed by the villagers at different points of times during the last several years in both villages under study. Besides, old structures were repaired and made effective for conserving rainwater. Different sites were chosen within the same village for sampling of vegetation keeping in view topographical differences that existed in each village. This served two purposes. First, the analysis showed how floral diversity varied between different sites with different topography within the same village, and second, how the same differed between upstream and downstream villages taking all sites together in each village after water conservation initiatives. Both had hilly, undulated and periplain terrains. Three sites were chosen in Bhanwata-Koliyala for sampling, including hilly, undulated (inside the village) and plantation site (periplain). Similarly, nine sites were chosen in Samra village with some of them located nearer large water-harvesting structures than the others. These included Khazuriwala Dhani, Sandya ki Dhani, Dev ka Devra, Tigria ki Dhani, Gyarah ki Dhani, Dhayawala Dhani, Samra (main village), Nizara Nichli Dhani, Van ki Dhani.

Sampling of vegetation was conducted in the months of April and September following the methodology as shown in Annexure 2.

Although several measures were developed to analyse floral diversity, the present study focussed only on four measures that included: (1) importance value index; (2) index of diversity and dominance; (3) species richness index; and (4) similarity index.

Importance Value Index

Trees and Shrubs

Importance value index (IVI) was estimated for examining the hierarchical order of all species present on the selected sites. This measure must show whether such hierarchical order of species contributes to the stability of the ecosystem. It depends on whether the order of species is friendly to the ecosystem in terms of its suitability to the local climate and ability to maintain synergism. It is in this context that one is interested to know whether 'keystone species' assumes greater value in the hierarchical order. For these species have a greater role to play in maintaining synergy even in the event of adverse climatic conditions. It is worthwhile to mention here that the community of the upstream village became more conscious about the conservation of flora after tne intervention was made. They not only stopped the practice of cutting plants but also started collecting only dry twigs for fuelwood. Of course, the village institution was renovated to manage the process of rainwater conservation in a way that anything that went against it was immediately checked. This seemed to have directly contributed to the expansion of floral diversity of Bhanwata-Koliyala.

Bhanwata-Koliyala

The IVI analysis of tree and shrub species on the hill site shows that *Anogeissus pendula* had the maximum IVI of 44.06 followed by *Adhatoda vasica* (41.83), *Rhus misurensis* (35.62) with *Capparis decidua* (7.55) at minimum (Table 3.27). Despite the fact that Bhanwata-Koliyala faced severe drought for around three years, some of the native plants, such as the ones mentioned above, continued to dominate and did not allow the regenerating capacity of the ecosystem to collapse.

These plant species played a significant role as keystone species. This seemed to have been facilitated by the construction of water-harvesting structures, which increased the soil moisture content. Discussions with the villagers revealed that the vegetation cover, even in the hilly areas of the village, greatly expanded post intervention. With the measures taken up by the village community for conserving wildlife, the regenerated vegetation cover remained unimpaired even during the prolonged drought period. The entire conservation effort was given social sanctity of highest order by naming it after God as 'Bhairon Dev Abhyaranya'.

Table 3.27

Importance Value Index for Trees and Shrubs Species at Different Sites in Bhanwata Koliyala

IVI	*Hilly Side*	*Inside Village*	*Plantation*
1-10	A. leucophloea Prosopiscinereria Wrightial inctoria Azadirachta indica Dalbergia sissoo Capparis decidua Calotropis procera Euphorbia caducifolia Casaearia elliptica Boswellia serrata Tecomella undulata	A. leucophloea Cassia fistula Ficus religiosa F. bengalensis Flacourtia indica Phoenix Anogeissus pendula Z. mauritiana Capparis decidua Casaearia elliptica Grewia tenax	
10-25	Acacia nilotica Cassia fistula Flacourtia indica Butea monosperma Zizyphus nummularia Commifora wightii	Acacia nilotica Prosopis cinereria Holoptelia integrifolia Azadirachta indica Dalbergia sissoo Butea monosperma Adhatoda vasica Rhus misurensis Ziziphus nummularia Nerium indicum Calotropis procera Ipomoea pestigrides Euphorbiacaducifolia	Acacia nilotica A. leucophloea Ficus religiosa F. bengalensis Phoenix Rhus misurensis Euphorbia caducifolia Hardwickia binata Commifora wightii
25-50	Adhatoda vasica Rhus misurensis Anogeissus pendula	P. juliflora	Azadirachta indica Anogeissus pendula Zizyphus nummularia
50-75			P. juliflora

Source: Prepared on the basis of field survey.

Thirteen trees and shrub species were recorded on the plantation site (periplain) with *Prosopis juliflora* having the highest IVI (72.06).

A. pendula (34.70), *Zizyphus nummularia* (33.27), *Azadirachta indica* (28.85) and *Phoenix* (11.10) with the least followed this. *P. juliflora,* however, did not record as high IVI (32.82) inside the village (undulated) as it did on the plantation site, though it was the highest. The minimum was 7.76 (*Capparis decidua*). The higher value of *P. juliflora* on plantation site was because this plant was allowed to grow exclusively for meeting fuelwood need of the village community. The lower value of IVI of the same plant inside the village (undulated) shows that the community was not interested in this species for it suppressed growth of other species. In any case, this plant species was not native one, and, as the field survey revealed, the villagers were more conscious about protection and growth of native species. Some of these were *Acacia nilotica, Prosopis cinereria, Rhus misurensis, Capparis decidua* and *Zizyphus nummularia*. This might have been the reason why only native species were found largely on the hillside also.

The enormous ecological sensitivity displayed by the village community seemed to be embedded in their culture. Their strong preference for protection and conservation of those native species that were well adapted to the local climatic and geographical conditions was but a legacy of their past practices. This appeared to have led the ecosystem, as Table 3.27 shows, to gain substantial floral diversity inside the village as compared to hilly and plantation sites. Apart from the favourable terrain inside the village, concentration of large number of small earthen structures along with soil conservation measures made it possible to gain comparatively more diversity with a large number of native species.

Samra

Table 3.29 shows distinct difference in the order of importance of plants between the sites that were located near the water harvesting structures and at a distance. *Lithophilic* shrubs such as *Rhus misurensis* and *Ziziphus nummularia,* with a tendency to grow on wastelands, were found to have higher IVI at Sandya ki Dhani, Gyaran ki Dhani and Van ki Dhani. All these places were located far from the conservation sites and had lesser number of plants with lower IVI. These were all xerophytic plants. Except these three *dhanis* (hamlets), all other *dhanis* that were endowed with comparatively larger water-harvesting structures had a large number of trees that grow comparatively better where soil is deep and has high moisture content. The greatest advantage of these is their utility in fulfilling human and livestock needs, besides having immense ecological importance.

Table 3.28

Importance Value Index for Grass and Forbs Species at Different Sites in Bhanwata-Koliyala

IVI	Hilly Side	Inside Village	Plantation
1-10	Indigofera cordifolia Lepidagathis trinervis Boerhavia diffusa Borreria stricta Lasiurus sindicus Dactyloctenium sindicum Trichodesma indicum Tephrosia apolliniea Fagonia Cretica Ipomoea eriocarpa	Cenchrus ciliaris Lepidagathis trinervis Polygala crioptera Tribulus terrestris Panicum turgidum Euphorbia hirta Heliopropuim- bengholensis Vicoa indica Melanocenchrus- jacquemontii Borreria stricta Boerhavia diffusa Indigofera cordifolia Cenchrus penutiferner Ocimum americanum Dichanthium annulatum Fagonia cretica Cyprus kyllinge Cenchrus setigerus Apluda mutica	Trichodesma indicum Euphorbia hirta Ocimum americanum Polygala erioptera Vicoa indica Indigofera cordifolia Heliopropuim- bengholensis Ipomoea eriocarpa Melanocenchrus- jacquemontii Zornia gibbosa Boerhavia diffusa Glossocardia bosvelea Commelina benghalensis Tribulus terrestris
10-25	Polygala erioptera Heliopropuim- bengholensis Eragrostis ciliaris Euphorbia hirta	Bulbostylis barbata	Aristida funiculata Borreria stricta Cyprus kyllinge Dichanthium annulatum Lepidagathis trinervis
25-50	Bulbostylis barbata	Eragrostis ciliaris	Bulbostylis barbata Eragrostis ciliaris Cynodon dactylon
50-75	Cynodon dactylon		Cenchrus ciliaris
75-100		Cynodon dactylon Aristida funiculata	
100-125	Cenchrus ciliaris		

Source: Prepared on the basis of field survey.

Table 3.29

Importance Value Index (IVI) for Trees and Forbs on Different Sites in Samra

IVI	Khazuriwala Anicut	Sandya ki Dhani	Dev ka Devra	Tigria ki Dhani	Gyaran ki Dhani	Dhayawala Anicut	Samra Main	Nizara Nichli	Vanki Dhani
1-10	Prosopis cinereria Cassia fistula F. bengalensis Mangifera indica Wrightia linctoria Zizyphus nummularia		A. leucophloea F. bengalensis Wrightia linctoria Flacourtia indica Azadirachta indica Phoenix Butea monosperma Bauhinia racemosa Z. mauritiana Casaeria elliptica Belanites roxburghii	Nerium indicum	Azadirachta indica Boswellia serrata	Acacia nilotica A. leucophloea Wrightia linctoria Phoenix Zizyphus nummularia Hardwickia binata Terminalia arjuna Eucalyptus S.persica	Prosopis cinereria Cassia fistula Ficus religiosa Wrightia linctoria Flacourtia indica Dalbergia sissoo Leucena leucophloea F. glomerata	Prosopis cinereria Cassia fistula Azadirachta indica Dalbergia sissoo Capparis decidua Moringa cancanesis Albizia lebbock F. glomerata	Phoenix Salvadora oleoides
10-25	Acacia nilotica A. leucophloea Ficus religiosa Flacourtia indica Holoptelia integrifolia	Anogeissus pendula Zizyphus nummularia Salvadora oleoides	Acacia nilotica Prosopis cinereria Mangifera indica Holoptelia integrifolia Zizyphus	Acacia nilotica Cassia fistula Ficus religiosa Wrightia linctoria Flacourtia indica Holoptelia	Acacia nilotica Zizyphus nummularia Hardwickia binata	Z. mauritiana	A. leucophloea Mangifera indica Holoptelia integrifolia Azadirachta	Acacia nilotica Ficus religiosa Holoptelia integrifolia	Holoptelia integrifolia Azadirachta indica Leucena leucophloea

contd. ...

contd. ...

VI	Khazuriwala anicut	Sandya ki Dhani	Dev ka Devra	Tigria ki Dhani	Gyaran ki Dhani	Dhayawala anicut	Samra Main	Nizara Nichli	Vanki dhani
	Phoenix Dalbergia sissoo Bauhinia racemosa Anogeissus pendula Z. mauritiana Calotropis procera Euphorbia caducifolia Aegle marmelos		*nummularia Calotropis procera Ipomoea pestigridis Boswellia serrate L. leucocephala*	*integrifolia Bauhinia racemosa Rhus misurensis Capparis decidua Euphorbia caducifolia Casaeria elliptica Grewia tenax Leucena leucophloea*	*Grewia tenax*		*indica Z. mauritiana Capparis decidua Ipomoea pestigridis A. catechu Tecomella undulata Eucalyptus*	*Zizyphus nummularia Ipomoea pestigridis A. catechu*	*S. persica Emblica officinalis*
25-50	*Adhatoda vasica*	*A. leucophloea Tecomella undulata*	*Anoegeissus pendula*	*A. leucophloea Azadirachta indica Butea monosperma*	*Butea monosperma Casaeria Commifora wightii*	*Cassia fistula Rhus misurensis Anogeissus pendula F.glomerata*	*Rhus misurensis*	*Phoenix Butea monosperma Rhus misurensis Z. mauritiana*	*Acacia nilotica Boswellia serrate*
50-75		*Acacia nilotica*	*Rhus misurensis*		*Adhatoda vasica Rhus misurensis*	*Azadirachta indica*	*Acacia nilotica*		*Rhus misurensis Zizyphus nummularia*
75-100									
100-125		*Rhus misurensis*							

Source: Prepared on the basis of field survey.

However, the hierarchy of plants exhibited a different order of importance on those sites where big water harvesting structures were constructed. The community analysis of these sites, as shown in Table 3.29, did not reveal a hierarchical order of species required for maintaining stability of the ecosystem. Of course, plant species such as *Rhus misurensis, Zizyphus nummularia, Anogeissus pendula, Acacia nilotica, Butea monosperma* were observed to have higher values at some sites. They were also keystone species. The occurrence of *Acacia nilotica* and *Rhus misurensis* was found to be slightly better than that of all other species mentioned above. The occurrence of the rest, as Table 3.29 shows, was not consistent on all sites, including the distant ones, thus putting the stability of the ecosystem at risk. For example, *Anoegeissus pendula* was found present with relatively higher values in the four *dhanis* of Khazuriwala *Dhani*, Sandya ki *Dhani*, Dev ka *Devra* and Dhayawala *Dhani*. However, it was missing from the rest of the *dhanis*.

In other words, the growth pattern of native tree and shrub species was found to be more inconsistent and disorderly downstream. Although both upstream and downstream villages experienced a change in the habitat, water and soil conservation efforts seemed to have brought about a more conducive hierarchical order of plant species in the upstream village. It may largely be attributed to the linkage the community institution established between conservation efforts and wildlife protection. However, no such attempt was made in the downstream village. Once the characteristics of the habitat change, conservation efforts must be extended to biodiversity also. While Bhanwata-Koliyala succeeded in doing so, Samra failed. This might be the reason why the hierarchical order of species in Samra was not found as much as in Bhanwata-Koliyala.

Grass and Forbs

Bhanwata-Koliyala

Seventeen grass and forb species were recorded at Bhanwata-Koliyala hillside and 23 species each on the site inside village (low land) and the plantation. The IVI of grass and forbs showed that *Cenchrus ciliaris (103.34)* was the dominant species on the hillside (Table 3.28). Grass species like *Cyanodon dactylon (68.21), Aristida funiculata, Eragrostis ciliaris,* etc. had higher IVI. It may be noted that all species except a few such as *Aristida funiculata,* which grew inside the village, were edible. While this might be considered as an improvement in the ecosystem that facilitated growth of livestock, quality of forage was not considered to have improved by the villagers. As revealed by the field survey, grass species such as *Sundardi, Bidardi, Motiburi* and *Satburi* (local names), nutritious for milch animals, used to grow in abundance in these places. However, they had all become extinct. Overgrazing and drought may have hindered the growth of these species.

It must be underlined that the low land inside the village exhibited the presence of large concentration of grass and forb species with lower IVI (Table 3.28). Obviously, this area achieved more diversity than other sites, which had different geomorphological features.

Samra

Samra presents an almost similar pattern of growth of grass and forbs. In this village again, the grass species of *Cyanodon dactylon* was found to have higher IVI on all sites except Gyaran ki Dhani (Table 3.30). It also appears that almost the same grass and forb species with high IVI (25-50) were found to have occurred even across sites in the same village. These included *Cynodon dactylon, Eragrostis ciliaris, Bulbostylis barbata,* and *Lasiurus sindicus,* all edible and adaptable to the local climatic conditions of the area.

It is important to note that Dhayawala Dhani had the largest number of grass and forb species as compared to all other sites of the same village. The main reason, as one could see, was the construction of a big anicut on this site. The concentration of a large number of *dhanis* was almost similar to the site inside the upstream village, which was endowed with large number of traditional water harvesting structures. No significant difference was found in the hierarchical order of importance of grass and forb species of both upstream and downstream villages as confirmed by the similarity index (Table 3.31).

Table 3.30

Importance Value Index (IVI) for Grass and Forbs at Different Sites of Village Samra

IVI	Khazuriwala Anicut	Sandya ki Anicut	Dev ka Devra	Tigria ki Dhani	Gyaran ki Dhani	Dhayawala Anicut	Samra Main	Nizara Nichli	Van ki Dhani
0-10	*Lepidagathis trinervis* *Indigofera linnari* *Cyprus kyllinge* *Tephrosia apolliniea* *Borreria stricta* *Euphorbia hirta* *Polygala eriaptera* *Vicoa indica* *Cleome viscosa* *Dichanthium annulatum* *Mollugo cerviana* *Zornia gibbosa*	*Dactyloctenium sindicum* *Cenchrus ciliaris* *Heliopropuim benghalensis* *Mollugo cerviana* *Tephrosia apolliniea*	*Aristida funiculata* *Indigofera linnari* *Heliopropuim benghalensis* *Vicoa indica* *Trichodesma indicum* *Mollugo cerviana* *Ocimum americanum* *Cenchrus penutiferner* *Borreria stricta*	*Melanocenchrus jacquemontii* *Ocimum americanum* *Dichanthium annulatum* *Mollugo cerviana* *Indigofera linnari* *Evolvulus alsinoides* *Euphorbia hirta, Borreria stricta Aristida funiculata* *Fagonia cretica* *Heteropogon contortus* *Ipomoea eriocarpa* *Tribulus terrestris*	*Ipomoea eriocarpa* *Tephrosia apolliniea* *Dactyloctenium sindicum* *Trichodesma indicum* *Euphorbia hirta* *Boerhavia diffusa* *Borreria stricta* *Heteropogon contortus* *Melanocenchrus jacquemontii* *Ocimum americanum* *Dichanthium annulatum* *Evolvulus alsinoides*	*Trichodesma indicum* *Lepidagathis trinervis* *Tephrosia apolliniea* *Evolvulus alsinoides* *Ipomoea eriocarpa* *Dichanthium annulatum* *Euphorbia hirta* *Mollugo cerviana* *Glossocardia bosvelea* *Fagonia cretica* *Indigofera linnari* *Apluda mutica*	*Trichodesma indicum* *Indigofera linnari* *Cyprus kyllinge* *Tephrosia apolliniea* *Polygala eriaptera* *Commelina benghalensis* *Ocimum americanum* *Vicoa indica* *Heliopropuim benghalensis* *Dichanthium annulatum*	*Lasiurus sindicus* *Borreria stricta* *Ipomoea eriocarpa* *Tephrosia apolliniea* *Ocimum americanum* *Aristida funiculata* *Dichanthium annulatum* *Heliopropuim benghalensis* *Trichodesma indicum* *Euphorbia hirta* *Tribulus terrestris* *Apluda mutica*	*Ipomoea eriocarpa* *Aristida funiculata* *Mollugo cerviana* *Tephrosia apolliniea* *Euphorbia hirta* *Cenchrus penutiferner* *Vicoa indica* *Dichanthium* *Heteropogon contortus* *Indigofera linnari* *Cenchrus ciliaris*

contd. ...

contd. ...

IVI	Khazuriwala Anicut	Sandya ki Anicut	Dev ka Devra	Tigria ki Dhani	Gyaran ki Dhani	Dhayawala Anicut	Samra Main	Nizara Nichli	Van ki Dhani
10-25	*Boerhavia diffusa Dactyloctenium sindicum Aristida funiculata Trichodesma Melanocenchrus jacquemontii*	*Aristida funiculata Polygala eriaptera Heteropogon contortus Borreria stricta Lepidagathis trinervis Cyprus kyllinge Melanoce-nchrus jacquemontii*	*Lepidagathis trinervis Melanocen-chrus jacquemontii Cenchrus setigerus*	*Lasiurus sindicus Cyprus kyllinge Lepidagathis trinervis Polygala erioptera*	*Cenchrus ciliaris Lepidagathis trinervis Polygala eriaptera Aristida funiculata Cyprus kyllinge*	*Aristida funiculata Lasiurus sindicus Dactyloc-tenium sindicum Polygala eriaptera Melanoc-enchrus Borreria stricta Boerhavia diffusa Cyprus kyllinge*	*Eragrostis ciliaris Cenchrus ciliaris Borreria stricta Aristida funiculata Lepidagathis trinervis Dactyloc-tenium sindicum Melanoce-nchrus jacquemontii*	*Boerhavia diffusa Heteropogon contortus Melanoce-nchrus jacquemontii Cyprus kyllinge Lepidagathis trinervis*	*Lepidagathis trinervis Indigofera cordifolia Melanoce-nchrus jacquem-ontii Ocimum americanum Polygala eriaptera Boerhavia diffusa*
25-50	*Cynodon dactylon Eragrostis ciliaris Bulbostylis barbata Lasiurus sindicus*	*Cynodon dactylon Boerhavia diffusa Eragrostis ciliaris Lasiurus sindicus Bulbostylis barbata*	*Boerhavia diffusa Eragrostis ciliaris Bulbostylis barbata Polygala eriaptera Cyprus kyllinge*	*Cynodon dactylon Bulbostylis barbata Boerhavia diffusa Eragrostis ciliaris Dactyloctenium sindicum Trichodesma indicum*	*Bulbostylis barbata Lasiurus sindicus Eragrostis ciliaris Cynodon dactylon*	*Bulbostylis barbata Heteropogon contortus Eragrostis ciliaris*	*Bulbostylis barbata Boerhavia diffusa Lasiurus sindicus*	*Eragrostis ciliaris Bulbostylis barbata Dactyloc-tenium sindicum*	*Cynodon dactylon Eragro-stis ciliaris*
50-75			*Cynodon dactylon*			*Cynodon dactylon*	*Cynodon dactylon*	*Cynodon dactylon*	*Cyprus kyllinge*

Source: Prepared on the basis of field survey.

Table 3.31

Similarity Index between Tree, Shrub, Grass and Forb Species of Village Bhanwata-Koliyala and Samra

	Trees and Shrubs	*Grass and Forbs*
Similarity index	0.73	0.90

Source: Prepared on the basis of field survey.

This index was prepared while taking the entire village into consideration. Table 3.31 shows that there was more similarity in the case of grass and forb species than tree and shrub species. Since both the villages were located within a distance of 15 kms, aeolian, fluvial and biotic factors played a crucial role in pollination of similar grasses in both the villages. However, this did not work for the tree species because even if pollination of same species does take place, the regeneration process takes a much longer time. Besides, it must be complemented by conservation programme. While deforestation took place in both villages before water and soil conservation efforts were made, conservation of forest was carried out consistently in Bhanwata–Koliyala. However, such an effort was missing in Samra.

A diaggregated analysis of the tree species showed that the dissimilarity index of tree and shrub species between upstream and downstream villages was more (0.45 similarity index, not shown in Table 3.31) among dominant species above IVI of 25 (earlier shown in Tables 3.27 and 3.29). This was equally true for grass and forbs that had similarity index of 0.40 and above (not shown in Table 3.31). However, tree and grass species were found to have more similarity between the two villages at the lower IVI level. Although both the villages fall under the same geographic region and have same climatic features, dissimilarity of such a higher order in dominant species of both trees and grasses shows a distinct change in their respective habitats with different microenvironments. However, such a difference may not be a cause for too much concern, for both villages had species that were largely found to be native and edible.

Index of Diversity and Dominance

Analysis of floral diversity is carried out at two levels. While at the first level, the diversity of both the villages has been analysed in its entirety, equal attention has been paid to examine how floral diversity differed within the village. As mentioned earlier, topography even within the village, *vis-à-vis* location of water harvesting structure, differed widely within the village. As it appears from Table 3.32, Samra had a higher level of floral diversity (2.17) than Bhanwata-Koliyala (2.12). Samra, however, was expected to have achieved a much higher level of diversity than observed. For, as it flows downstream, it carries not only water but also minerals and pollen; it provides alluvium through fluvial and aeolian aftermath.

Table 3.32

Index of Diversity for Tree, Shrubs, Grass and Forbs Species in Bhanwata-Koliyala and Samra

	Tree and Shrubs	*Grass and Forbs*	*Total*
Bhanwata-Koliyala			
Hillside	1.18	0.85	2.04
Inside village (low land)	1.35	0.91	2.27
Plantation site	1.01	1.08	2.10
Average	1.18	0.94	2.12
Samra			
Hilly sites			
Van ki Dhani	0.94	1.08	2.03
Dev ka Devra	1.16	1.03	2.19
Sandya ki Dhani	0.72	1.13	1.85
Undulated			
Samra main	1.19	1.14	2.33
Tigria ki Dhani	1.17	1.15	2.32
Nizra Nichli Dhani	1.18	1.10	2.28
Periplain			
Khazuriwala anicut	1.25	1.10	2.35
Dhayawala anicut	1.06	1.17	2.24
Gyaran ki Dhani	0.95	1.13	2.08
Average	1.06	1.11	2.17

Source: Prepared on the basis of field survey.

Besides, there were other streams and channels, such as Jhiri, through which water flew to this village from upstream and enriched the quality of soil as reported by the villagers. Despite being in an advantageous position, Samra did not achieve floral diversity to the

Table 3.33

Index of Dominance for Tree, Shrubs, Grass and Forbs Species in Village Bhanwata-Koliyala and Samra

	Tree and Shrubs	Grass and Forbs	Total
Bhanwata-Koliyala			
Hilly site	0.08	0.20	0.28
Inside village (low land)	0.04	0.19	0.24
Plantation site	0.11	0.10	0.22
Average	0.076	0.163	0.24
Samra			
Hilly sites			
Van ki Dhani	0.13	0.11	0.24
Dev ka Devra	0.10	0.11	0.21
Sandya ki Dhani	0.2	0.08	0.32
Undulated			
Samra main	0.08	0.09	0.17
Tigria ki Dhani	0.07	0.08	0.16
Nizra Nichli Dhani	0.07	0.10	0.18
Periplain			
Khazuriwala Dhani	0.06	0.09	0.16
Dhayawala Dhani	0.10	0.09	0.17
Gyaran ki Dhani	0.13	0.95	1.00
Average	0.10	0.18	0.29

Source: Prepared on the basis of field survey.

extent it should have. It must be remembered that the community institution in Bhanwata-Koliyala witnessed renovation over the years and played a crucial role in enforcing rules of conservation of water, soil and wildlife. However, this was not the case in Samra, which was why it witnessed large incidence of deforestation. Of course, there had been continuous rise in the number of water harvesting structures and *medhbandi* for soil conservation on private lands. Hence, whatever little changes in the ecosystem, in terms of floral diversity, there were mainly because of private conservation efforts. However, Samra at 2.90 had a much lower species richness index than Bhanwata-Koliyala at 3.45 (Table 3.34).

Based on the result of this index, it can be said that Samra witnessed a greater growth in the variety of species, but the individual number of each of these species was much less than that of Bhanwata-Koliyala. In other words, the coverage of the Basal area of the species in Samra was much less. Bhanwata-Koliyala achieved more richness in floral diversity mainly due to institutional renovation. Perhaps the religious and social sanction attached to wildlife conservation efforts paid off in ensuring total community commitment.

Diversity of Hilly and Non-Hilly Sites

It is worthwhile to note that there is little difference in floral diversity between the hilly and periplain areas within each village under study. As can be seen from Table 3.32 floral diversity in the low land areas of Bhanwata-Koliyala was higher at 2.27 than the other sample sites with values of 2.04-2.10. However, what is important, is that the diversity index of the hillside was not significantly lower in this village. One may observe almost a similar situation in Samra. Of the nine sampling sites, the periplain and undulated sites (where relatively large water harvesting structures were available) had much more floral diversity than the hilly sites. The availability of better moisture and edaphic features due to fluvial and aeolian aftermath created more conducive environmental conditions for better growth and survival of large number of tree and shrubs in the low lands of the village. A similar study that was conducted in Japan suggests that wetlands are created through water-harvesting structures, and thus become habitats for a variety of flora and fauna (Kitakami River Integrated Dam Management Office, 1999).

Table 3.34

Species Richness Index for Bhanwata-Koliyala and Samra

	Trees & Shrubs	*Grass & Forbs*	*Total*
Bhanwata-Koliyala	2.86	0.59	3.45
Samra	2.18	0.71	2.90

Source: Prepared on the basis of field survey.

However, the existing water-harvesting structures on the hilly sites also contributed to the improvement of floral diversity. Had there been no water-harvesting structures in hilly sites whatsoever, the level of diversity would have been far lower than observed even after three years of drought. Further, the real difference between the lowlands and the highlands lay in the diversity of grass and forbs (rather than trees and shrubs) found in the former. On the hillsides, the growth of trees and their survival rates were limited due to comparatively less moisture in the soil. Besides, higher lamination might have resulted in higher evapotranspiration loss.

Higher dominance lowers complexity in the ecosystem. Assessment of dominance index in both villages showed higher dominance in village Samra (0.29) as compared to Bhanwata-Koliyala (0.24) (Table 3.33). Despite having a large number of tree shrubs, grass and forbs species, Samra did not gain dilution in dominance. For, there was a large number of few species and mere presence of some species with very low strength. Intra-village analysis indicated higher dominance on hilly sites in both villages, not surprising considering moisture content was low on the rocky terrain allowing only tolerant species to grow.

To Sum Up,

- There was no significant difference between upstream and downstream villages in terms of floral diversity. However, species richness index shows the upstream village achieved more floral diversity as compared to downstream village. The number of individual species declined considerably downstream, due to deforestation. Villagers upstream were found to be more sensitive to forest conservation.
- The importance value index analysis indicates some disturbing trends upstream as well. Some plants, which had very high IVI, were neither palatable nor eco-friendly. *Prosopis juliflora*, for example, was neither palatable nor eco-friendly. It was used only as fuelwood. Similarly, *Adhatoda vasica* was eco-friendly in terms of its ability to improve the moisture retention capacity of the soil. However, it was not palatable.
- In the case of downstream village, Samra, plant species with higher IVI showed declining moisture content of the soil. It suggests that the aridity in the environment was increasing. The growth of plants such as *Rhus misurensis*, *Zizyphus nummularia*, *Acacia nilotica*, etc. fairly indicated this.

IV

Conclusion

In a drought prone state like Rajasthan, there appears to be no discernible alternative to rainwater harvesting for ensuring rural livelihood. The revival of traditional practices may improve agro-ecology interaction, which an interventionist seeks to achieve. However, the operational aspects that shape the intervention, impact the sustenance of such practices. One may, therefore, be curious to know, in the present context, whether an integrative and comprehensive land and water management of the complete catchment area was ever considered (Pangre *et al.*, 2002). For the study area is a single watercourse system in that natural resources such as soil, water and vegetation are interconnected. The impact of the intervention on one resource affects the others suggesting non-seperability of their externalities (White and Runge, 1994). In view of this, habitats and breeding environment of plants and animals must not be damaged and requires coordination between landholders upstream and downstream. In the absence of this, land productivity of downstream farmers declines (White and Runge, 1994).

Unless the need for integrated management, as explained above, is explicitly recognised the revival strategy of rainwater harvesting is bound to be discrete and village-centric, no matter where the village is located. The present study shows that the approach followed by the intervening agency towards land and water management was essentially village-centric. The social costs that such an approach entails, in terms of the inequitable development gains between upstream and downstream, may go unnoticed in the immediate terms. Even an upstream village like Bhanwata may fail to escape such an eventuality ultimately, because of the mismatch between production and consumption of groundwater despite comparatively higher levels of groundwater recharge. The present study shows that the village gradually entered into the unsustainable zone in the late 1990s, when groundwater draft far exceeded its recharge.

This case exemplifies a case of 'induced' expectation among the village community upstream for more agricultural gains without rooting them in sustainability. The availability of groundwater at a manageable depth (in terms of costs incurred) upstream stood out as the principal source of expectations. However, community shortsightedness, in terms of intensification of irrigation and its fallout on groundwater availability, may push it back to a state of underdevelopment in the foreseeable future. The potential social hazards may prove invincible in such a situation—especially when the *Gram Sabha*, the local institution developed by the villagers at the initiative of TBS—was found to be practically defunct and riddled with conflicts. It is not known whether such a *Gram Sabha* can ever be revived in a situation where private initiative for rainwater harvesting gained has overriding importance at the sponsorship of the intervening agency. Inequality may have sufficient ground to breed even upstream then.

This, however, must not reduce the importance of water harvesting through revival of traditional practices. Then, one must have clear understanding about the extent of the availability of irrigation in the area concerned before intervention is sought. The present study shows that irrigated area as a percentage of the total cropped area was 50 per cent and 60 per cent in the upstream and downstream villages respectively during the starting points of intervention (late 1980s, the initial years of intervention). These do not go well with the guidelines as stipulated by the government of India for planning watershed interventions. Hence, one may have reasons to question selection of the areas for intervention especially when one recognises the national priorities and intends to draw policy implications from the present study.

In other words, one must recognise that the starting points of intervention are critical in that resource allocation in terms of public energy and funds must not be utilised in the areas, which are relatively well off. However, in the present context, since resources are already allocated in these selected areas and rainwater is harvested, the real issue is sustainability of water use. This is essentially a problem of demand management, which is no less important than supply augmentation of groundwater. One can address such problem by way of following basin-level perspective as explained earlier. Such perspective might help in avoiding conflict, if any, across the stream.

It is true that breakdown of village institution contributed immensely to degeneration of status of floral diversity in the downstream village. The absence of basin-centric approach to rainwater harvesting in the study area was not less important to aggravate in the years to come.

4

Development of Farm Forestry

Introduction

The promotion of farm forestry in the 'wasteland' gained considerable importance as a strategy to implement the forestation programme that was launched in the country in pursuance of the National Forest Policy. Farm forestry is but growing trees by farmers individually or in groups on privately owned or leased land. The practice of planting trees by the farmers on their lands is an old one. However, the need for government intervention to promote farm forestry was articulated by the National Commission of Agriculture (NCA) in 1976. The National Forest Policy (NFP) of 1988 also emphasised the role of farm forestry as a resource for forest-based industries and advocated that degraded land should be utilised for tree plantation (Singh, 1994b).

In 1988, when Rajasthan was reeling under severe drought, the Indian Farmers' Fertiliser Corporation Ltd. (IFFCO) initiated a pilot project to promote farm forestry on wasteland in the state. It had a threefold objective including improvement of degraded land, generation of employment and providing fuel and fodder to the rural community. *Tehsils* namely Mavli and Girwa of Udaipur district were chosen to ground the project with cooperativisation as an organisational approach. As many as 10 primary farm forestry cooperative societies (PFFCs) were formed in collaboration with 10 village *panchayats* in these *tehsils* and registered under the State Cooperative Act, 1965. The villages included (1) Sangwa, (2) Rakhyawal, (3) Chandesara, (4) Bijanwas, (5) Mahuda, (6) Sindu, (7) Palana Khurd, (8) Peepalwas, (9) Jawad and (10) Nai.

Initially, the villagers were enthusiastic about the promotion of farm forestry on their village pastures through formation of cooperative societies. On an average, more than 100 people came forward to become members in each of these societies. As much as 1,850 hectares of village pastures that were controlled by the village *panchayats* were brought under the supervision of these societies for the development of farm forestry. The Wasteland Development Board and Department of Non-Conventional Energy, Government of India joined IFFCO in providing financial assistance to achieve the goal.

However, the initial enthusiasm of most villagers gradually waned as the years passed; inefficient management deterred these societies from rooting in the community life of these villages. As a result, they failed to evolve on their own, largely because no serious attempt was made to effectively involve the villagers in the management of the societies. This pushed almost all the societies into dormancy after three years of operation in 1990-91 when IFFCO also stopped providing financial assistance and guidance.

There isn't any problem with the formation of cooperative societies as an organisational form of community institutions while developing and managing a resource base such as the one examined here. However, troubles begin to creep in if stakeholders fail to develop a sense of communal responsibility for developing and managing such a resource base. In Udaipur, perhaps, the problem lay in the perception. As it was found in the field survey, the villagers thought the ownership of the land was with the government, and, therefore, free and theirs to use without any responsibility towards regenerating it. The old practice of using the pastures with callous disregard for their rejuvenation resulted in their degradation to a point of no return. It was the cooperative society's failure to engender that perception that the land was community-owned and, hence, the responsibility of the community to manage it and manage it well.

The cooperative societies, formed in these villages of Udaipur district, did not go through such a process, never had it made participatory, which was why most of the societies became virtually non-existent. When IFFCO learnt about the problems, it raised a sister organisation, the Indian Farm Forestry Development Corporation Ltd. (IFFDC), to bridge the organisational gaps. IFFDC took note of the missing links and began its activities on a pilot basis in the 10 villages mentioned above. Subsequently, the programme was extended to other areas of Udaipur, Rajsamand and Chittorgarh districts in 1995 with financial assistance from the Indo-Canada Environment Facility (ICEF) and IFFCO (IFFDC, 2001-02).

IFFDC reframed the objectives of promotion of farm forestry to include employment generation as a potential benefit that could accrue to the local community. It chose 12, 15 and 18 villages of Udaipur, Rajsamand and Chittorgarh districts, respectively, with a total area of 7,049.6 hectares of village pastures, with the following objectives:

- to develop national, state and village level cooperative institutions that promote farm forestry;
- to assist farmers, specially women, to promote forestation on wastelands by organising PFFCs and making it a people's participative programme;
- to help in improving soil and water conservation and bring about ecological balance; and
- to generate employment opportunities for the rural poor (IFFDC, 2001-02).

The present study concentrates on one such PFFC located at Bijanwas of Mavli *tehsil* of Udaipur district and revived by IFFDC on a pilot basis. This was one of the few societies that functioned relatively better, as claimed by IFFDC officials, even before IFFDC entered the scene.

The case study is divided into eight sections. While section II looks into the historical context in which the intervention was sought, Section III overviews the resource base of the village including land, livestock and human resources that had direct and indirect bearing on the development and conservation of natural resources. Section IV examines how cooperative society evolved as an institution over the years. Subsequently, Section V describes the steps taken by the institution for the conservation and growth of farm forestry. Section VI examines the impact of the institutional intervention on the economic well-being of the village community. The prospect of institution in terms of sustainability within the framework of Cooperative Act is an important issue that has not received much attention. An attempt is made to examine it in Section VII while Section VIII presents a few conclusions.

II

HISTORY OF INTERVENTION

The farm forestry intervention in Bijanwas started with the *sarpanch* of Sangwa *panchayat* advising the *sarpanch* of Bijanwas to develop farm forestry in his *panchayat* with IFFCO support. Having perceived the potential benefit in promoting farm forestry earlier, especially when the villages were reeling under drought, the *sarpanch* persuaded IFFCO to intervene. The proposal found spontaneous acceptance with the residents of Bijanwas and Bikarni (both under the same *panchayat*) at a meeting held subsequently where the decision to develop farm forestry on their pastures was taken. This came against a backdrop of the continuous drought that brought in its wake a growing livelihood crisis with its limited employment opportunities and resources essential for maintaining livestock.

With the proposal gaining the approval of both the villages, the Bijanwas Primary Farm Forestry Cooperative Society was registered in 1988 under the State Cooperative Act, 1965. Pasturelands were leased out to the society initially for a period of five years by the village *panchayat*, though, on completion of the term, the contract was further extended for a period of 25 years. The important terms of the contract between the society and the *panchayat*, on the one hand, and IFFDC, on the other, were as follows:

Society and *Panchayat*

- The ownership of the pastureland would remain with the *panchayat*. The society would be responsible for its development for the lease period.
- Forty per cent of the net profit of society would go to *panchayat*, with 30 per cent to the society and 25 per cent to IFFDC.
- The *panchayat* would have to respect the terms of lease and it would have no right to take back the land from the society unless the latter became defunct.
- The period of contract would have to be renewed once the lease period ends.

Society and IFFDC

- In case the society failed to function, it would have to return all its assets to IFFDC.
- Society would have to submit its annual account to IFFDC regularly.
- IFFDC would have to be taken into confidence on decisions on dividend if any activities other than plantation, like fertiliser distribution, were taken up by the society.
- The rights to scarp the agreement remained with the society.
- Society must accept any change of arrangement that may take place in the light of the contract that IFFDC and ICEF entered.

While Bijanwas PFFCs Ltd. had its own by-laws that governed its functions, the role of IFFDC in influencing the decision making process of the latter has to be critically examined. The role of the external agency in promoting collective action for development may turn out to be inconsequential or adverse to achieving self-sustainability, if the latter continued to be directed by the former. It must also be examined whether a cooperative society that had been in existence for nearly 14 years could generate sufficient momentum for collective action on its own.

Village Profile

Located around 40 km away from Udaipur town, Bijanwas and Bikarni together form the Bijanwas village *panchayat* (Map 9) though Bijanwas is relatively larger, in both geographical area and population (1,021 hectares and 2,601, respectively) than Bikarni (701 hectares and 1,118, respectively). At the time of the study, half the households in both the villages were settled in hamlets around the main village. The noticeable feature of this settlement pattern was that not a single household from the scheduled tribe (ST) of *Gameti* lived in the main village. The villages also had similar landholding patterns with the relatively affluent *Dangis*, belonging to the other backward castes (OBCs), occupying the pre-eminent position (Table 4.1) socially with their larger landholdings and better educational background. Though the *Gametis* were stronger numerically, their small landholdings and poor education put them lower than the *Dangis* in the power hierarchy. Besides, their settlement at the village periphery reinforced their estrangement from the mainstream development process. As many as

44 of 57 medium farmers belonged to the Dangis (Table 4.1). Most of the development initiatives were taken by the households of this community only.

Table 4.1

Caste-wise Landholding Pattern of All Households in Bijanwas and Bikarni

Category	*OBC*		*General*	*SC*	*ST*	*Total*
	Dangi	*Others*				
Landless	—	5	5	3	—	13
Marginal	111	96	40	63	217	527
Small	68	19	18	4	12	121
Medium	44	6	7	—	—	57
Total	223 (31.06)	126 (17.55)	70 (9.75)	70 (9.75)	229 (31.89)	718 (100.0)

Note: 'Others' in OBC category included *Kumhar, Suthar, Teli, Vaishnav, Nai,* etc.
Figures in parentheses are percentages.
OBC: Other backward castes; SC: Scheduled Castes; ST: Scheduled Tribes.
Source: Field survey.

The farm forestry initiative was one such example where the *sarpanch,* who was a *Dangi,* took the lead along with some members of his community. The rest of the villagers supported the move in the hope that such an initiative would ensure employment and fodder. The drought-induced scarcity had created a homogeneous base for all to promote farm forestry in the degraded village pastures. However, the leadership continued to be in the hands of a powerful few from the *Dangi* community.

III

RESOURCES

The status of household livelihood in rural Rajasthan is determined by the village resource base and its availability. Apart from private agricultural land, village commons (other than village pastures) generate livelihood in the form of fuelwood and fodder. An enriched educational background also forms another resource base that expands the scope for livelihood generation in terms of

Map 9

LOCATION MAP OF BIJANWAS AND BIKARNI

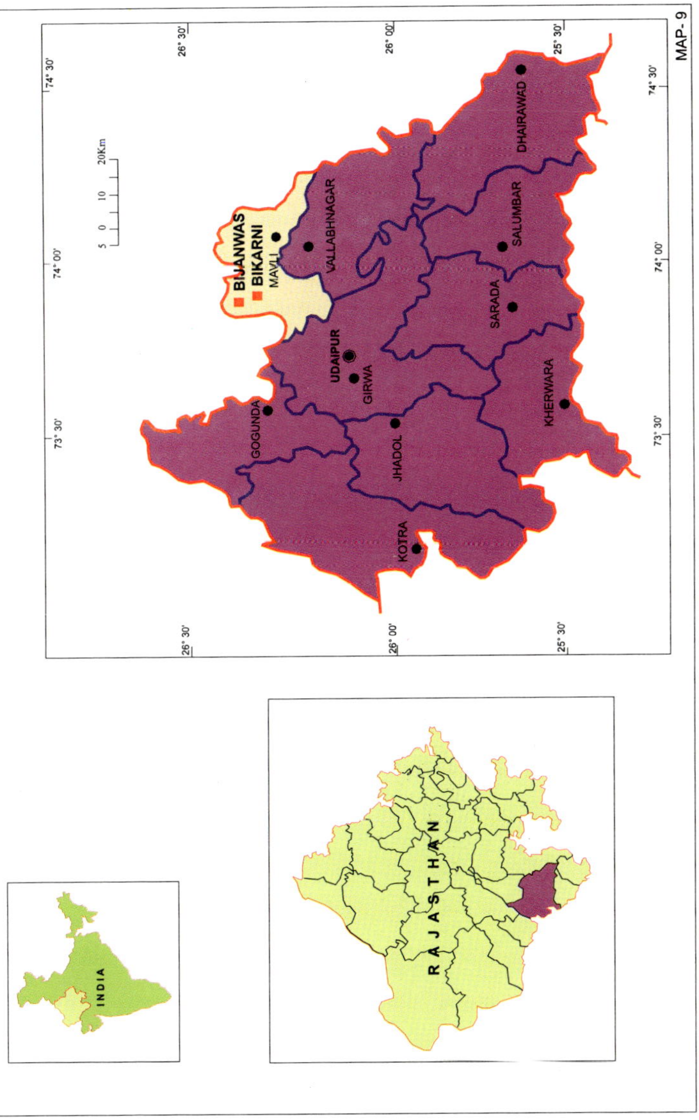

employment and income. In Bijanwas, given the limited prospect of livelihood generation from educational resources and gradual decline of all land-based resources, the villagers found no other option than to jointly seek restoration of the village pastures.

The worst that the village witnessed in the last decade was the decline of agricultural prospect. Only 22 per cent of the available land was cultivated in Bijanwas in 2001-02. It was a drought year. It was worse in Bikarni at 10 per cent. However, it was 34 and 15 per cent, respectively, in 1990-91 when rainfall was normal. While one might explain such a difference in terms of rainfall received, its trend even otherwise was not encouraging (Table 4.2).

Table 4.2

Land Use Pattern of Both Villages

(in hectares)

	Bijanwas		*Bikarni*	
	1990-91	*2001-02*	*1990-91*	*2001-02*
A. Land unavailable for cultivation	326 (32.0)	326 (32.0)	452 (64.0)	452 (64.0)
• Permanent pastures	134	134	62	62
• Others (forest, mountain, barren land, etc.)	192	192	390	390
B. Land available for cultivation but not cultivated*	347 (34.0)	470 (46.0)	141 (20.0)	179 (25.53)
C. Cultivated land	348 (34.0)	227 (22.0)	108 (15.0)	70 (10.0)
• Irrigated	177	45	70	23
• Un-irrigated	171	182	38	47
Total	1021 (100.0)	1023 (100.0)	701 (100.0)	701 (100.0)

Note: * (1) Currently fallow, (2) Fallow 2-5 years, (3) Culturable waste, (4) Tree crops and groves.

Figures in the bracket are percentages of the total land available to the villages.

Source: *Tehsil* office, Mavli.

The year-wise analysis of the land use pattern from 1990-91 (not shown in Table 4.2) shows that the gross cropped area and the area under irrigation were the most in both the villages in 1991. However, after this year, the villages witnessed a decline in both. In other words, the agricultural prospects were gradually limited and, therefore, also livelihood generation from agriculture. The topographical condition of the land being rocky, barren and undulating was also responsible for the constrained agriculture prospects, which could not keep up with the growing population. Table 4.3 shows that the population grew annually @1.42 per cent in Bijanwas, and by 1.80 per cent in Bikarni during 1981-2001. In the absence of agricultural prospects, the only options left to the villagers for livelihood generation were migration for wage employment and rearing of livestock.

Table 4.3

Population of Bijanwas and Bikarni (1981-2001)

Year	*Bijanwas*		*Bikarni*	
	No of HH	*Population*	*No of HH*	*Population*
1981	388	1,961	164	784
1991	458	2,219	181	926
2001	478	2,601	239	1,118

Source: Population Census 1981, 1991 and 2001.

There was a corresponding rise in the livestock population as well, especially of large ruminants, in these villages over the years

Table 4.4

Animal Resources Per Household in Both Villages

Animal	*1992*	*1997*	*2003*
Cow	683	1025	779
(per HH)	(1.07)	(1.43)	(1.09)
Buffalo	993	1004	796
(per HH)	(1.56)	(1.40)	(1.11)
Sheep	590	107	57
(per HH)	(0.92)	(0.15)	(0.08)
Goat	2024	1800	1553
(per HH)	(3.17)	(2.51)	(2.16)
Bullock	756	846	629
(per HH)	(1.18)	(1.18)	(0.88)

Note: Data on livestock pertaining to 1992 and 2003 was collected from the patwari office, Bijanwas and for 1997 from field survey. To estimate livestock per household in 1992, data on the number households was collected from the population census of 1991. In the same way, data on the number of households for 1997 and 2003 was collected from the population census of 2001.

Source: Patwari office, Bijanwas.

(Table 4.4). While both cows and buffalos increased during 1992-1997 (both these years received normal rainfall), the rise in cow population from 683 to 1025 was comparatively much higher than that of buffalo (1.1 per cent). However, the stock of large animals declined sharply in 2003 by 23.34 per cent (671 animals). Of course, there was decline of small ruminants also. While drought in the consecutive three years previously may be a reason for the declining livestock—it was challenging for households to even maintain a minimum stock of animals—the pressure on the village common lands must not be underrated either. In the absence of animal migration from these villages, households had to depend on the local resource base that reached a critical stage in its carrying capacity, especially during drought.

Some might see fluctuations in stock position of animals as an outcome of the household adjustment mechanism being followed against any economic odds that the households encountered. While not going against this view, it is necessary to consider other aspects that normally go unnoticed. Most of the people, especially the poor, are inclined to sell their animals during drought as required by the financial adjustment mechanism of the household. However, they sell at a distress price since supply of animals for sale far exceeds demand at this time. Under such circumstances, those, who can, retain as many animals as possible depending upon the condition of the local resource base. Besides, those who are capable of buying animals also do so. It was found during the survey that sample households not only sold their animals but also bought them even during drought (Table 4.5).

Table 4.5

Animals Bought and Sold by the Sample Households in Bijanwas & Bikarni during 2002

Category	*Animals Dought*		*Animals Sold*	
	Small	*Large*	*Small*	*Large*
Landless	0	0	0	0
Marginal	8	17	25	8
Small	0	3	5	5
Medium	2	14	2	3
Total	10	34	32	16

Source: Field survey.

It appears that villagers preferred to sell small ruminants more than the large animals and buy more large animals as compared to the small ones. More importantly, the volume of such transaction was concentrated among the marginal farmers. In other words, poor farmers tended to make use of the opportunity of falling prices of the large animals by buying them to the extent they could do so. However, fodder availability in the local resource base had to be ascertained.

Another notable aspect of the status of livestock was that there was continuous decline of the stock of small ruminants over the years. The field survey revealed that steady encroachment and enclosure of common lands by private households motivated the village community to stop rearing small ruminants. Enclosure of the encroached land prevented the small ruminants from uninterrupted grazing. This increasingly limited availability of common land was another reason why villagers felt it necessary to improve the conditions of the village pastures where no such encroachment, as the villagers perceived, could take place. However, the encroachment was not limited to the common lands alone. Even pasturelands of both the villages were encroached.

Encroachment

Encroachment of common land was found to be an accepted practice among the villagers. This was what largely ensured fodder availability to the villagers, especially the poor farmers. Village pastures were also encroached for quarrying of soapstone. Quarrying began inside the pastureland in the 1970s when there was no plan for promotion of farm forestry with a 15-year lease from the revenue department. The lease period was ended in the early 1990s. However, the existing *sarpanch* did not take any step to scrap the contract, even when the development of farm forestry began. Rather, another quarry was opened up in the same period inside the pastureland without permission from the village *panchayat* or license from the mining department. There was also extensive quarrying of stone for house construction; private individuals not only quarried stone from their own lands but also from pasture and revenue that were encroached. These lands were levelled after mining was complete.

No definite pattern of encroachment of lands was found in terms of the social and economic background of the villagers. In other words, although the *Dangis* were the most powerful group, households of other less powerful social groups also encroached upon both revenue and pasture lands. However, no landless household encroached upon land (Table 4.6)

Bijanwas had 951 *bighas* of revenue land (common open land) and 837.5 *bighas* of pastureland; 82.98 *bighas* of revenue land (8.72 per cent) and 70.35 *bighas* of pasture land (8.40 per cent) were encroached (Table 4.6). The incidence of encroachment in Bikarni was slightly higher—of 1,574 *bighas* of revenue land 402.40 *bighas* of land was encroached upon. An aggregate 24 per cent of all households together of both the villagers occupied the common land resources. The act of encroachment, gradually, seemed to have gained social legitimacy with a mutually supportive relationship among the villagers. There was no discernible conflict between the encroachers, although social tension seemed to have gradually built up as common land became scarce.

Table 4.6

Incidence of Encroachment of Both Revenue and Pasture Land in Both Villages in 2003-04

(*in bighas*)

Caste	*Bijanwas*				*Bikarni*			
	Revenue Land	*No of HH*	*Pasture Land*	*No. of HH*	*Revenue Land*	*No of HH*	*Pasture Land*	*No of HH*
Dangi	30.15	26	10.30	3	147.10	17	7.60	2
Rajput	13.75	7	20.00	4	—	—	—	—
Bhil (Gameti)	5.50	7	25.75	7	232.80	45	18.50	6
Khateek	20.75	6	—	—	—	—	—	—
Brahmin	0.40	1	—	—	0.50	2	27.50	4
Others	12.43	16	14.30	10	22.00	6	4.00	1
Total land-encroached	82.98	63	70.35	24	402.40	70	57.60	13
Total available-land	951.0		837.5		1574.0		384.0	
Percentage of-encroached-lands to the-total available-land	8.72		8.40		25.57		15.0	

Source: Patwari office, Bijanwas.

It is also equally important to underline that the act of encroachment had nothing much to do with the pattern of land ownership. In other words, it was not necessarily true that those who had large landholdings were more active in occupying common lands. On the contrary, the poor farming households encroached upon more common lands than the richer ones. All together 91.51 per cent of the total encroached land in Bijanwas and Bikarni was shared by marginal and small farmers, and the rest by the medium farmers (Table 4.7). Here, the incidence of encroachment was guided more by availability of common land in view of the topography and settlement pattern of the households.

Table 4.7

Encroachment by the Poor Farmers vis-à-vis the Rich Farmers

(in bighas)

Farmers	*Bijanwas*		*Bikarni*	
	Revenue	*Pasture*	*Revenue*	*Pasture*
Poor farmers	69.03 (83.19)	70.40 (100.00)	366.90 (91.18)	55.00 (95.49)
Rich farmers	13.95 (16.81)	0.00 (0.00)	35.50 (8.82)	2.60 (4.51)
Total	82.98 (100.0)	70.40 (100.0)	402.40 (100.0)	57.60 (100.0)

Note: Figures in parentheses are percentages of the land encroached. Poor farmers include marginal and small farmers. No landless farmers encroached either revenue or pastureland.

Source: *Patwari* office, Bijanwas.

Most of poor households, who belonged to the ST community, lived in hamlets that were located far from the main village. Maps 10 and 11 show that agricultural lands were mostly concentrated near the main settlement, where the scope for encroachment was much less. However, such scope was higher nearer to the lands of poor households, which were located around hamlets.

The act of encroachment of the village commons received the indirect approval of the government too. The revenue department was regular in collecting penalty charges from the encroachers depending upon the condition and use of the land. For example, if the land was utilised for cultivation by the encroacher, agriculture produces were supposed to be auctioned by the government. However, in practice, this hardly happened. Even otherwise, it failed to make any impact on the farmers in stopping them from crop cultivation. Penalty varied from Re. 0.30 to Rs. 15 per *bigha* for each harvest of grass for those types of lands that were not cultivable at all. However, field survey

Map 10

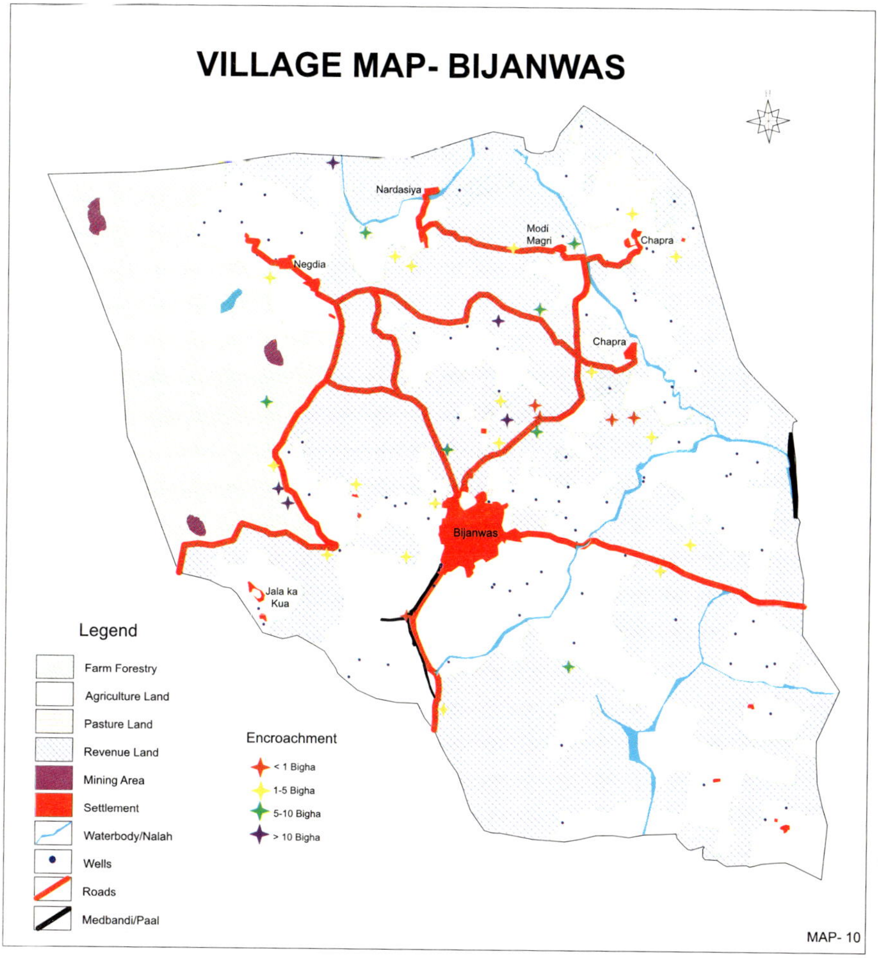
VILLAGE MAP- BIJANWAS
Nardasiya
Modi Magri
Chapra
Negdia
Chapra
Bijanwas
Jala ka Kua
Legend
Farm Forestry
Agriculture Land
Pasture Land
Revenue Land
Mining Area
Settlement
Waterbody/Nalah
Wells
Roads
Medbandi/Paal
Encroachment
< 1 Bigha
1-5 Bigha
5-10 Bigha
> 10 Bigha
MAP- 10

revealed that most of the encroachers of this type of lands were charged Rs. 15 per *bigha*.

In case of cultivable lands, there was a penalty ranging from Rs. 2.50-125 per *bigha* for each harvest of crop. This was in addition to crops, if grown, being auctioned by the revenue department. In practice, auctioning of the crops under the present condition was meaningless. For crops were auctioned to the same farmers who cultivated such crops. Strangely enough, the land was freed each time to be occupied again by the same encroacher when it would be freed again in a repetitive cycle. It must be remembered that the revenue collected from those who encroached pasturelands regularly went to the *panchayat* department.

It is unclear yet how the government should address the act of encroachment on its lands. It is true that it can impose penalty on the encroachers as per the Revenue Act, 1956. But, the manner in which encroachers are penalised is tantamount to no penalty and giving tacit approval to encroachment. This is also a poor reflection of government policy towards using common land and incidence of encroachment on it. This also throws up an important development issue that one must not ignore—especially in the context of the Rajasthan economy that has a bearing on its vast stretches of common land. The issue is whether it is desirable to seek improvement of land that has witnessed continuous deterioration over the years. If so, what role does the state conceive for itself? The tacit state approval of encroachment in these villages only indicates that it favours privatisation of the commons simply because it has failed to bring about improvement.

However, it may be asked whether encroachment should be allowed to continue on state sponsorship. Even if such an action enjoyed political patronage, there would be reason to be concerned about the prospects of the community-owned natural resource base and the serious implications for the development of the village economy, especially of the poor. Besides, there are lessons to be learnt from what happened to the livestock economy as a sequel to encroachment and enclosure, which resulted in the rise of grass production. It may be argued that, in the absence of any regulatory system for improving the conditions of these lands, it was better were private households allowed to grow grass. In other words, it would have been impossible to achieve increased grass production had common lands been allowed to remain in the same condition as before. The rise in grass production in Bijanwas facilitated the growth of the informal grass market, which was completely absent earlier. It is, of course, a different story how such market came into being.

The field survey revealed that villagers learnt to grow grass on the encroached land and sell it only after the cooperative society began its activities. The practice of encroaching was prevalent even before the society came into being though the purpose was not to grow grass. Enquiries made with the villagers revealed that around 1 quintal of grass (average yield rate of both normal and drought years) were available to the encroachers per *bigha* of land in a year. Hence, it can be roughly estimated that 613 quintals of grass was available per year in 613 *bighas* of encroached land in both villages. In order to examine the extent to which villagers increasingly depended on grass and dry fodder from the encroached land, data was collected from the sample respondents (Table 4.8).

Table 4.8

Dependence of the Sample Households on Private Informal Grass Market in Both Villages

(no of households)

	Bijanwas		*Bikarni*	
	Before Farm Forestry	*After Farm Forestry*	*Before Farm Forestry*	*After Farm Forestry*
Green fodder (including grass)	10	22	4	3
Dry fodder (including dry grass)	34	70	13	23
Tree lopping	4	10	2	3

Source: Field survey.

Table 4.8 shows that after farm forestry came into being and the grass market developed, the number of dependent households, especially on dry fodder from private sources, increased sharply. Green fodder included fodder grown on private lands and the green grass available mostly during the rainy season. During rest of the years, demand for dry fodder had a tendency to rise. Where they had earlier depended on the neighbouring villages for dry fodder, the residents now had the informal private grass market to fall back on. While the growth of the grass market contributed substantially to the sustenance of livestock, the compositional shift of livestock that these villages witnessed over the years must not be ignored. The burgeoning grass market seemed to have lent credible support to rearing of large animals at the expense of small ruminants (Table 4.4).

Another reason for which villagers became dependent on the grass market for livestock production was that the villages witnessed a change of cropping pattern from water-intensive crops like wheat to mustard, gram, etc. which required lesser water. Apart from the change in the cropping pattern, the irrigated land, as observed earlier, continuously declined over the years. The increasingly unavailability of land for cultivation, accompanied by the change in cropping patterns, caused a serious gap in the supply of by-products that were useful for livestock consumption. Such a gap was fulfilled by the grass market.

It may be noted that grass was available from three different sources including protected pasturelands, private lands and common open lands that were encroached. The grass available from village pastures was sold to the villagers during the months of October-November in each year. However, farmers sold grass normally before summer set in or when *rabi* crops were harvested. In other words, they sold at the time of scarcity when the price of grass went up to Rs. 2 per kg, which was higher than the price at which society sold (estimated to be Re. 1 per kg).

In view of the constrained agricultural and livestock prospects, wage employment was seen as a reliable option for livelihood generation of many households. An attempt was made to examine which group of farmers migrated more in two different drought years for wage employment. It was found that more marginal and small farmers migrated than medium farmers (Table 4.9).

Table 4.9

Incidence of Migration of the Sample Households

Options	*1987-88*					*2002-03*				
	Landless	*Marginal*	*Small*	*Medium*	*Total*	*Landless*	*Marginal*	*Small*	*Medium*	*Total*
(A) For employment										
(i) Daily	0	10	3	1	14	0	23	6	5	34
(ii) More than one month	0	19	6	3	28	0	28	7	13	48
(B) For animal grazing	0	1	1	0	2	0	0	0	1	1
(C) No migration	3	62	13	20	98	3	41	10	5	59
Total	3	92	23	24	142	3	92	23	24	142

Source: Field survey.

It is clear that the incidence of migration from the same households increased in 2002-03—going up to 58 per cent from 31 per cent in 1987-88. It may be noted that the migration was mainly for obtaining wage employment in cities like Bombay, Ahmedabad, Surat and Udaipur. Other members of the households looked for employment opportunities in quarrying and other activities, commuting daily. However, there was virtually no migration of animal for grazing outside the village. The increased availability of fodder inside the village over the years had reduced the necessity for livestock migration.

The formal educational background of the sample households did not appear to be high enough for it to enhance their employability and generate livelihood. Almost half of the population of the sample households was illiterate (Table 4.10).

Table 4.10

Educational Background of the Sample Households

Educational level	*Male*	*Female*	*Total*
Illiterate (above 5 years of age)	132	251	383 (48.42)
Primary	138	83	221 (27.94)
Sr. Secondary	126	39	165 (20.86)
Above Sr. Secondary	19	3	22 (2.78)
Total	415	376	791 (100.00)

Note: Figures in parentheses are percentages.
Source: Field survey.

IV

COOPERATIVE SOCIETY AS AN INSTITUTION

As mentioned earlier, Bijanwas Cooperative Society came into existence mainly at the initiative of IFFCO, which had considerable influence on the formation of the executive committee of the society and its regular functioning. Since this society was registered under the

State Cooperative Act of 1965, it had to follow the guidelines formulated at the state level. However, the most important aspect was how rules of the institutions were framed and enforced, especially in context of the benefits to be accrued to the people at large from institutional growth without any scope for discrimination. This could come only through promoting the spirit of equal participation of all sections of the community. The Bijanwas Cooperative Society, however, failed to succeed in involving the village community. Even the selected members of the executive committee were hardly taken into confidence. The irony was that almost the same set of people continued to be the members from 1992 onwards (Table 4.11). Though elections took place twice, in 1992 and 1996, they were in name only, having been conducted to maintain the organisational *status quo*.

Table 4.11

Composition of the Members of the Executive Committee of the Cooperative Society, Bijanwas

Name of the Members	*Category of Farmers*	*Social Background*	*Year*	
			1992	*1996*
Bheru Lal	Middle	Dangi	Yes	Yes
Panna Lal	Marginal	Suthar	Yes	Yes
Madhav	Small	Brahmin	Yes	Yes
Lila Devi	Small	Brahmin	No	Yes
Fatta	Marginal	Gameti	Yes	Yes
Radhu	Marginal	Gameti	Yes	Yes
Shyam Lal	Landless	Jain	Yes	Yes
Bhawarlal	Marginal	Dangi	Yes	No
Omkar Lal	Marginal	Kalal	Yes	Yes
Shankar Giri (from Bikarni village and employee of the society)	Marginal	Gusai	No	Yes
Mohan Lal	Marginal	Luhar	Yes	No
Ganesh	Marginal	Khatik	No	Yes
IFFDC			—	—
Cooperative			—	—

Source: Audit reports, Bijanwas Primary Farm Forestry Cooperative Society Ltd.

It is clear that the composition of the executive committee of the society did not change significantly. Truly speaking, Ganesh was the only new member to be inducted in the committee in 1996, who was neither a society employee nor a relative of any other member. However, other cases gave different signals. For instance, Leela Devi was inducted as a new member only to replace her husband, Madhav.

While an almost unchanged organisational *status quo* of the cooperative society gives wrong signals about its representative character, its continuation had much to do with the control exercised by president of the society ever since the society came into being. Here was a case of how leadership, emerging out of a powerful social group, could take advantage of indifferent members with poor social and economic background and remain uncontested. Table 4.11 shows that a majority of the members hailed from poor farming categories and marginalised social background. The only exception was Bheru Lal Dangi, a medium farmer with strong social backing. President of the Society since 1988, ensured that his position remained unchallenged by other powerful persons even from the same *Dangi* community even as he retained monopolistic control over organisational leadership.

He neither invited competition nor was inclined to accept opposition, ensuring total control through selection of those people, as committee members, who were never vocal and remained outside the mainstream development initiatives taken in these villages. It was definitely a progressive attempt to bring the subordinate to the centre of the development activities, though with a very different motive from what appeared. Discussions with members revealed that they were hardly informed about meetings, decisions, etc. Only the secretary was taken into confidence by the president. In other words, institutional growth and stability for promoting farm forestry failed to be rooted in the popular perception of the village community. The monopolistic functioning dictated the terms of evolution of the society. Scepticism about its sustainability was, therefore, hardly surprising.

The cooperative society had another organisational limitation since its inception that of intra-village rivalry between Bijanwas and Bikarni over management never resolved by the president. The big brother attitude by the Bijanwas-dominated society management created a rift between the two villages. There was little representation in the executive committee from Bikarni from day one of its inception; the only person from Bikarni, appointed as a guard, could never voice any resentment since he was an employee of the society. The growing disenchantment in Bikarni with the society did not allow them to develop a sense of community ownership of the farm forestry. This was also perhaps why there was more incidence of freeriding here.

Fortunately, this failed to gain momentum due to the initiative taken by Bikarni residents themselves. Apart from the existence value of the forest, its importance to their livelihood generation was realised by the villagers when they understood the loss 'they' would suffer if tree felling continued. Coming together on their own, they engaged themselves to rebuild their forest in defiance of the rules of the cooperative society headquartered at Bijanwas. The context in which village community of Bikarni exhibited their eagerness to rebuild their farm forestry illustrated the potency and resilience of the informal institution.

V

STEPS TAKEN

As the Bijanwas cooperative society evolved as an institution over the years, it took several steps for the conservation and growth of farm forestry. The society's activities can be grouped into two major areas of land improvement and plantation, and grass and fuelwood distribution. As a first step, the society invited several agricultural scientists, who advised it on the construction of contour trenches, staggered trenches, box-cum-pits, stone check dams, etc. These were all required to improve the moisture content of the soil and improve the quality of land. Initially, 100 hectares of land were earmarked for plantation of both tree and grass species. This was followed by another 100 hectares of land in the subsequent year.

An absolute care, as claimed by the secretary and the president of the society and the office bearers of IFFDC, was taken in the choice of the species for plantation. The selection was based on the trees' suitability to the local agro-climatic conditions and their usefulness in terms of fuelwood, tree lopping, and timber and fruit production. A proportion of 40:40:20 was followed for plantation of tree species for fuelwood, timber and fruits. The plantation of grass seeds, of different varieties including *Cenchrus ciliaris* (*dhaman*) sourced from Central Arid Zone Research Institute (CAZRI), was an equally important activity taken up by the society, since grass was useful for not only grazing but also arresting soil erosion. Grass cover was also necessary for the survival of the young saplings planted on barren hillocks. The society did not have a nursery of its own and had to purchase species from elsewhere. However, a nursery was developed in Bikarni in 1991-92 that helped tree plantation a great deal in the village pastures. Besides, a large number of households also purchased saplings for planting them in their own lands.

Contour trenches of 0.3x0.3 metres (m) were dug on hill slopes and tree saplings planted in the space between two trenches. The method of 'box-cum-pit' was also followed to grow plants on the slopes with each box-cum-pit constructed separately for each plant in order to save and use water judiciously. With boxes sized at 1.25x.45x.45 m and dug at a distance of 2.5 m, this method was particularly useful for growing plants on rocky surfaces. In places where the land was not so steep, staggered trenches were dug with plantation by their side. Apart from this, stone check dams were constructed to arrest the rainwater run-off and stop soil erosion.

It may be noted that as many as 400 check dams were constructed by the watershed department at different points of time. In fact, their construction began much before the farm forestry programme was initiated on these lands. However, once the society came into being and the programme was started, these check dams proved extremely beneficial. In addition to these, five check dams were constructed by IFFDC and the *panchayat*. The watershed department also made six feet deep trenches some 2,000 m in length.

Another important feature of the plantation was that the society chose the indigenous method because of its appropriateness to soil conditions and other climatic factors (*see box*).

Indigenous Technique

From the beginning, plantation techniques that could contribute greatly to improve vegetation cover on barren hills were adopted. The pits were filled with cow dung mixed with black soil collected from the village tank. This was conducive to good growth of not only the saplings but also of grass simultaneously because of the percolation of water and manure provided to the plants. Wherever it was difficult to dig pits on the slopes, seeds, such as *Prosopis juliflora* (*vilayati babool*), *Acacia nilotica* (*desi babool*), *Anogeissus pendula* (*dhokda*), *Azadirachta indica* (*neem*), *Acacia senegal* (*kumta*), etc. were sprayed. Normally, this was performed before the monsoon set in. This too helped grass grow around these plants during the monsoon. In this way, nearly five lakh plants and trees were recorded to have grown in both the village pastures.

In order to protect the pasturelands of both the villages, walls were constructed on all sides with locally available stone. Besides, thorny plant species such as *Euphorbia caducifolia,* locally known as *thor* was planted along the boundary as another means of protection.

Management of Grass Distribution

Once grass started growing, after a year or so, the task ahead of the society was to evolve a management regime for distribution of grass among the villagers at administered prices. It had to decide on the quantity of the grass to be distributed to each household and the rate at which it was to be done so. These decisions were based on the rainfall received and yield rate of the grass during the year. While these criteria were acceptable to both the villages, the Bikarni community resented the management rules for grass distribution. For, as they argued, the topography of their village pastures was undulating, steep and rocky, which posed serious difficulties for them in harvesting grass. Bikarni residents felt that the society management refused to understand their problems and appropriately design management rules for harvesting grass from their village pastures. It was found during the field survey that residents felt the society's approach was discriminatory and inequitable.

Initially, the system of harvesting grass allowed villagers to harvest grass on their own for 3-4 days a year. Each household had to deploy one person to harvest as much grass as he or she could from 10:00 am to 3:00 pm. Each household had to deposit Rs. 30 before harvesting grass. However, residents of Bikarni refused to pay this amount because collection of grass from the sloping terrain of its pastureland was more difficult. This led to the introduction in both villages of issuance of a coupon against the ration card of the household and a fee of Rs. 15. This too did not work in the following year with villagers failing to follow the timings stipulated for harvesting. The society then began to engage wage labour at its own cost to harvest grass. The coupon system, however, remained intact. Labourers now made bundles of 1 kg for distribution at the rate fixed by the society. The rate per bundle increased from Re. 0.60 during 1994-95 to Re. 0.70 in 1995-96 and Re. 0.80 in 1998-99, finally going up to Re. 1, which prevailed until recently. But Bikarni inhabitants refused to pay at this rate for they found the charges still higher. It may be noted that the labour had to be paid higher wages to harvest grass on sloping terrain.

In view of this non-cooperation by Bikarni, it was now decided to introduce the auctioning of harvested grass. However, this too failed to yield desirable results. For some powerful people in Bikarni formed a cartel and jointly bid much lower. The irony was that they re-sold the same grass at a relatively higher price to the villagers. Villagers were also willing to purchase at that price because it was the ongoing rate in the informal grass market. It has to be borne in mind that it was the informal grass market on which villagers depended more than their own source; collection of grass from village pastures remained a secondary source for them as long as the informal grass market met their requirements. As observed earlier, the incidence of encroachment was much higher in Bikarni. Adequate availability of grass on the encroached land made them somewhat indifferent to access grass grown on the village pastures.

In Bijanwas, however, the coupon system of grass distribution continued uninterruptedly (Table 4.12). It may be noted that the production in the village pasture here was more than the distribution. The field survey revealed that lack of proper management, especially at harvest time, led to considerable wastage. Despite this limitation, the system proved to be equity driven.

Table 4.12

Distribution of Grass in Bijanwas through Cooperative Society

Year	*No. of Households Benefited*	*Average No. of Bundles Received by Each Benefited Household*	*Rate per Bundle*	*Total Quantity of Grass Distributed (in Quintal)*
1994-95	270	300	0.60	810
1995-96	230	150	0.70	345
1996-97	452	199	0.75	900
1997-98	400	200	0.75	800
1998-99	368	381	0.80	1402
1999-00	400	100	0.80	400
2000-01	579	60	1.00	345
2001-02	395	150	1.00	593
2002-03	463	100	1.00	463
2003-04	434	150	1.00	651

Source: Bijanwas Primary Farm Forestry Cooperative Society Ltd.

It appears that the per household availability of grass from the society was much lower during the drought years, varying from 60-150 kg per household. Importantly, however, the villagers were certain about the availability of grass even in drought whatever the quantity. Hence, in terms of systemic efficiency of grass distribution, irrespective of how it was run, the cooperative society seemed to have made a positive dent as an institution at least in Bijanwas.

The second most important issue was whether the management system of grass distribution ensured self-sustainability for the society. The data available shows that gross income from the sale of grass alone was to the tune of Rs. 3.65 lakh against the total expenses of Rs 2.69 lakh incurred in harvesting and distribution of grass from 1995-96 to 2000-01 (Table 4.13). Of course, this is not a comprehensive and credible analysis to show how had the society achieved self-sustainability over the years without taking into consideration the income and expenditure status of other activities. Nonetheless, it suggests that the institution that evolved even within the cooperative framework of the type under the present study had the potential to ensure sustainability through its own efforts.

In fact, it had to excel further to utilise its full potential so that it did not have to depend on IFFDC even for the payment of salaries to the secretary and forest guards that it recruited. Table 4.13 shows that around Rs 3.4 lakh was paid as salary during 1995-2001, which the society received as development grant from IFFDC.

Table 4.13

Income and Expenditure of the Cooperative Society, 1995-96 to 2000-01

(in Rs.)

Sources	*Income*	*Expenditure*
Nursery	2,61,373	1,37,265
Grass	3,64,792	2,68,662
Fuelwood	9,594	3,720
Interest on bank deposit	64,853	
Revolving fund	10,696	
Salary of secretary & forest guards		3,40,790
Other activities	26,885	49,775
Total	7,38,193	8,00,212

Note: Other activities included the sale of cattle feed, dry fodder, urea, maize seed, sugar, etc. The society constructed its office building at a cost of Rs. 1,42,099, which was spent from its own income. This is not shown in the table.

Source: Bijanwas Primary Farm Forestry Cooperative Society Ltd.

Table 4.13 shows that the sale of grass and plant species from the nursery was a major source of income for the society. Sale of fuelwood was not a significant source since, as explained later, villagers could procure it from private and common lands. In any case, the balance sheet of the society, mainly in terms of income and expenditure, roughly indicates that the society was quite capable of ensuring sustainability and scaling up its activities. However, it all finally depends on how it could succeed in making the organisation more participatory.

Apart from promoting farm forestry, the society performed several other activities, as shown in Table 4.13, for the benefit of the villagers. One such welfare activity was the opening up of the grass depot in the village during all the drought years in the 1999-2003 decade. The society was subsidised by IFFDC to sell grass to the villagers at a lower price. This helped the villagers significantly and cooperative society in turn received, as the field survey revealed, incredible support. However, this was not true of other activities that the society performed without any real commitment. For instance, the goat breed improvement programme, which had no tangible benefit. It was found in the field survey that such an initiative had no appeal for the villagers. Similarly, the society's sale of blankets, tea, urea, cattle feed, maize seed, etc. at lower prices also flopped. The gap in the supply and the demand precluded equal access to all. The sale of fertiliser through the society failed to yield encouraging results. Villagers could purchase the same from the cooperative society, which was managed by the village *panchayat* at the same rate. Besides, they had the option of buying from the open market on credit although at a higher price.

It is hard to understand why the society undertook activities that did not have continuity, consistency and wider coverage. The idea of the society, established for a specific purpose, expanding its role to gain the confidence of the villagers in it as a forum for sustainable management of natural resources and village development was definitely laudable. But, in practice, the services provided discretely, as narrated above, failed to evince a sense of commitment to the society among the residents. Rather, the management's indecisiveness in penalising free riders—although it declared its intention of imposing penalty through the local police station, there is no evidence to suggest it—only added to the lack of commitment.

VI

IMPACT OF THE INTERVENTION

The growth of farm forestry in the once denuded village pastures brought about several changes that directly and indirectly benefited the economy of both villages. Even its existence value was very high for the villagers. They may have been, by and large, indifferent to the management of the farm forest, but they remained uncompromising towards the conservation of the one that they had lost a few years ago. This was keenly registered in a village meeting during the survey. It was significant that farm forestry had created space in the social life of the village community. As they learnt its importance over the years, the villagers became more conscious of conserving it, although they were not involved in management due to reasons explained elsewhere. The tangible benefits of farm forestry were varied in magnitude. The impact was discernible, largely, in three areas, including the availability of grass, fuelwood and tree lopping and villagers' access to them; floral diversity and creation of employment opportunities.

Access to Resources

Access to improved resource base by all groups of farmers, especially the poor, significantly determines institutional stability and prospects. If the rules of cooperative society are enforced to the disadvantage of the poor, there is every possibility of conflict over ownership of resources. This may be reason enough to debase a cooperative institution that has been in existence for over one decade. However, at the time of the field survey, the society had no such threat perception, for very little of the villagers' requirements of grass, fuelwood and tree lopping are met by the village pastures. Their needs were largely fulfilled from private lands (Figure 4.1). It may be mentioned here that private land also includes encroached lands. It was found that, of their total requirement, villagers could collect only 6.16 per cent grass, 9.58 per cent fuelwood and 0.70 per cent tree lopping in a year from the village pastures or common protected land. Obviously, villagers would be less concerned about the availability of these resources in the village pastures as long as their requirements were met from the other two alternative sources. Hence, one does not see these common property resources as an immediate potential source of conflict. Nevertheless, apprehensions exist of conflict emerging in the future when competing claims on the common open land (revenue land) for these resources would increase. This may aggravate further if encroached lands are decolonised.

Figure 4.1

Availability of Grass, Fuelwood and Tree lopping from Difference Sources

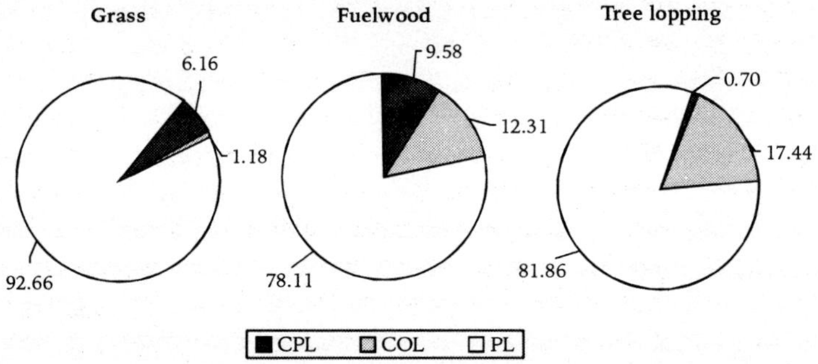

Note: CPL: common protected land; COL: common open land; PL: private land.
Source: Field survey.

However, it is important to note that the present management regime of the cooperative institution has been able to successfully enforce its rules of allocation of these resources across all categories of farmers, including the poor, without perpetrating discrimination (Table 4.14).

Table 4.14

Average Collection of Grass, Tree Lopping and Fuelwood Per Household from Protected Pastureland vis-à-vis Common Open and Private Lands

(in quintals)

Category of Landholdings	*Grass*				*Fuelwood*				*Tree Lopping*			
	CPL	*COL*	*PL*	*Total*	*CPL*	*COL*	*PL*	*Total*	*CPL*	*COL*	*PL*	*Total*
Landless	1.13	0.00	0.00	1.13	0.00	0.00	0.00	0.00	0.00	0.00	0.00	0.00
Marginal	1.27	0.26	18.85	20.38	1.70	2.17	10.00	13.87	0.10	2.08	9.53	11.71
Small	3.07	0.78	25.39	29.24	1.57	0.83	11.11	13.51	0.00	0.78	2.13	2.91
Medium	1.34	0.00	42.37	43.71	0.00	1.22	16.69	17.91	0.00	0.08	2.69	2.77
Average	1.57	0.30	23.62	25.49	1.37	1.76	11.17	14.3	0.06	1.50	7.04	8.60

Note: CPL: common protected land; COL: common open land; PL: private land.
Source: Field survey.

Ecological Impact

Wasteland development through commercial plantation assumes considerable importance not only in terms of improvement of ecology but also in that of the economic status of the people. Plantation, which is sought to transform wastelands into fertile lands, facilitates a mature interactive process to take place between abiotic and biotic factors of the ecosystem. Of course, it produces tremendous existence value (Bromley, 1998). Once the plantation of trees, along with grasses, takes place, the natural process of ecological recovery is activated through reduction in soil erosion, continuous litter fall, improvement in the microenvironment, and, finally, growth of biomass. However, much depends upon the nature of the plant species selected for wasteland development.

The selection of plant species plays an important role in achieving a desirable plantation programme. Several instances have shown how flawed selection of species adversely affects the ecosystem. The reclamation of soil in tea orchards near the Kaziranga national park is a case in point, where the luxuriant spread of *Mimosa sp.* hindered the growth of native grass species, which were important for the survival of animals such as the rhinoceros and the elephant, in adjacent wildlife park.

The farm forestry development in Bijanwas and Bikarni was one such attempt to convert wasteland into fertile land. Villagers were involved from the primary stage of planning and implementation of programme for planting an almost similar type of species in both the villages although both differ topographically. The terrain in Bijanwas is a mix of undulating and periplain with yellow-brown gravely soil. Bikarni, on the other hand, has a hilly terrain although the quality of the soil is the same.

Another notable feature of the intervention is that the Bijanwas site is a continuum one or available in one single plot whereas the Bikarni site is divided into three plots based on the topography of the area. Plantation on all plots took place one after the other at intervals. To protect them, all the sites were fenced in and grazing disallowed. However, it will take a few years for the trees to mature and benefits to accrue to the villagers.

The present exercise has been undertaken to assess the extent to which the conservation programme contributed to the improvement of the local ecosystem. It is in this context that the floral diversity is measured based on standard indices.

Floral Diversity

The field team did the sampling twice, in August 2003 and February 2004. The sample size of both grass and tree species taken together was 80 for both the villages (40 samples in each village).

Importance Value Index

Tree and Shrubs

The importance value index (IVI) indicates the hierarchical position of the plant species. Analysis of the species in terms of their hierarchical position also indicates the direction in which the ecosystem swings. If an unsuitable species or groups of species assume higher IVI, there is great potential for disastrous consequences for the ecosystem. It may be noted that the pattern of plantation and the species chosen for both trees and grasses was the same for both the villages. The only difference, as mentioned earlier, was the topography. The question is, can there be a difference in the status of floral diversity of both sites in terms of the IVI and other measures of diversity. A species that dominates the hierarchy at the initial stage of development of an ecosystem may lose its supremacy in successive phases. It all depends on how the species adapts to the environment. The number of tree and shrub species was 15 in Bijanwas, while it was 28 in Bikarni. With its capacity to grow on wasteland, *Leucaena leucocephala* became the dominant species on both sites.

The society planted *Leucaena leucocephala, Dandrocalamus strictus, Azadirachta indica, Prosopis cinereria, Albizia lebbeck, Dalbergia sissoo, Butea monosperma, Acacia nilotica, Acacia catechu* and *Emblica officinalis* on both sites. In Bijanwas, *Leucenea leucocephala* showed the highest IVI (62.9) followed by *Acacia nilotica* (56.31), *Ziziphus mauritiana*(47.54) with *Prosopis cinereria* (2.3) at the lowest (Table 4.15). *Leucaena leucocephala* is an important plant for forage and timber. At the time of the field survey, it was found to be in the growing stage and not ready for use as timber. However, there were some plants, such as *Prosopis cineraria,* that were being used for forage for small ruminants in both the villages. This was why its extensive growth was stalled; this can be gauged from its lower IVI on both sites.

Table 4.15

Importance Value Index for Tree and Shrub Species

IVI	*Bijanwas*	*Bikarni*
0-10	*Calotropis procera* *Cassia auriculata* *Dalbergia sissoo* *Eucalyptus tereticornis* *Prosopis juliflora* *Prosopis cinereria* *Azadirachta indica*	*Prosopis juliflora* *Mayteurus emarginata* *Cassia auriculata* *Albizia lebbeck* *Prosopis cinereria* *Calotropis procera* *Barleria priorites* *Derris indica* *Grewia flavescens* *Butea monosperma* *Acacia nilotica* *Grewia flavescens* *Dalbergia sissoo* *Anogeissus pendula* *Acacia senegal* *Boswellia serrata* *Emblica officinalis* *Cassia fistula* *Ailanthus excelsa*
10-25	*Hibiscus microranthus* *Rhus misurensis* *Dandrocalamus strictus** *Mayteurus emarginata*	*Ziziphus mauritiana* *Hibiscus microranthus* *Acacia catechu* *Rhus misurensis* *Annona squamosa* *Eucalyptus tereticornis* *Azadirachta indica*
25-60	*Acacia nilotica* *Ziziphus mauritiana* *Euphorbia caducifolia*	*Leucaena lucocephala* *Euphorbia caducifolia* *Dandrocalamus strictus**
60-100	*Leucaena lucocephala*	

Note: *Gramineae.
Source: Prepared on the basis of field survey.

In Bikarni, *Leucaena leucocephala* (53.85) had the highest IVI and *Ailanthus excelsa* (1.34) had the lowest (Table 4.15). A significant aspect of the IVI analysis was that species that were planted on both sites did not grow equally well. While IVI of some plants declined, some were wiped out in the process of succession in Bijanwas. It was different in Bikarni. For example, Table 4.15 shows that *Leucaena leucocephala, Dandrocalamus strictus, Eucalyptus teretecornis* and *Acacia catechu*, which were commercially important, had higher IVI on Bikarni sites. More importantly there was a large number of species that were absolutely missing even at lower IVI.

Grass and Forbs

The IVI analysis of grass and forbs shows that *Cenchrus setigerus* that was planted on both sites had lower IVI (within the range of 10-25, Table 4.16) than those grasses, which grew naturally. The latter included *Sehima nervosum, Cyprus kyllinga, Cynodon dactylon, Aristida funiculata, Cenchrus ciliaris and Boerhavia diffusa,* etc. *Cenchrus setigerus* was not found to be successful despite the fact that it exhibited adaptive behaviour even in adverse climatic conditions. The prolonged three-four year drought seemed to have constrained its growth. It may be noted that grasses such as *Sehima nervosum* and *Cyprus kyllinga* (more than 25 IVI) were dominant (Table 4.16) in both villages and were native and palatable. The analysis indicates that native species succeeded in occupying the dominant status in the hierarchy by virtue of having more capacity to survive under harsh conditions as

Table 4.16

Importance Value Index for Grass and Forbs Species

IVI	*Bijanwas*	*Bikarni*
0-10	*Borreria stricta* *Ipomoea eriocarpa* *Lepidagathis trinervis* *Ocimum americanum* *Dactyloctenium sindicum* *Lasiurus sindicus* *Cyperus compressus* *Vernonia cinerea* *Boerhavia diffusa*	*Dactyloctenium sindicum* *Lepidagathis trinervis* *Ipomoea eriocarpa* *Mollugo cerviana* *Boerhavia diffusa* *Ocimum americanum* *Indigofera cordifolia* *Cyperus compressus* *Melanocenchrus jacquemontii* *Vernonia cinerea* *Glossocardia bosvelea* *Lindenbergia muraria*
10-25	*Polygala erioptera* *Cenchrus ciliaris* *Melanocenchrus jacquemontii* *Cenchrus setigerus*	*Cynodon dactylon* *Sehima nervosum* *Polygala erioptera* *Cenchrus ciliaris* *Dichanthium annulatum* *Borreria stricta* *Lasiurus sindicus* *Cenchrus setigerus*
25-60	*Cyprus kyllinga* *Aristida funiculata* *Cynodon dactylon*	*Cyprus kyllinga* *Aristida funiculata*
60-100	*Sehima nervosum*	

Source: Prepared on the basis of field survey.

compared to species like the exotic *Cenchrus setigerus*. It is clear from the preceding IVI analysis that Bikarni had more variety of species than Bijanwas.

Table 4.17 shows that, at 2.40, Bikarni achieved relatively more floral diversity than Bijanwas at 1.97. At the disaggregated level, although the diversity of both tree and grass species was more in Bikarni, it was more in respect of the former. The reason, which has been explained earlier, was the suitability of the plant species to the Bikarni site. Contour bunding and construction of trenches transformed this site into a better habitat with higher moisture content in the soil along the slopes and across them. In addition to this, evapo-transpiration was low here because of the vegetation cover on the slopes; there is lower solar radiation on a sloping topography, which reduces the heating of the upper surface of the crust. Low evapo-transpiration maintains pH of the soil. The consequent improvement in the microenvironment along with improved edaphic factors increased the stability of the ecosystem.

Table 4.17

Index of Diversity and Dominance of Trees and Grass Species

	Tree		*Grass*		*Total*	
	Bijanwas	*Bikarni*	*Bijanwas*	*Bikarni*	*Bijanwas*	*Bikarni*
Index of dominance	0.13	0.07	0.13	0.09	0.26	0.16
Index of diversity	0.97	1.27	1.00	1.13	1.97	2.4

Source: Prepared on the basis of field survey.

The index of dominance was 0.13 for tree and shrub species in Bijanwas and 0.07 in Bikarni (Table 4.17). Bijanwas showed a higher index of dominance for grass species also at 0.13; in Bikarni it was 0.09. Bikarni had a higher number of tree and shrub species, which were more evenly present. Higher IVI was recorded for *Cenchrus setigerus*, which was planted at both the sites, at Bikarni. This was due to the favourable microenvironment that developed here.

Table 4.18

Species Richness Index for Tree and Grass Species

Village	*Tree*	*Grass*	*Total*
Bijanwas	0.91	0.31	1.22
Bikarni	1.50	0.40	1.90

Source: Prepared on the basis of field survey.

Species Richness Index

The species richness index was recorded at 0.91 and 1.50 for tree and shrubs species in Bijanwas and Bikarni villages, respectively (Table 4.18). However, the values for grass and forbs species were 0.31 and 0.40, respectively. Higher index was recorded for all species (trees and shrubs, and grass and forbs) at Bikarni due to diversity in morphological features at this site as well as better edaphic features.

Similarity Index

The similarity between the two sites was more among the grass species at 0.87 as compared to the tree species at 0.68 (Table 4.19). This was mainly because the survival rate of the natural grass species appeared to be much higher than the planted ones such as *Cenchrus setigerus*. Even though a similar pattern of plantation was followed on both sites, as noted earlier, the tree species adapted comparatively better to the Bikarni habitat with an increased survival rate. The relatively harsher Bijanwas environment was not conducive to a good survival rate resulting in a lower tree similarity index. Three factors led to the higher survival rate at Bikarni—better moisture and humus in soil and the 'natural' protection offered by the difficult terrain that restrained human and livestock movement.

Table 4.19

Similarity Index of Two Protected Sites

Tree	*Grass*
0.68	0.87

Source: Prepared on the basis of field survey.

In any plantation programme, it is the survivorship of the planted species that indicates the technical efficiency of the intervention. Here, technical efficiency essentially refers to the appropriateness of the selection of species in terms of climatic conditions, topography and edaphic characteristics. If these aspects are taken into consideration at the time of selection of species, the survival of the plants is largely ensured. In view of this, the Bijanwas plantation programme did not yield enough even after a decade to say that it succeeded. Only 34.5 per cent of the planted species on an average could survive till the end of 2004 (Table 4.20). The declining trend of survival is shown in Figure 4.2. The total number of species that survived at the end of 2004 was 1,25,545 as against the 3,63,796 species that were planted initially. Interestingly, 98,290 native species were regenerated, contributing more to achieving floral diversity than the planted species.

Table 4.20

Stock of Planted Species in 2004

Species	*Botanical Name*	*Planted*	*Plants Survived*	*Plants Damaged*	*Plants Survival Rate %*
Bamboo	*Dandrocalamus strictus*	1,435	816	619	56.86
Desi babool	*Acacia nilotica*	520	207	313	39.8
Eucalyptus	*Eucalyptus tereticornis*	742	277	465	37.33
Kanji	*Derris indica*	1,207	721	486	59.73
Neem	*Azadirachta indica*	1,170	461	709	39.4
Prosopis	*Prosopis species*	18,032	3,878	14,154	21.5
Siris	*Albizia lebbeck*	1,445	379	1,066	26.22
Shisham	*Dalbergia sissoo*	3,233	1,465	1,768	45.31
Su. Babool	*Leucaena lucocephala*	45,156	14,567	30,589	32.25
Others		2,90,856	1,02,774	1,88,082	35.33
Total		3,63,796	1,25,545	2,38,251	34.5

Source: IFFCO & IFFDC office, Udaipur, 1988-2004.

To sum up, Bikarni achieved more floral diversity due to all the reasons mentioned elsewhere. In the case of grass species, IVI analysis shows that native species succeeded in occupying the dominant status in hierarchy by virtue of having greater capacity to survive harsh conditions as compared to the exotic species. Grass species that grew naturally were also palatable. Importantly, the inter-species association between natural and planted species that grew stronger over the years contributed a great deal to the achievement of floral diversity whatever its extent. The lower survival rate of plant and grass species shows inappropriate selection of species against the backdrop of edaphic characteristics and climatic conditions.

Employment Opportunities

As mentioned earlier, farm forestry was taken up during drought when villagers had to face severe livelihood crisis. In fact, it was the creation of employment opportunities through farm forestry, which prompted the villagers to accept the programme readily. In every possible sense, it was a well-conceived programme.

Figure 4.2

Survival Trend of Planted Species of Bijanwas & Bikarni

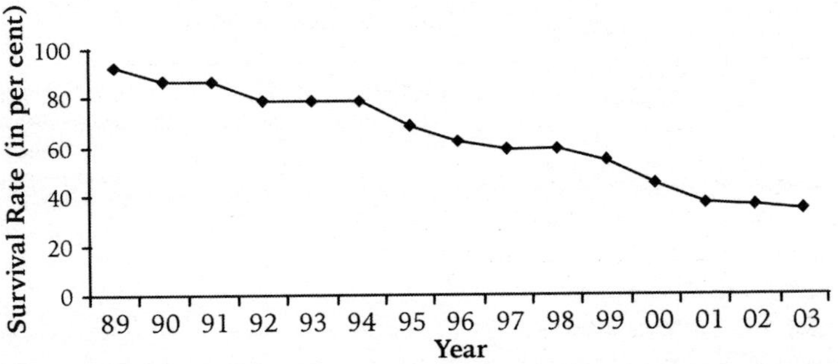

Source: Estimated on the basis of data collected from IFFCO & IFFDC office, Udaipur, 1988-2004.

The village economy gained in terms of creation of employment opportunities inside the village on the one hand, while on the other, its barren land was transformed into a farm forest that provided fodder and fuel to the villagers. Table 4.21 shows the extent to which employment was generated by the cooperative society for both men and women in both villages in terms of man-days during 1988-2003. As many as 1,57,716 man-days of employment were created. At the initial stage, construction works related to soil and water conservation were taken up which was why employment generation was comparatively higher. Later on, it centred mostly around the maintenance of plantation and conducting other activities such as harvesting of grass and fuelwood.

Table 4.21

Man-days of Employment Created by the Society (1988-2003)

Activities	*Male*	*Female*	*Total*
Pit digging	36,149	47,918	84,067
Plantation	8,967	13,451	22,418
Maintenance	13,571	22,530	36,101
Others*	5,610	9,520	15,130
Total	64,297	93,419	1,57,716

Note: Data for 1988-2000 was collected from the society. Data for three years from 2001-2003 was estimated on the wages paid to the workers who were deployed by the society on daily wage basis and bundles of grass and fuelwood harvested as fixed by the society.

* Others included harvesting of grass, fuelwood, and other activities.

Source: Bijanwas Primary Farm Forestry Cooperative Society.

Table 4.22 shows that the society spent Rs. 37,61,046 in 15 years on 1,57,716 man-days of employment. In other words, roughly 10,514 man-days of employment were generated in a year for the villagers at an average wage of Rs. 23.85 per day. Although IFFCO/IFFDC made the largest contribution to the investment towards wage payment, the cooperative society, as seen earlier in Table 4.13, had displayed its ability to manage these activities on its own. However, it was yet to run independently of IFFDC.

Table 4.22

Activity-wise Wage Cost Incurred by the Society (1988-2003)

Activities	*Total Man-days*	*Average Wage Rate per Day*	*Total Wages Paid (Rs.)*
Initial stage (pit digging, plantation)	1,06,485	18.00 (Rs. 14-22)	19,16,730
After 1990 (maintenance & grass cutting & others)	51,231	36.00 (Rs. 22-50)	18,44,316
Total	1,57,716	23.85	37,61,046

Source: Bijanwas Primary Farm Forestry Cooperative Society.

In any case, the development implications of the programme cannot be denied. The lesson of the small illustration given above is that if the natural resource base is restored, critical to a state like Rajasthan, the rural economy will definitely achieve lasting improvement. For long-term cyclical effect of restoration of natural resource base will not only generate development impulses in other areas of productive activities, but it will also be a constant source of employment opportunity. Organisational efficiency of the local institution and technical efficiency of the activities taken up for restoration of the natural resource base are the required minimum to achieve this. The Bijanwas Cooperative Society is yet to ensure organisational efficiency to the extent required, not to talk about the technical efficiency.

Livestock

As discussed earlier, the livestock economy of the village did not undergo any significant improvement in the villages under study, remaining more or less stagnant over the years. The per capita holding of livestock remained almost the same even after farm forestry was initiated. The analysis of the sample households' opinion on growth of livestock in the period since the initiation of the programme did not differ significantly from the observations made earlier in Table 4.4 (Table 4.23). Although about 25 per cent of the sample respondents found an increase in the number of livestock, the rest said that the livestock population had either declined or remained stagnant after the intervention. The scenario was not encouraging also with regard to milk productivity; only a minuscule percentage observed enhancement in milk productivity (Table 4.23).

So how did the renovated village pasture contribute to the livestock economy after farm forestry was promoted on it? The interest in enhancing milk productivity is largely linked to income generation through milk trade. However, as the survey discovered, milk was yet to become a tradable good and a significant source of household income in these villages. At 18-19 per cent of the average production, there was virtually no difference in the quantum of milk being sold during the drought year and when monsoon was normal (Table 4.24).

In other words, the village economy did not depend significantly on livestock development for livelihood generation, although 'it was' considered of some importance for meeting financial exigencies. Hence, the primary concern of each household was to have a few animals, including milch, to fulfil household needs. This could account for livestock holding per household remaining almost the same, as observed earlier, in the normal rainfall years (Table 4.4). Of course, the aggregate stock of animals in the village did rise over the years, though more for subsistence than livelihood generation. The rise was really an outcome of the increase in the population. As part of the strategy for sustaining livelihood in crises, households tended to have a minimum stock of animals.

Table 4.23

Response of the Sample Households on Livestock Prospects

	Bijanwas	*Bikarni*
Livestock		
(a) Increase	28 (26.17)	8 (22.85)
(b) Decrease	31 (28.97)	17 (48.57)
(c) Same	48 (44.86)	10 (28.57)
Milk Productivity		
(a) Increase	15 (14.00)	2 (5.71)
(b) Decrease	22 (20.56)	12 (34.29)
(c) Same	70 (65.42)	21 (60.00)

Note: Figures in parentheses are percentages.
Source: Field survey.

Table 4.24

Average Production, Consumption and Sale of Milk Per Household

(in quintal)

	Bijanwas		*Bikarni*	
	2003	1998	2003	1998
Production	8.04	13.46	9.29	17.58
Consumption	6.59	10.77	7.85	14.66
Sale	1.47	2.69	1.44	2.92

Source: Field survey.

Households could also earn from sale of animals, dung, etc. Owing to the very low quantity of milk being sold in the market, the average gross income earned through livestock was much lower than other sources of income like agriculture and the non-farm sector. Table 4.25 shows the gross value of agricultural production per household in 2003, the last of the three years of consecutive drought. Its contribution to the total household income was the lowest. In fact, it was lower than the income generated from livestock. Wage income in the non-farm sector stood out as the major source of household income contributing 62 and 52 per cent to it, respectively, in Bijanwas and Bikarni.

Table 4.25

Average Annual Gross Income per Household from Different Sources

(in Rs.)

	Bijanwas	*Bikarni*
Agriculture	4,564.58 (11.80)	8,819.86 (20.61)
Non-farm activity	24,007.10 (62.07)	22,452.43 (52.48)
Livestock	10,105.51 (26.13)	11,514.86 (26.91)
Total	38,677.19 (100.00)	42,787.15 (100.00)

Note: Figures in the bracket are percentage of gross average income of the household. Gross values of agricultural products were estimated on market prices of each produce as prevailing in the village during the field survey. Values of agricultural by-products were added to agricultural income.

Source: Field survey.

As mentioned earlier, members of a large number of households regularly migrated to seek wage employment for a few months in a year. Besides, some were also engaged in petty trade and business. Wage income from agriculture and village pastures was comparatively very low in both villages. However, it should not be forgotten that data from sample respondents pertained to the third year of consecutive drought. Hence, the picture obtained from Table 4.25 may not represent the volume of income that the sample respondents could earn during normal rainfall years. At the same time, one cannot ignore the insights gained from the analysis, for agricultural prospects were constrained in both villages.

It is in this context that one seeks to understand the crucial role-played by livestock in drought prone villages. Table 4.25 shows that the gross value of output from livestock was 26-27 per cent in both villages. It was made possible mainly because the availability of grass was largely ensured through the informal grass market emerging out of the grass production on the encroached lands. In any case, aggregate stock of animals was bound to increase whether livestock as a household unit went commercial or not.

In view of this, one may visualise the gradual decline of the carrying capacity of the village pastures, open or protected, especially during drought years. It would be reasonable, therefore, to seek an increase in the grass production in the village pastures managed by the cooperative society. While it is true that the sale of timber from the farm would generate income for the society in the future, at present, no one knew how villagers would gain individually from it. All that they looked forward to was a better management of the village pastures so that grass production increased. This may ease the pressure on grass availability, likely to intensify in future and jeopardise the prospect of livestock as a larger source of livelihood generation. However, it all depends on how the cooperative society evolves.

VII

INSTITUTIONAL PROSPECTS

Before examining the possible scenarios that can emerge under different conditions, it may be worthwhile to appreciate the present status of the institution and how it is perceived by the villagers. The analysis of the executive committee formation, as described earlier, has sufficiently indicated that the society failed to effectively involve the villagers in its management. However, the latter were involved at different levels of activities launched by the society.

Table 4.26 shows that, initially, a large number of households were morally committed to the development of farm forestry; as many as 51 per cent of the sample respondents in Bijanwas and 43 per cent in Bikarni worked voluntarily for the development of the farm forestry without wages. However, this level of commitment gradually declined in the subsequent years. What remained, for majority of the villagers, was to work for the society as wage labour in the name of participation. Another significant aspect of the bad management that became a source of discouragement was the concentration of power in the president and the secretary. This can be gauged from the fact that 55 per cent of the sample respondents in Bijanwas and 85.7 per cent in Bikarni did not know how management decisions were taken (Table 4.26).

Table 4.26

Response Elicited from the Villagers on the Functioning of the Cooperative Society

	Bijanwas	*Bikarni*
(1) Participation of the villagers in supervision of grass and fuelwood distribution (as unpaid labour)		
(a) Initially	54 (50.5)	15 (42.9)
(b) At present	34 (31.8)	3 (8.6)
(2) Participation of the villagers (as paid wage labour)	53 (49.5)	12 (34.3)
(3) Appropriateness of intervention	84 (78.5)	24 (68.6)
(4) Adequacy of intervention	24 (22.4)	10 (28.6)
(5) Decision making process		
(a) Not aware	59 (55.1)	30 (85.7)
(b) Taken by the president and secretary	22 (20.6)	4 (11.4)
(c) Taken by the members	17 (15.9)	3 (8.6)
(d) Taken by the villagers together	5 (4.7)	3 (8.6)
(6) Freeriding	26 (24.3)	5 (14.3)
(7) Penalty imposed	18 (16.8)	1 (2.9)

Note: Figures in the bracket are percentages of the households responding to the questionnaire to the total sample households.

Source: Field survey.

The incidence of freeriding was also too big to be ignored. In fact, it was much higher, especially in Bikarni, than reported by the sample respondents; this was observed by the secretary of the society. However, in practice, no penalty was imposed on the free riders. Interestingly, despite their indifference to the management of the society, 70-80 per cent of the sample respondents felt that the promotion of farm forestry was appropriate but inadequate. They felt the management should be made more people-centric; its activities should be geared so that the institution was not reduced to a mere grass distribution centre and forum for creating employment for a few. It must also be a forum to resolve conflicts, likely to grow with increasing pressure on the village pastures and in case encroached lands are decolonised.

Other Probable Institutional Outcomes

The issue is what are the other alternative probable scenarios depicting different institutional outcomes; better or worse than the present one. If we concentrate only on the better ones, which is desirable, the task is then to examine under what conditions are such institutional outcomes produced. Here, institutional success refers to positive outcomes consequent to intervention. Indicators of positive institutional outcome of collective action are based on the perceptions of the village community. These include: improvement in livestock, improvement in the natural resource base, facilities created by the cooperative society, availability of dry fodder from the village pastures and decline in migration.

In view of this, data on different important variables that have a bearing on the estimation of the institutional outcomes have been analysed (Table 4.27). Before estimating the results of the probability (Y=1), the values of the independent variables (predictors), found significant in Table 4.27, have been estimated and shown in Table 4.29.

As explained earlier, the objective is to indicate alternative scenarios of institutional outcomes that the cooperative society may witness under different conditions. Different conditions are represented by different combinations of independent variables as shown in Table 4.29. Such an exercise helps to identify the best possible combination of independent variables (predictors) that yield highest achievable institutional success. Table 4.28 shows goodness of fit of the parameters, which are estimated in Table 4.29. It shows that the parameters that are finally included in the model are statistically significant, thereby implying that the predictors are chosen correctly for the estimation of the probability of institutional success.

Table 4.27

Results of Significant Test of the Parameters

Variables	*Type of Response*
Awareness about the cooperative society*	Yes=1, No= 0
Regular attendance of the meeting *	Yes=1, No= 0
Participation in the activities undertaken by village *samiti* for natural resource management *	Yes=1, No= 0
Effective participation in the decision making process*	Yes=1, No= 0
Regularity of meeting of cooperative society *	Yes=1, No= 0
Decision taking process in the meeting *	With majority of members=1 Only a few =0
Awareness of rules *	Yes=1, No= 0
Performance of cooperative society ##	Satisfactory=1 Not satisfactory =0
Association with cooperative society##	Voluntary labour =1Part-time labour =2 Fully paid labour =3
Whether denied participation in the management of the society *	Yes=1, No=0
Access to the improved natural resource base *	Yes=1, No=0
Freeriding* (non-compliance)	Yes=1, No=0
Whether satisfied with the distribution pattern of resources ##	Yes=1, No=0
Conflict between villagers and the members of the cooperative society *	Yes=1, No=0
Landholding pattern##	Landless =1 Marginal/ small =2 Medium/large =3

Note: * *Significant* ## *Not Significant*

Table 4.28

Results of Goodness-of-fit Test of the Model

Estimation terminated at iteration number 4 because log likelihood decreased by less than 0.01 per cent	
Cox & Snell $-R^2$	0.42
Goodness of fit	136.33
$x_{(8)}^2$	21.69 ($p>0.05$)

Table 4.29

Estimated Value of the Significant Parameters

Parameters	*Estimates*
Awareness of the cooperative society	-0.608
Participation in the activities undertaken by village *samiti* for natural resource management	0.984
Denied participation	2.039
Regularity of meeting	0.563
Attending the meeting regularly	0.528
Decisions making process in the meeting	-1.081
Effective participation in decision making process	-2.012
Awareness of the rules	-0.018
Freeriding (non-compliance)	0.676
Conflict between villagers and the members of the-cooperative society	-0.829
Access to the improved natural resource base	-0.505

Scenarios of institutional success in terms of its outcomes are worked out within the probability of 0 to 1 (Table 4.30).

Table 4.30

Probable Sscenarios of Institutional Outcome of the Cooperative Society Under Different Conditions

Variables	*Scenarios*								
	A	*B*	*C*	*D*	*E*	*F*	*G*	*H*	*I*
Awareness about the co-op society	1	0	0	0	0	**1**	0	0	1
Participation in the activities	1	1	1	1	1	**0**	1	1	1
Denied participation	1	1	1	1	1	**0**	1	0	0
Regularity of meeting	1	1	1	1	1	**0**	1	1	1
Attendance of the meeting-regularly	1	1	1	1	1	**0**	1	1	1
Decisions making process in the meeting	0	0	0	1	0	**1**	0	0	1
Effective participation in decision-making process	1	1	1	1	1	**0**	1	1	1
Awareness of the rules	0	0	0	0	0	**0**	0	0	0
Freeriding (non-compliance)	1	1	1	1	0	**0**	1	1	0
Conflict	1	1	0	1	1	**1**	1	1	0
Access to the improved natural resource base	0	0	0	0	0	**1**	0	0	1
Probability of Success	**0.399**	**0.399**	**0.508**	**0.631**	**0.643**	**0.705**	**0.736**	**0.921**	**1.000**

Table 4.30 presents nine alternative scenarios of institutional outcomes and the cooperative society may opt to achieve any one of them. The scenario that exists at present is represented by F (0.705) indicating that the performance of the society was a little higher than average although majority of the respondents were never part of the decision making process. What mattered significantly was the equal access to the resources that was provided to the villagers. However, this was not sufficient to ensure the stability of the cooperative institution in these villages. For at this level of performance, as one can see, the conflict between the management and villagers and between the villagers that exists presently might push back the society to a lower level of performance in future and destabilise it further.

Figure 4.3

Probable Scenarios of Institutional Performance

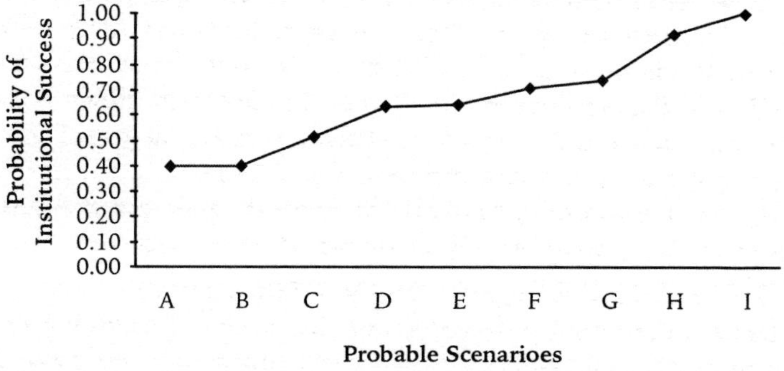

Besides, this level of performance was achieved without intensifying the villagers' participation. With the given level of performance, the ideal alternative institutional outcome is indicated by 'I'. It shows that the cooperative society might be able to achieve institutional stability if it succeeds in creating a condition for conflict resolution by promoting villagers' participation in the management of the society. This needs to be achieved while providing the poor equal access to the improved resource base. In other words, the combination of the factors as indicated by alternative 'I' has to be noted if the society chooses to make a transition from 'F to I' in Figure 4.3.

VIII

Conclusion

The basic trouble with the cooperative approach was the absence of collective action. It may be argued why collective action is required as a precondition for sustainable resource management when a cooperative institution, as the one under study, had survived for nearly a decade without it. However, one is not certain whether the co-operative society will continue to sustain in the absence of participation of the villagers especially the poor ones. While the villagers were yet to develop a sense of community ownership, IFFDC must seek complete exit from the scene. The village community must no longer nurture the idea that the farm forest was not owned by what they called 'company' (IFFDC).

However, one must look at the survival odds against the backdrop of the material conditions of grass production that underwent change once the village commons were privatised at the sponsorship of the state. The large availability of grass in these lands had never allowed scarcity condition to deepen, or dormant inter-village or intra-village conflicts to sharpen over control of grass and fuelwood. However, with a growing human and livestock population, commercial gains from the sale of trees could threaten the present social equilibrium in the near future. It is not certain how the latter would gravitate if the state discontinues its approval of privatisation of the village commons.

All these together suggest that the village community must come to terms with rebuilding the cooperative institution by institutionalising the rules of conservation of their farm forest on their own. The cooperative institution has already displayed its potential to sustain on its own—even as it generated large employment opportunities during the agricultural production crisis, it could meet its expenses through the sale of grass and fuelwood at administered prices. This reflects the community's potential to conserve and manage its resource base, including the privatised commons, through new institutional practices as they may evolve. It is here that the state must define its role *vis-à-vis* local *panchayat* in terms of complementing and supplementing community performance for growth and conservation of its own natural resource base.

5

Revival of a Degraded Forest

Introduction

Nothing can better exemplify how the lives of the tribals are interwoven with the forest than Dagal of Banswara district in south Rajasthan (Map 12). The name Dagal derives from *dagla,* which, in local parlance, stands for the roof of a house on stilts. Once wood was abundant in the thick forest around this village and collected freely by the villagers along with other forest produce. But that was over a hundred years ago. Gradually, the old forest thinned out as the traditional symbiotic relationship between the forest and the people withered away. Things came to such a pass that the villagers' very survival was put to stake. The forest they had grown up with, that had sustained generations before them, was on the verge of extinction. Threatened, they looked to the forest resource base of neighbouring villages for subsistence only to be turned back humiliated. But in this lay a great lesson for the residents of Dagal. They learnt they must bring back their own forest for their survival.

The village was first inhabited by 50 households of the *Dabi* family line of the *Meena* tribe. As the oral history of the village goes, the area had a thick forest of *Dandrocalamus strictus* (locally known as *baans*) and a few other species such as *Tectona grandis* (*sagwan*). Although the families were sparsely settled, their dependence on a common resource base for survival did not allow the community to disintegrate. However, this was severely threatened by a devastating epidemic that caused many deaths about a century ago. A sense of insecurity made the inhabitants bring over relatives from other places to settle down in Dagal. The new settlers—belonging to different family lines, including *Kala Sua, Ninama, Gonda, Mairha, Baroda, Charpota, Dindore, Bargot, Bagora, Bamaniya* and *Katara* besides, of course, *Dabis*—struck roots at various locations on the forestland. The habitations, or hamlets (*pada* in local parlance), came to be called Rohena *pada,* Hijudi *pada,* Hindula *pada,* Masalda *pada* and Timba *pada*

(Map 13). The influence of forest on the lives of the villagers was so deep that some of these hamlets were named after tree species that grew predominantly in the area. For example, Rohena pada was named after the species locally known as *rohena* and Hijudi pada after *Hijudi* plant.

Although villagers were content with their living conditions, their household equilibrium was increasingly disturbed as the forest cover thinned out unable to take the pressure of a spiralling human and livestock population. This was despite the fact that the inhabitants were culturally conditioned to not extract more than what they required from the forest. If they had food at home, they preferred not to go out and earn. Like other tribal communities in other parts of the country, the *Meenas* of Dagal were imbued with the traditions, customs and beliefs of, if one may say so, a static society.

It was static in the sense that there was no advancement of its material base and therefore, no switchover to a lifestyle contrary to the one culturally defined and accepted. However, even in such a situation, the mismatch between the increasing demands made on the forest and its comparatively slow regenerative capacity began to threaten the *Meenas'* survival base. The situation was similar in the neighbouring forest of Amarpura prompting stiff resistance from its inhabitants when Dagal residents began to collect produce from it. Having been hurt and insulted, and left with no choice, the inhabitants of Dagal decided to replicate the Amarpura initiative for the restoration of their own forestland.

The study has eight sections. Section II traces the genesis of institutional development and examines how social capital was instrumental to it. Section III analyses the steps taken by the village institution for the regeneration and conservation of its forest. Section IV gives a brief account of the changes witnessed by the village resource base While section V looks into how poor tribal households gained access to the improved resource base, section VI examines the ecological impact of the forest regeneration. Institutional prospects remained a matter of serious concern from the point of view of growth of local initiative. This is examined in section VII. Section VIII concludes the case study.

Map 12

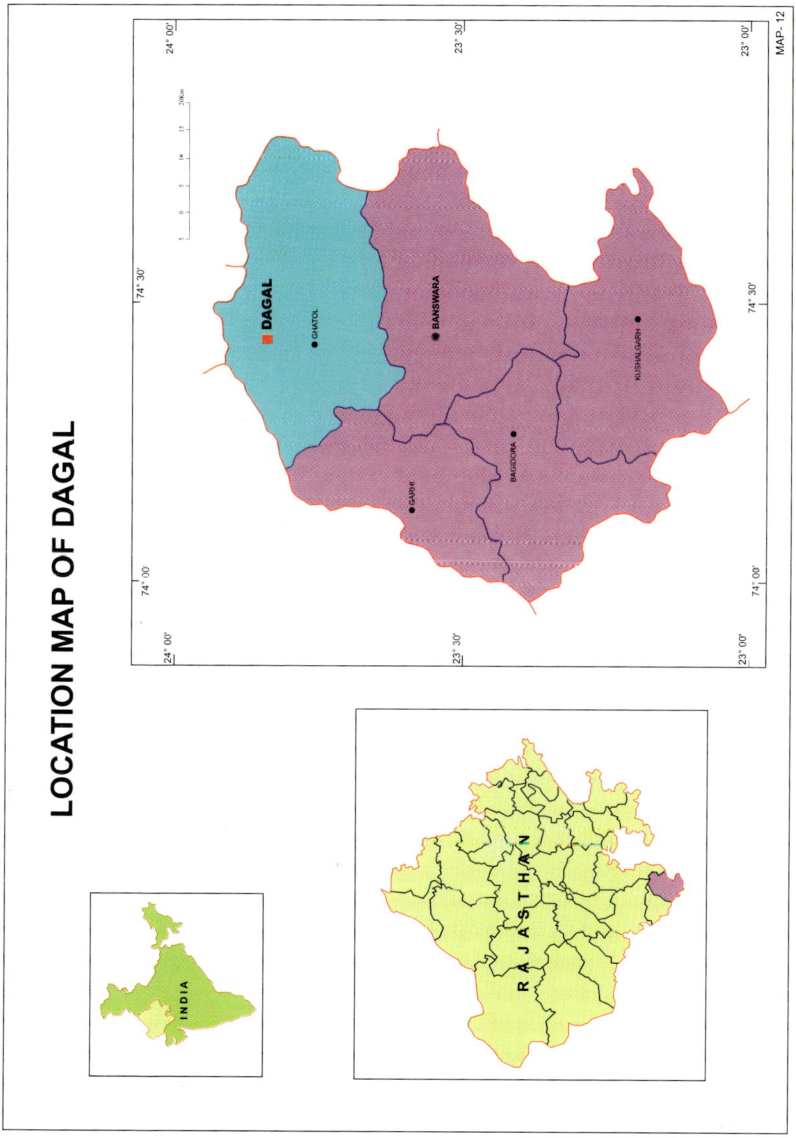
LOCATION MAP OF DAGAL
MAP- 12
24° 00'
23° 30'
23° 00'
74° 00'
74° 30'
DAGAL
GHATOL
BANSWARA
KUSHALGARH
BAGIDORA
GARHI
INDIA
RAJASTHAN

Map 13

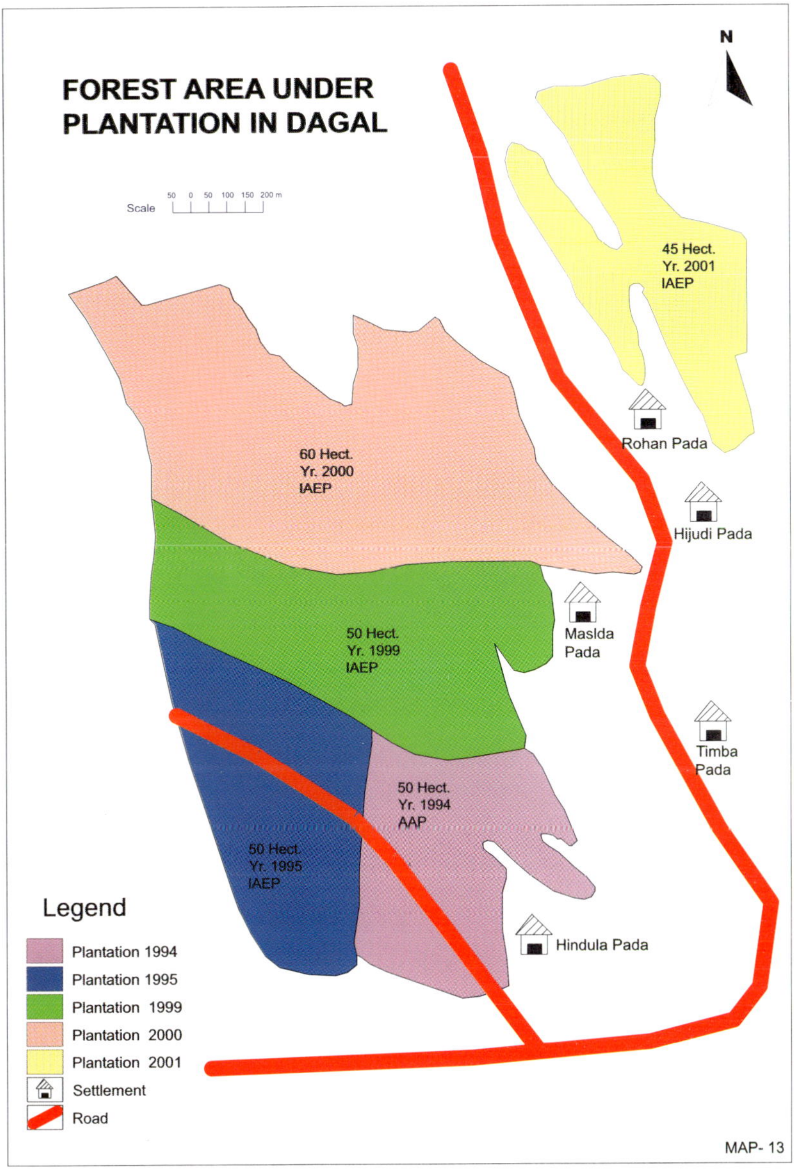
FOREST AREA UNDER
PLANTATION IN DAGAL
N
Scale
50 0 50 100 150 200 m
45 Hect.
Yr. 2001
IAEP
60 Hect.
Yr. 2000
IAEP
Rohan Pada
Hijudi Pada
Maslda
Pada
50 Hect.
Yr. 1999
IAEP
Timba
Pada
50 Hect.
Yr. 1994
AAP
50 Hect.
Yr. 1995
IAEP
Hindula Pada
Legend
Plantation 1994
Plantation 1995
Plantation 1999
Plantation 2000
Plantation 2001
Settlement
Road
MAP- 13

II

Genesis of Institutional Development

Sometime in the mid-1970s, the villagers of Dagal came together under the leadership of Chogaji and decided to regenerate their forest collectively. As a first step, they formed a frontal group consisting of five representatives from each *pada*. A forest guard was appointed to look after the forest. All households agreed to contribute five kg of cereals after each harvest as payment towards wages of the forest guards. However, this proved too inadequate to significantly affect regeneration of the degraded forest. As the villagers recounted in the field survey, the defective management practices failed to ensure compliance of the villagers. For example, half the forestland (around 100 hectares) was to be looked after by the forest guard during the rains while the other half was to be kept open. In practice, this did not work. Many villagers failed to comply with the rules they made themselves. The problem was compounded with the entry of people from neighbouring villages inside the forest. One may ask why there was such a high incidence of non-compliance or why the villagers were indifferent to traffic from neighbouring villages when the restoration of their forest was top priority for Dagal.

Perhaps the reason lay in the presence of the forest guard appointed by the forest department. A feeling was created among the villagers that the forestland was owned by the government while they owned its trees and produce. Hence, they thought, they had reason to assert their rights over the forest produce. It may be noted that this kind of dichotomy regarding the ownership rights over the forest did not exist before the entry of the government into the forest. Once such a dichotomy received social legitimacy, it eased the thoughtless exploitation of the forest resources. This continued uninterrupted until the late 1980s, when the forest cover was reduced to the lowest level that the village had ever witnessed.

It was at this stage, when the forest was left with only critical minimum reserve, that the same village headman brought the villagers together again and formed a committee of four persons to work out ways and means to revive the dying forest. The interesting part was that the meeting in which the committee was formed was chaired by the *sarpanch* of the Dagal village *panchayat*. This was the first time that the village community went beyond its own system of administration to seek redress of the problem. Perhaps, the *sarpanch* would not have been involved had the government remained absent. Absolute inaction on the part of the government-appointed forest guards had proved that no protective mechanism could succeed unless the local people were involved. Villagers were more enthusiastic now about the revival of their forest as compared to earlier, because the intense drought had reduced the scope for livelihood generation. Now that their lives had become more uncertain, the villagers realised the relevance to their lives of the forest. Steady depletion of their livestock resources threatened their household economy. Many had to migrate to survive and indebtedness to moneylenders and traders was at an extraordinary high.

This precarious condition led the villagers to redesign their institution with a new set of rules and enforcement characteristics:

- Two forest guards must be appointed, even if government-appointed forest guards were in operation. Each one must be given five kg of cereal by each household after each harvest;
- Nobody should be allowed to cut trees;
- Nobody should be allowed to enter the forest with animals;
- If anyone from outside the village was found inside the forest and cutting trees, the village committee should be immediately informed to initiate action;
- Cutting of grass was allowed only for one month in a year after Deepawali;
- Anyone found entering the forest with a sickle (*dantli* in local parlance) should be fined Rs. 25 (in case the person had an axe, the fine would be Rs. 51); and lastly
- Anyone found entering the forest with animals should be fined Rs. 25 and 50 for a small and large animal, respectively.

It was also decided that the money collected through penalty would be utilised mainly for mending the fence. No one, however, was penalised financially as stipulated above. What gained overriding importance for ensuring compliance was community pressure created by those who were inclined to comply with the rules of management. This 'social fencing' worked more than anything else, especially in the following years when there was normal rainfall and the forest began to regenerate. The substantial production of grass, thereafter, helped villagers to improve their livestock position, devastated by the years of drought.

More importantly, they gained confidence in themselves when they found the dying forest begin to regenerate because of their initiative and commitment. The village *sarpanch* continued to be their inspiring force. Several meetings held at regular intervals under his chairmanship, directly and indirectly lent credible support to enforcement of the rules of conservation. Each household was allowed to harvest the same quantity of grass from the forestland free of cost. In this way, the informal village institution developed by the villagers for forest conservation gradually evolved. The rallying point was how to strike a balance between the regeneration of biomass produced in the forest and its consumption. It was more or less an endogenous change that unleashed a new form of social embeddedness of the village community. Before it matured, however, Aravali, a government-sponsored NGO stepped into the village in 1994 to implement joint forest management (JFM) programme through the forest department. The evolution of institutional renovation at the local initiative was halted. It then took a different route with the local initiative being subsumed into that of the forest department.

Village Institution: Transition from Informal to Formal

A cluster of nine villages, including Dagal, of the Umarjhala Pathara cluster of the Ghatol forest range of Banswara district were selected for the implementation of the JFM programme. The purpose was to restore the degraded forest through plantation. Although there were several evidences to suggest that the degraded forest could have been regenerated without plantation through the protection and practice of silvicultural operations, the case of Dagal was different. Plantation here was unavoidable despite the critical minimum stock of plant species being present. There were several arguments for and against the entry of JFM in villages situated on different types of degraded forestlands. Extremely rich in their content, these arguments had the potential to create the appropriate policy framework for making the programme more people-centric. However, the issue here was that of the choice of implementing a programme like JFM, irrespective of its significance and appropriateness to the existing conditions.

Without tinkering much with the existing committee, engaged in conservation for last several years, some additional members were inducted into the committee at the suggestion of Aravali. The name of the village organisation was changed to Van Suraksha Evam Prabandhan Samiti. This was also done at the suggestion of Aravali. Many other changes were made in the functioning of the village institution subsequently by Aravali without any confrontation from the villagers. The members of the village committee and the rest of the village community supported the move for two reasons that they thought were extremely important. First, they saw great potential for generation of wage employment in the forest development activities to be initiated by Aravali and second, Aravali would take the entire responsibility of plantation in the degraded land in addition to fencing. This involved huge expenditure, which the villagers were unable to bear.

One may argue that forest regeneration was the responsibility of the state and, therefore, a step undertaken in the right direction by the state in collaboration with the community. However, the JFM programme here was conceived so that the initiative of the state was appropriately blended with the community participation; the state was expected to create space for local initiative to grow further to ensure that regeneration of the forest sustained. However, in practice, if the mechanism and the process of implementation magnified the role of the state at the expense of the local initiative, alienation of community was bound to occur. Dagal was one of a few villages in the state to have witnessed the growth of local initiative for regeneration of forest. The issue is whether such local initiatives should be eventually subsumed into state action, usually found to alienate the village community from the management of its forest.

As discussed earlier, the first step that the forest department took was to rename the village committee Van Suraksha Evam Prabandhan Samiti, uniformly called so in all places where Aravali and forest department intervened for the development of forests. Maybe, it was right for the forest department to maintain this uniformity. But it is also true that this marked the beginning of the alienation of the Dagal village community from the management of the forest. While the previous members were retained under the leadership of Bhagwanji, eight new members were inducted as proposed by the forest department, including three women and one nomination by the department. One of the members was made vice president of the committee.

It must be remembered that the president of the committee selected members, of course, with the full approval of the village community. The notable feature of the constitution of the committee was that the majority of members were drawn from the *Dabi* family line—for it had as much as 43 per cent share in the total population of the village with the rest shared by 12 other family lines. The *Dabis* also dominated by virtue of being the original settlers on the land. However, despite the numerical and other advantages enjoyed by them, the institutional approach towards enforcement of management rules was never discriminatory against the minority groups of other family lines. This was contrary to what was observed in the case of Singhariya where the stronger family line discriminated against the relatively disadvantaged family lines (Ray and Sharma, 2004).

Content of Social Capital

It was, in fact, the content of the social capital that made the difference in managing natural resources in Dagal and Singhariya. The content of social capital in a tribal society like the one that existed in Dagal was more equity driven. The villagers' social embeddedness was inherited from a distinct cultural identity that continued to nurture a system of self-determination. This was why they had their own *panchayat* parallel to the state *panchayat* system; it was their own forum and the primary body (rather than the state *panchayat*) for seeking solution to any conflict. The sanctions imposed by this *panchayat* were binding on all; non-compliance meant excommunication. It was this cultural embeddedness within which villagers learnt how to make choices about individual actions that affected the community. A sense of togetherness and reciprocity that shielded the community from any disturbance, social, economic or natural, was central to such cultural embeddedness. In fact, the strength of collective action and its sustainability for forest regeneration was drawn from the social forces that constituted the social capital. The sustainability of collective action for forest regeneration and its management deserved to be examined in this cultural context only.

One may not have a better example than *nautra* to illustrate how a traditional cultural practice was instrumental in averting the economic crisis faced by a household. In this traditional tribal custom, a poor householder, left with no means of subsistence, would invite the entire community for a feast. All who came to take part in the feast would in return gift substantial amounts of money to the impoverished household to help it out.

However, the tribals' attitude towards women was as chauvinistic as that of male members of any other community. The only exception was that women enjoyed the independence of communicating freely with the male members of the village. Hence, they never felt secluded. However, they did not have this space in any collective action and they should have been treated at par with the males. They continued to remain excluded in this regard since they were thought less able to think and act. However, contact with outsiders, awareness campaigns launched by NGOs like Bhartiya Agro-Industries Foundation (BAIF) and the increasing mobility of the villagers thanks to the new road seemed to have induced some change in their attitude towards women. This was further boosted with the improvement in the education of women.

It was in this context that one must see how useful it was to create space for women in the village forest protection committee (VFPC). As per the rules of the government, three women were inducted into the committee. It was definitely a step in right direction for variety of reasons. They could now articulate their perceptions about forest development in terms of their advantages and disadvantages. Besides, they stood out as a source of inspiration for other women members of the community to join collective action for collective goods.

The women were Soni Bai, Phula Devi and Mani Devi; all from one particular hamlet (Masalda *pada*) and from the same *Dabi* family line. Discussions with these women revealed that they were culturally constrained to speak before the elders, to do so was considered a mark of disrespect. Notwithstanding this difficulty, women members exhibited their sense of involvement with the forest development at every stage and ensured their participation repeatedly. However, the selection of all the three women members from the same hamlet and family line reflected biased action on the part of the leadership, though this did not create inequitable access to the improved resource base. The formation of the women's advisory committee in 1994 was more of an eyewash, lip service to the forest department requirements. Although enthusiasm was extraordinarily high initially, it lost momentum and eventually became defunct in a few months.

The conversion of the informal institution into a formal one as explained above had some African parallels observed by Dia (Dia, 1996). Dia observed that formal institutions are not rooted in the local culture and are at odds with societal behaviour, expectations and incentive systems, and therefore, face a crisis of legitimacy and enforcement. Local informal institutions, on the other hand, despite having advantages over formal institutions in this regard, are not inclined to renovate by shedding dysfunctional traditional practices and always evolve in response to the changes outside. The issue, therefore, is how the informal institution built by the villagers reconciled to the formal institutional approach as planted by the forest department. If it were a reconciliation that lent credible support for local initiative, it might ensure institutional sustainability for forest conservation. However, sustainability was uncertain if local initiative was subsumed to Aravali/forest department. One may gain more clarity while examining how rules were framed and reframed as a part of the strategy followed for the revival of the degraded forest.

III

STEPS TAKEN

Framing of Rules

While the rules of management of forest development and its conservation that were framed in 1988 remained the same, villagers witnessed a basic change in them so far as collection of grass from the forest was concerned. The earlier practice of grass collection was guided by the requirement. Each household was allowed to collect as much as they required, not as much as they could collect. This system of management was maintained for some years because pressure on the land was not too high, meaning, thereby, that the carrying capacity of the protected land was high. However, after the entry of Aravali, VFPC was clearly instructed to seek permission from the forest department to harvest grass from the forestland. This imposition of a new practice was not a contentious issue for the villagers. Hence, they did not voice their resentment against it. So long as they could access the forest, however limited, they were content with the new institutional arrangement.

Discussions with the forest department revealed that it was needed on their part to enforce new practices since they apprehended that newly planted species were likely to be damaged, if entry was not restricted. Whether imposition of new practices was necessary or not was a matter of debate. But, it was true that such an action did engender some alienation among the villagers with regard to the ownership of the forest; the dominant opinion was that the forestland was owned by the government.

Plantation Programme

Growth and conservation of the forest was the main activity for which Aravali entered the village and renovated the informal institutional arrangements. It was decided to bring 255 of the 393 hectares of land (65 per cent) under plantation or regeneration programme successively each year (Table 5.1). This included the 100 hectares of land being earlier conserved by the villagers.

Table 5.1

Area Brought under Plantation Programme in Dagal (1994-2001)

Year	*Area (Hectares)*
1994	50
1995	50
1999	50
2000	60
2001	45
Total	255

Source: Field survey.

It is worthwhile to mention that the villagers were duly consulted while selecting the tree species for plantation. The tree species included *Dandrocalamus strictus* (bamboo), *Tamarindus indica (imli)*, *Emblica officinalis (aonla)*, *Acacia catechu (desi kher)*, *Azadirachta indica (neem)*, *Albizia lebbeck (siras)*, *Echinochloa colona (sawan)*, *Churel*, *Terminalia belarica (behra)*, *Zizyphus mauritiana (desi ber)*, *Leucaena lucocephala* (*su-babool*), *Tectona grandis* (*sagwan*), *Kegar kher, Ailanthus excelsa* (*ardu*), Eagle marmelose (*beel*), etc.

The villagers relied on plants' religious, medicinal, feed and cultural value for selecting them for plantation. For instance, *Tamarindus indica* (*imli*) was chosen for *Aamli Gyaras*, a local religious festival on which the tree was worshipped. Therefore, the people wanted to have thc plant in their own forestland so that they did not have to search for it on important day. The tree was also important for its medicinal value. *Ailanthus excelsa* (*ardu*) was needed for feeding goats. Similarly, *Echinochloa colona* (*sawan*), was required for making local musical instruments such as the *dholak*. *Acacia catechu* (*desi kher*) was equally important for them to make ploughs. They also needed *Pithicolobium dulsi* (*jungle jalebi*) for its fruit apart from the shade it provided. Of course, other species were planted over the years in different plots.

Besides plantation, several other activities were taken up related to the preparation of the land and its development. These included staking; stone fencing and ditch fencing (locally known as *khai*); land preparation for planting species; contour bunding; staggered trenches; V-ditch; check dams; continuous trenches and sowing of seeds for growing trees. The total expenditure that was incurred by the forest department during 1993-94 to 2001-02 was Rs. 30,36,385 for all the activities that were taken up during this period (Tables 5.2 and 5.3 for activity and plot-wise expenditure incurred each year)

Table 5.2

Activity-wise Expenditure Incurred during 1993-2002

Particulars	*Total*
Plantation	10,50,886
Fencing	6,91,990
Soil and water conservation	3,88,753
Security	80,609
Maintenance	4,28,904
Others (Miscellaneous)	3,95,243
Total	30,36,385

Source: Forest Department, Ghatol Range District, Banswara.

Table 5.3

Plot-wise Expenses Incurred for Forest Regeneration in Dagal during 1993-2002

Year	Plot A	Plot B	Plot C	Plot D	Plot E	Total
1993-94	1,77,311					1,77,311
1994-95	1,94,853	1,80,831				3,75,684
1995-96	42,120	3,16,655				3,58,775
1996-97	19,993	44,911				64,904
1997-98	12,000	9,731				21,731
1998-99	10,750	10,750	3,35,800			3.57,300
1999-00		6,372	1,77,588		2,39,540	4,23,500
2000-01			80,667	4,11,932	3,60,519	8,53,118
2001-02			15,790	2,66,776	1,21,496	4,04,062
Total	4,57,027	5,69,250	6,09,845	6,78,708	7,21,555	30,36,385

Source: Forest Department, Ghatol Range, Banswara District.

These activities were taken up to develop all plots that were successively adopted for plantation. They provided employment opportunities to a large number of people in the village. The field survey revealed that as many as 56 out of 78 sample households (72 per cent) obtained wage employment during the period under study. A three-pronged strategy seemed to have been followed to create conditions that were conducive for forest regeneration. The first was to create a condition for improving soil moisture of the forestland. The second was to stop trespassing of the neighbouring villagers to the forestland and the third was to restrict free access of the villagers to the forest. It was the third one that the forest department was quite concerned about. For, as mentioned earlier, they felt that damage might be inflicted on the regeneration process of the plants in the absence of the restrictions imposed on free access. Besides, they feared that tree felling might continue. This was the reason why a specific period was earmarked for harvesting grass. Accordingly, the overall organisational management was changed from time to time for grass collection depending upon how the situation evolved especially in terms of growing maturity of the plants. This is shown below:

1995 Two persons from each household were allowed to cut and carry grass only for seven days a year with one cutting and the other collecting. No payment was stipulated.

1996 Grass collection was made through introduction of bundles (*pule* in local terms). The weight of each bundle was around 1.5 kg of grass. Two persons of each household were permitted to collect grass not exceeding 105 bundles a day for 7 days a year. Of these 105 bundles, 5 bundles were given to VFPC. After having collected five bundles from each household, VFPC, in turn, distributed the same to those who volunteered to look after the forest. It was an incentive designed to promote the interest of the villagers in directly contributing to forest protection. In addition to the restriction imposed on collection of limited numbers of the bundles, villagers were asked to pay Rs. 25 per household for grass collection.

1998 Two persons from each household were allowed to collect grass from the plot in which plantation was done in 1994, for 7 days with 55 bundles in a day for 8 hours. Each household was charged Rs. 20 for this. This system of grass distribution did not appeal to the villagers any longer. Discontent brewed and gave rise to conflict between VFPC and villagers. The committee found that the introduction of contractual system was the only way out to resolve conflict. Around 25 households selected from Hijudi *pada*, Masalda *pada* and Timba *pada* were given the contract upon payment of Rs. 12,000. They were given absolute freedom to collect grass from the plot developed during 1994.

1999 Having found the contractual system to be working efficiently and solving inter household conflicts related to procurement of grass from the forest, VFPC decided to extend it to the next plot that was brought under plantation during 1995. In fact, these two plots (of 1994 and 1995) were exclusively earmarked for harvesting grass until 1999. The plot of 1994 was given on contract to the villagers of three hamlets including Timba *pada* , Rohena *pada* and Hindula *pada* with the introduction of slight change in the management. Similarly, plot of 1995 was given on contract to two other hamlets, including Hijudi *pada* and Masalda *pada*. In each of these plots, two persons from each household were allowed to cut grass on a collective basis. Once cutting of grass was over, 100 bundles were distributed to each household. Of course, in addition to the above, each household had to pay Rs. 50 for grass collection.

2003 While the same system continued in respect of the plots of 1994 and 1995, no contractual system was introduced in the plots of 1999, 2000 and 2001. In the case of these plots, the system of grass collection was similar to the one that was in operation before contractual system came into being.

One of the important features of VFPC functioning, as it appears, was that it introduced the incentive system for better management of forest regeneration programme. Five persons from each hamlet were identified to check whether grass was harvested as per the norms. In addition to it, they had to locate and report those people who were found felling trees or lopping trees without seeking permission from the village committee. As mentioned earlier, this group of people was given five bundles of grass each by each household everyday, while grass was collected in the month of November. This system of incentive promoted direct participation of the villagers in forest management.

The second aspect was related to the incidence of non-compliance of the villagers to the rules framed for fuelwood collection. The earlier rules, prior to Aravali's entry, were more concerned with cutting or felling of trees. They had nothing to do with the collection of dry wood from the forest as fuel; a traditional practice long followed by the villagers. Although the forest department knew about it, it insisted that the village committee must seek permission from the department for collection of fuelwood in as much as it was done in respect of grass collection. This, however, went contrary to the villagers' practice of fuelwood collection. Hence, in practice, the regulatory behaviour on fuelwood collection imposed by the department did not materialise. While the position taken by the forest department might be convincing from the viewpoint of forest protection, it reduced the importance of what one called 'social protection' into insignificance.

The village committee formed before the entry of Aravali emphasised on this aspect. Instead of imposing such regulatory behaviour, some other means of collection and distribution could have been tried in order to meet the requirement of the villagers for fuelwood. Since no other alternative means was available, villagers collected dry wood from the forest with the knowledge of volunteers and forest guards, who were sympathetic. They allowed the villagers free access for collection of fuelwood only. However, the problem began to multiply when such an informal system eventually encouraged incidence of freeriding to rise. A detailed analysis of the proceeding of the meeting held in each year shows that members of VFPC expressed their concerns about such incidents repeatedly. Decisions were also taken to penalise the free riders. However, no single evidence showed that anybody was ever penalised. The village FPC seemed to have been caught in a situation that neither could they say no to the villagers nor could they afford to ignore the formal rule imposed by the forest department for regulating villagers' access to the forest for fuelwood collection.

Transition from the informal to the formal institution had serious implications for the sustainability of collective action by the village community and the restoration of the degraded forest. While it was difficult for the villagers to bear the cost of plantation and meet other expenses associated with forest regeneration, they increasingly undermined their own community strength. As noted earlier, the system of incentive and punishment continued to be a significant cultural trait that the village community had nurtured for long. Paradoxically, such a system was effective in other socio-cultural affairs of the village community even now after the government intervention. The use of such informal traditional system was also found when Chogaji, the earlier village head of Dagal, took initiative for restoration. Collective social pressure was then the most reliable means through which the incidence of free riding could be controlled.

However, after the entry of Aravali/forest department, they shifted their dependence on government. Slowly the idea that the forest was no longer their child to protect and grow took root; it was the responsibility of the government. So much so that, in one of its meetings, VFPC decided that freeriders must be punished by the police administration. Although nothing happened to this effect, it indicated how social space had shrunk in that the village community could no longer impose social sanction on its own. Once villagers began to think that conservation and management of the forest was essentially an affair of the government, they were not inclined even to comply with the restrictions imposed, especially for fuelwood collection.

IV

Resource Base and its Changing Status

It might be imperative to examine the prospect of forest regeneration in terms of its relevance to the household economy of the villagers. Here, relevance must be understood in relative terms; in the sense that other resource bases might be paying more to the village community. This might motivate the latter to give less importance to the conservation of forest resource base, and, therefore, discourage them to sustain collective action for forest regeneration. Hence, one must look into the sustainability issue of collective action for forest regeneration in broader correlates of different resource bases.

While availability of forest produces had direct implications for livestock development, agricultural prospect over the years, in all probability, expanded the scope for its trade-off to take place with livestock in Dagal. In other words, relative gains and strength of supplementary relationship between agriculture and livestock had much to determine households' interest in improving upon community resource base. Of course, availability of fuelwood from the forest was a factor that constantly generated villagers' interest in forest regeneration. In any case, purely from the viewpoint of material value (as compared to the existence value), it was necessary to account for other resource bases of the village, which changed over the years. It was equally important to know their relative gains that accrued to the village community. A critical understanding to the other changing resource bases might give insight into the sustainability of collective action within the institutional framework followed for growth and conservation of forest resource base of the village.

Land Resources

Table 5.4 shows that around 53 per cent of the total geographical area were under forest. Cultivated land varied between 29-36 per cent of the total geographical area. The encouraging aspect of the cultivable land was that the share of irrigated land of the total cultivated land increased indicating agricultural prosperity during the last one-decade or so. Around 82 per cent of the total geographical area was shared between agricultural and forestland during 2003-04. Of course, villagers were dependent on other types of lands, including barren, uncultivable and cultivable waste, for livelihood generation. The most interesting feature of this village was that only 15 hectares of land were under irrigation during 1991. It increased to 82 hectares in 1996-97, 147 hectares in 2000-01 and 130 hectares in 2003-04. The share of non-irrigated land of the village declined from 94 per cent in 1991 to 38 per cent in 2003-04 of the total cultivated area of the village, thanks to the Mahi Canal that was extended to this village. There had been a significant change in cropping pattern and improvement in crop yields that enhanced agricultural prosperity in the village. Besides, availability of grass and fodder increased manifold in the private lands. Villagers no longer had to excessively depend on forestland for grass collection. However, it did not mean that restoration of degraded forestland was irrelevant for them. It continued to be an important source for them for grass and fuelwood collection, especially in the face of the increasing human and livestock population of the village.

Table 5.4

Land Uses Pattern in Dagal

(in hectares)

	1990-91	1996-97	2001-02	2003-04
Total area	733	733	733	733
Uncultivable land				
(a) Land not available for cultivation	393	467	410	410
• Forest	393	393	391	391
• Others	0	74	19	19
(b) Available for cultivation but not cultivated	78.91	53.0	111.0	111.0
Cultivated land	261	213	212	212
• Irrigated	15	82	147	130
• Unirrigated	246	131	65	82

Note: Others indicate mountain, barren and unculturable land, land put to non-agriculturable uses, etc.

Source: Data for 1990-91 from Population Census, Government of India and data for rest of the years was collected from the Revenue Department, Government of Rajasthan.

Encroachment

The forestland of Dagal was not free from the ongoing debate on the ownership rights of the land. A patch of forestland (around 50 *bighas* or 8 hectares of land) was, as discovered by the forest department, while resettlement was going on during 1992, encroached by some tribal households in this village. However, according the households of this village, these plots of lands were under the revenue department in 1967 when they encroached upon it. Since then, they had been cultivating these lands. During 1991-92, when resettlement of forestland was done at the instance of Aravali, the forest department incorporated these plots of lands into forestland. However, it was not fenced.

These tribal households did not even know about it. In the years to follow, they were asked by the forest department to free the land from 'encroachment'. They, however, refused to do so, for they had been cultivating these lands for more than two decades. It created an impasse that was resolved only when they were persuaded to pay a penalty of Rs. 25 per *bigha* to the forest department for a year for all 50 *bighas* of land that were encroached. In other words, in actual the forest department had 387 hectares of land under its disposal. Subsequently, a deal was struck between the forest department and these households by which around two hectares of land were allowed to remain with them, while around six hectares of land were returned to the forest department. The decline of forestland by two hectares from 393 to 391 hectares was an outcome of this deal.

Livestock and Human Resource

The population of the village increased by 2.51 per cent annually from 1981-2001; from 905 in 1981 to 1,486 in 2001. This was associated with the rise in the number of households from 162 in 1981 to 270 in 2001, at the rate of 2.58 per cent per annum. Along with the rise in the number of households, the aggregate stock of animals of the village increased, although the stock per household did not change substantially. The exception was the stock of small ruminants, especially goats, which witnessed a steep rise in 2000. This is contrary to what was observed in other case studies.

Other case studies showed that the availability of grass increased wherever land was protected leading to a rise in the number of large animals. In Dagal, the production condition of small ruminants, especially goats, improved considerably over the years. First, only 255 of 391 hectares of forestland were fenced leaving 136 hectares of forestland open for grazing. Besides, the adequate availability of tree lopping in the protected forest and grass in the private agricultural lands motivated the rearing of more numbers of goats (Table 5.5).

Table 5.5

Changing Livestock Resources

Animal	*1990-91*			*1999-00*			*2002-03*		
	Livestock per HH	*No*	*Grazing units*	*Livestock per HH*	*No*	*Grazing units*	*Livestock per HH*	*No*	*Grazing units*
Cow	1.69	372	372	1.60	432	432	1.47	399	399
Buffalo	1.24	273	546	1.19	321	642	1.31	354	708
Sheep	Nil	Nil	Nil	0.24	65	16	0.36	97	24
Goat	2.00	440	110	3.08	832	208	2.46	664	166
Bullock	1.75	385	770	1.64	443	886	1.65	446	892
Total			1798			2184			2189

Note: Data on livestock for 1990-91 was collected from Micro planning for plantation programme of Dagal conducted by the forest department. Data on average holding of each type of animal was estimated based on field survey. The number of households of the village was multiplied by this. The data of number of household collected from the Population Census. The grazing unit in terms of cattle was estimated based on the norm which was 1 Cow = 1 Cattle, 1 Buffalo = 2 Cattle, 1 Sheep = 0.25 Cattle, 1 Goat = 0.25 cattle, 1 Bullock – 2 Cattle

Largely, three factors normally tend to influence households to increase their stock of animals—commercial gains, agricultural prosperity and availability of grass in the village pastures. Despite the fact that Dagal witnessed increasing agricultural prosperity and improvement in the conditions of its forestland from where they collected grass, why was it that there was no substantial improvement of livestock, especially large animals? The complete absence of a dairy market appeared to be largely responsible for this. Several studies show that wherever the sale of milk was promoted through dairy development it provided an assured market for milk and households reared milch animal, especially buffalos for their milk had higher fat percentage (Ray, 2004). In Dagal, neither was such a market outlet available nor were the villagers willing to sell milk outside the village. In other words, milk production remained outside the sphere of commercial transaction. It continued to be guided by the law of requirements of the households. This was the reason, as we will see latter, why village community never preferred any market dependent feeding packages for large animals.

It should, however, be noted that rationale for rearing cows was never linked to milk production at the household level. Rather, it was linked to agricultural operation that was boosted, as mentioned earlier, in the recent past. Unlike many other places, the village community of Dagal never intended to switchover to tractor from bullock for cultivation despite intensification of irrigation. They reared cows with the hope that the next calf born would be a male; this was the way they wanted to ensure availability of bullock at the household level.

Apart from the use of bullock for cultivation as a part of their traditional cultural practice, it was more economical for them. One might argue that mechanisation of agriculture might have brought more efficiency and growth in agriculture in Dagal that was endowed with more plain agricultural areas than many other places in Banswara district. However, Dagal presented a different scenario in its continued dependence on the bullock for cultivation. Here, the choice was not between efficiency *versus* inefficiency and growth *versus* stagnation, but it was between production *versus* consumption that had cultural compatibility. Besides, they did not allow their economic calculations of using bullock, as mentioned earlier, that had a bearing on natural resources, to be overtaken by the principles of capital accumulation. The simple calculation was that, they did not have to bear maintenance cost of bullocks since the latter could survive on available resource base of the village that produced grass; it might be private lands or the village commons.

Table 5.5 shows that the stock of both large and small animals increased steeply in 2000. Intervention for both forest regeneration and agricultural improvement seemed to have contributed to this rise. In order to examine its pressure on the forestland from where grass and fuelwood were collected, all categories of animals were converted in grazing unit in terms of number of cattle and shown in the same Table. Table 5.5 shows that there had been a continuous rise of grazing unit from 1798 in 1990, to 2184 in 2000 to 2189 cattle units in 2002-03. Consequently, availability of land per grazing unit showed a declining trend. It was estimated to be 0.22 hectares of forestland per cattle unit in 1990, 0.18 in 2000 and 0.17 in 2003. It was obvious that, as livestock grazing unit or livestock population grows, per capita availability of grass goes down. However, what was more important to see whether growth of livestock population was more than the rate of growth of grass production.

Figure 5.1

Year-wise Production of Grass Per Hectare

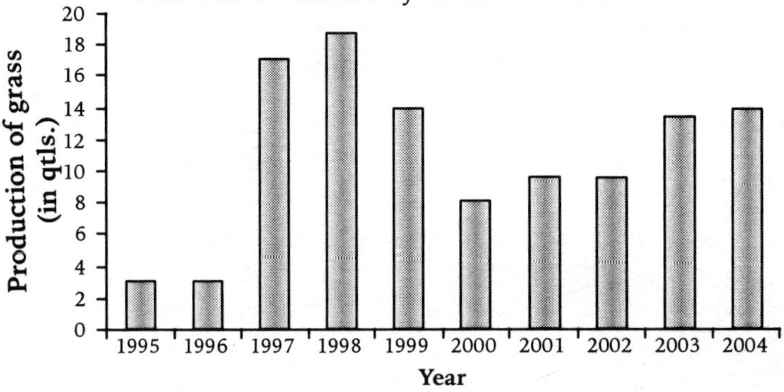

Source: Estimated based on the data collected from Forest Department, Ghatol Range, District Banswara.

In the context of Dagal, as explained earlier, 255 of 391 hectares of forestland were developed by having divided the same into six plots and developed them in each successive year from 1994 onwards. It means that the production of grass started becoming available increasingly to the villagers from 1995 onwards. Figure 5.1 shows that productivity of grass increased in 1997.

As mentioned earlier, the intervention took place in 1994 on only 50 hectares of forestland, which was why grass was available from 1995 onwards. The production of grass was estimated to be as low as 3.1 and 3.0 quintals per hectare during 1995 and 1996, respectively. However, the productivity of land increased steeply in the following years. Figure 5.1 shows that it was as high as 18.7 quintals per hectare in 1998. Although it showed a declining trend in the following years due to drought, it started exhibiting increasing trend during 2003 and 2004. Besides, more importantly, what all was produced was not harvested by the villagers. The FPC, as discussed earlier, in terms of number of days fixed the limit and hours per day for grass harvesting.

In any case, what needs to be underlined was that the carrying capacity of livestock in the forestland did not decline in view of the increasing productivity of land in terms of grass yield. It only shows that, if productivity of land in terms of grass production rises it may lend credible support to the sustenance of livestock especially in a situation like that of Dagal where village community never opted for expanding its size due to the reasons mentioned above. It was not the forest resources alone that supported sustenance of livestock, village community was equally or more dependent, as we will see later, on private land and common open land for this purpose. Dagal would have witnessed a different scenario if its village community would have inclined to increase the size of its livestock more than what existed at present.

In order to assess the perception of the villagers on the progress of stock of animals in the village, enquiry was made with the sample respondents about it. Responses elicited by them were analysed and shown in Table 5.6.

Table 5.6

Status of Livestock Over the Years as Observed by the Sample Respondents

Category	*Animals*			*No Response*	*Total*
	Increased	*Decreased*	*Same*		
Landless	0	0	1	0	1
Marginal	19	13	14	0	46
Small	6	6	8	0	20
Medium	6	0	4	1	11
Total	31	19	27	1	78

Source: Field survey.

As many as 59 per cent of the sample households felt that livestock in the village did not increase. It either declined or remained same. To whatever extent it increased, it did so more in respect of small ruminants. Since milk production was linked to household consumption, no extra efforts were made to increase milk productivity. This was why average production of milk per household per year was as low as 6 quintals. In other words, around 1.5 kg of milk was available on an average per day to each household. Table 5.7 shows that average production of milk per household in respect of small farmers was more than the average production of the village per household. It was around 9 quintals a year. It was possible because sample households of this category of farmers had on an average two buffalos, while others had one buffalo each.

Table 5.7

Production of Milk During 2003

(in Quintal)

Category	*Cow*	*Buffalo*	*Goat*	*Total*
Marginal	1.37	2.17	0.48	4.02
Small	1.84	6.35	0.70	8.89
Medium	1.69	3.98	0.71	6.38
Average	1.52	3.47	0.56	5.55

Source: Field survey.

The lower production of milk indicates low productivity of the animals that largely depend on breed and feeding pattern. Since breeds of both cow and buffalo were nondescript, there was no reason to expect more production of milk as compared to the present level, unless animals were also fed concentrate, cattle feed, etc. at a certain proportion (Ray, 2004). Table 5.8 shows that large animals were fed mainly grass and agricultural by-products that were available from both protected forestland and private agricultural lands. The use of concentrate, which was a recent addition to the feed basket of the animals, as the field survey revealed, was too low to incorporate into cost calculations. Table 5.8 further shows that there was no substantial difference between small and medium farmers in their feeding pattern except in the case of food grain that was used more by the medium farmers. Medium farmers could do so obviously due to more landholdings. However, it never provided extra productivity gains to this group of farmers.

Table 5.8

Feeding Pattern During 2003

(in Quintals)

Category	*Grass (Green + Dry)*	*Agricultural by product*	*Food grain*	*Tree lopping*	*Concentrate*
Marginal	19.9	27.68	1.61	8.56	0.01
Small	32.81	62.05	2.50	5.64	0.48
Medium	32.96	88.55	1.07	9.32	0.33
Average	24.8	44.72	1.74	7.80	0.18

Source: Field survey.

The performance of livestock in terms of milk productivity and feeding pattern, especially of the large animals followed by the farmers seemed to be more context specific than being a general trend. In other words, one must link the cultural context in which the livestock was reared to the feeding pattern that had implications for potential pressures on the local resources. The increased availability of grass both in the common protected and private lands in addition to agricultural by-products did not encourage the villagers to commodify milk. This happened despite the village was endowed with metal road and well connected with the mainstream. A very few households occasionally made *ghee* and sold it in the local market. What was produced, as shown in Table 5.7, was consumed by the households. A tribal village like Dagal was yet to transform its nature based livestock economy to a market-based one.

In the absence of commodification of the milk economy that continued to thrive mainly by being dependent on the natural system, the relative contribution of livestock, if expressed in monetary terms, to the household economy was the lowest in the village. It was estimated to be Rs. 7,519 on an average per household in a year. This is shown in Table 5.9. It was almost closer to the income generated by the non-farm sector. Some members of all farming households worked as wage labour in different non-agricultural avocations outside the village. Both these sources together contributed around 44 per cent to the gross annual income of the household. However, agriculture, as Table 5.9 shows, was the largest contributor to the household economy. It contributed to the tune of 56 per cent to the estimated gross income of the household.

One must bear in mind that comparatively very low contribution of livestock to the household economy in money terms could never reduce its worth. For, one can hardly discount the contribution being made, for instance, by milk to nutritional security of the household. Hence, one must account for it in real terms and seek to understand the relevance of regeneration of the forest to sustainable livelihood generation.

Education

It was not very far back that Dagal was almost in a state of dark zone on the literacy map. During 1991, only 17 per cent of the entire population were literate (Government of India, 1991). Much worse was the literacy rate of women, which was as low as 4 per cent.

Table 5.9

Annual Relative Share in Gross Income from Livestock vis-à-vis Agriculture and Non-farm Sector

(in Rs.)

Category of Landholdings	*Agriculture*	*Livestock*	*Non-farm*	*Total*
Landless*	Nil (0.00)	Nil (0.00)	96,000 (100.0)	96,000 (100.0)
Marginal	12,542 (48.67)	5,630 (21.85)	7,599 (29.49)	25,770 (100.0)
Small	29,113 (65.09)	11,579 (25.89)	4,032 (9.02)	44,725 (100.0)
Medium	40,946 (64.87)	8,722 (13.82)	13,454 (21.31)	63,122 (100.0)
Average per HH	20,635 (56.08)	7,519 (20.43)	8,644 (23.49)	36,798 (100.00)

Note: *There was only one landless household, which had a small grocery shop inside the village. He belonged to trading community and migrated from outside the village a few decades ago. He had neither agriculture nor livestock. His only source of income was this grocery shop. He neither belonged to the tribal community nor was involved with any activity related to forest regeneration.

Source: Field survey.

However, the scenario underwent a significant change during the last decade or so rising up to 27 per cent in 2001. What was noticeable was that literacy rate of the population of age group 5-14 was more in the recent past than that of the population of other age groups (Table 5.10a and 5.10b).

Table 5.10a

Educational Achievement of the Sample Households

Age Group	*Literate*		*Total*	
	Male	*Female*	*Male*	*Female*
5-14	51 (86.4)	26 (60.4)	59 (100.0)	43 (100.0)
14-30	59 (83.0)	26 (37.6)	71 (100.0)	69 (100.0)
Above 30	45 (67.0)	10 (16.3)	67 (100.0)	61 (100.0)
Total	155 (78.7%)	62 (36%)	197 (100.0)	173 (100.0)

Notes: Figures in the brackets are the percentage of total male and female population.

Source: Field survey.

Table 5.10b

Educational Status of the Sample Households

Age Group	*Illiterate*		*Primary*		*Middle*		*8 - 12*		*Above 12*		*Total*	
	M	*F*	*M*	*F*	*M*	*F*	*M*	*F*	*M*	*F*	*M*	*F*
5-14	8	17	37	19	13	7	1	0	0	0	59	43
14-30	12	43	15	9	13	8	23	6	8	3	71	69
30-60	18	47	22	7	9	3	9	0	1	0	59	57
> 60	4	4	4	0	0	0	0	0	0	0	8	4
Total	42	111	78	35	35	18	33	6	9	3	197	173

Note: M = Male; F = Female

Source: Field survey.

Dagal had the fortune of having relatively good infrastructure. It had a school (up to class eight) that was set up 20 years ago. Now, it had an English medium primary school, an *Aurvedic* dispensary, a metal road, etc. Besides, there had been an *aganwadi* centre for last few years. Thanks to the initiative taken by BAIF (Bharatiya Agro-Industries Foundation), villagers were now sending their children to school regularly.

Intervention for the restoration of the forest also made a significant contribution to the improvement in the educational level of the children. The design of intervention was such that grass was now available for seven days a year at a stretch. After having collected the same, it was stored inside the house and used as feed for livestock. This was contrary to the earlier practice in those children, mainly girl children used to be sent for collection of grass from the forestland as and when required. Now, two adult members of the household went to the forest for the stipulated number of days to collect grass. It was difficult for the children to carry back head loads of the collected grass. Since the quantity was much more, the adult members of the household did not want to take the risk of engaging the children in this task any longer. This factor also facilitated girls' regular attendance in school. Table 5.11 shows that an insignificant number of girl children were sent for collecting forest produce.

Table 5.11 also shows that, it was mostly the adult female members who collected the forest produce. However, males occasionally did so. Although the pressure on the womenfolk of the village for collection of forest produce was relatively more, the certainty about the availability of grass, fuelwood and tree lopping in the regenerated forest eased their difficulties they had to face earlier. Earlier, they had to travel a long distance for collecting the same. Now, they had more time to look after their children and attend other household works, which were more productive.

Table 5.11

Involvement of Children for Collection of Forest Produces

	Occasionally				*Mostly*			
	Male	*Female*	*Children* (M)	*Children* (F)	*Male*	*Female*	*Children* (M)	*Children* (F)
Fuelwood	13	3	1	1	11	67	0	2
Grass	18	3	1	1	14	65	0	2
Tree lopping	16	2	0	0	9	47	0	2
Total	47	8	2	2	34	179	0	6

Source: Field survey.

V

Access to Improved Resource Base

As discussed in the preceding section, the content of social capital in a tribal village such as the one under study was determined more by mutual trust and reciprocity. There was little room for power relations to breed and create social divisions within the village in terms of the powerful and the powerless. The continuity of such individual behaviour that was shaped by homogeneous conditions as determined by the historical forces seemed to have blockaded such divisiveness. This 'could' be reason, however, for some potential conflict between the village community and the external development agency (forest department) in the transition from the informal to the formal institution. However, there was no space for conflict while the village community was provided access to the protected forestland. The relatively more access gained by the marginal and small farmers to the protected forestland for forest produce showed this very clearly (Figures 5.2, 5.3 and 5.4). The only exception was tree lopping—small farmers collected only 11 per cent of their household requirements of tree lopping from the protected forest (Figure 5.4).

Figure 5.2

Average Collection of Grass Per Household of Each Category of Farmers from CPL,COL and PL

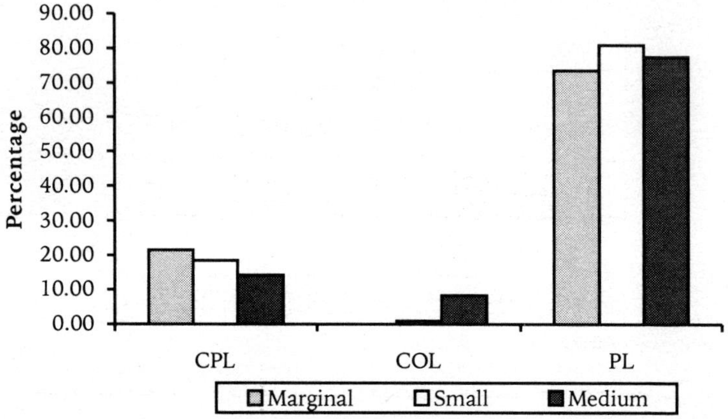

Source: Field Survey

Figure 5.3

Average Collection of Fuelwood Per Household of Each Category of Farmers from CPL,COL and PL

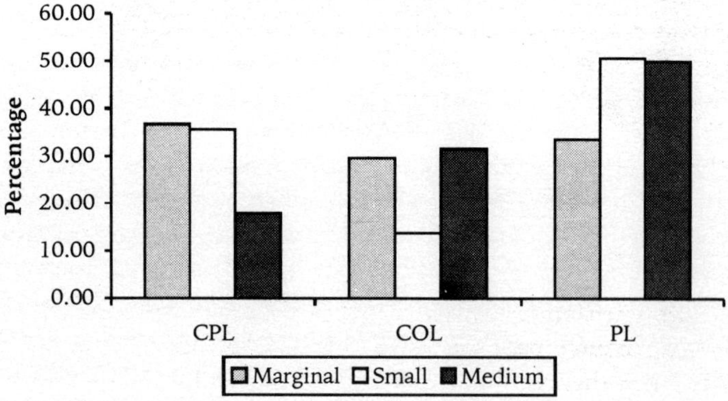

Source: Field Survey

Figure 5.4

Average Collection of Tree Lopping Per Household of Each Category of Farmers from CPL,COL and PL

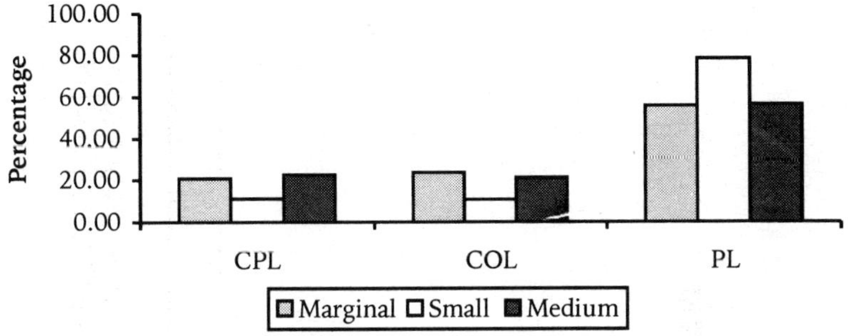

Source: Field Survey

Figure 5.5

Collection of Grass, Fuelwood and Tree Lopping from Different Sources

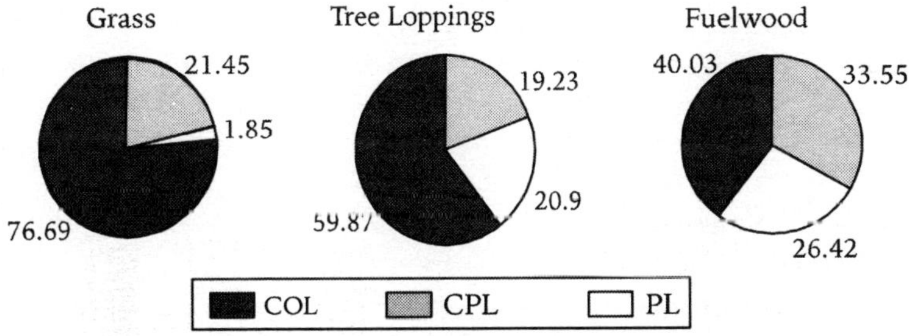

Source: Field Survey

While entitlement of the relatively poorer households to the improved resource base was never an issue to reckon with, one must bear in mind that the village had other sources through which it met household requirements. For example, the average requirement per household in a year was estimated to be 24.80 quintals of grass, 7.72 quintals of fuelwood and 7.80 quintals of tree lopping. Only 21.45 per cent of requirement of grass were met by the protected forestland, while 76.69 per cent met by private lands (Figure 5.5). Of course, around 34 per cent of the fuelwood requirement were met by the protected forestland. One may argue that the issue of entitlement might assume considerable importance in case the village community switched over to dairy economy and witnessed a rise in both human and livestock population. In such a situation, the pressure on the common land might increase disproportionately. One does not know whether a VFPC functioning under the shadow of the forest department would be able to ensure institutional stability under such critical conditions. There might not be any problem of accessing the forestland, but Dagal might witness degeneration of its forest once again.

VI

Ecological Impact

As mentioned earlier, forest department had the option of regenerating the forest either through fencing alone or implementing a plantation programme. In any case, the primary condition, which was implicit in the method of intervention, was the existence of the threshold limit of the forest. Regeneration, as observed by several scholars, was bound to be facilitated in case the critical minimum limit of plants was not destroyed. There had been degeneration without interruption until the late 1980s. It has been seen earlier how the revival strategy of the forest was locally built around local initiative, later overtaken by the formal government institution. Such a strategy for forest regeneration was followed in the entire forest range of Umarjhala Pathara of which Dagal was an important constituent.

The strategy that was followed by the forest department in consonance with the village community for plantation was both protective as well as regenerative. In order to protect the entire area taken for regeneration in each successive year, fencing was done. Of course, the village community was mobilised to provide social fencing of the forest. Apart from several steps being taken for soil and water conservation in the forestland, as described earlier, as many as 1,58,000 tree species were planted. Saplings were brought from nurseries located at different places of the district. Besides, seeds were sown for 1,27,000 plants in all plots developed in successive years. (Table 5.12)

Table 5.12

Plantation in Dagal Forests

(No.)

Year of Plantation	Through Planting Saplings	Through Seed Sowing
1994	35,000	31,000
1995	35,000	26,000
1999	35,000	16,000
2000	26,000	30,000
2001	27,000	24,000
Total	1,58,000	1,27,000

Source: Forest Department, Ghatol Forest Range, Banswara District, Government of Rajasthan.

This section examines the health of the forest in terms of the floral diversity and how both native and planted tree species contributed to it. Apart from the positive changes expected to take place in the forest ecosystem through regeneration, the likely commercial benefit that must accrue to the villagers was no less important. However, except few species such as *Dendrocalamus strictus* (bamboo) that matures in a year or two, the average age of the other trees planted to attain maturity was 25-30 years. Hence, it would take that much time to be harvested for commercial purposes. Therefore, it may not be an immediate concern for the present study. The topography of the forestland was hilly and undulating. The soil was sandy loam with black clay. Although the forest witnessed steady degeneration over the years, native plants such as *Tectona grandis* (*sagwan*), *Churel*, *Butea monosperma* (*khukra*) and *Diospyros melanoxylon* (*tendu*) had their rootstock in certain patches of the forestland.

Floral Diversity

Floral diversity is considered an important indicator of the health of the forest. The method of measuring diversity was the same as other case studies. The team used the Quadrate method (10x10m) for sampling trees. Sampling was done in the month of August after the rains. Three indices, including the importance value index (IVI), index of diversity and dominance and the species richness index, were followed to examine the impact of intervention on forest regeneration by means of increasing floral diversity. It is difficult to make any conclusions about any floral diversity achieved through plantation in the short term, as it was in the case of plots C, D and E. However, the emerging trends of diversity in these plots could indicate potential growth and its contribution in attaining maturity of the forest in future. Here, in the context of the present analysis, an attempt was made to examine diversity in all plots including A and B, which were initiated nearly 10 years ago.

The importance value index of each plot is estimated and shown in Table 5.13. In Plot B *Acacia catechu* had the maximum IVI of 71.45 followed by *Dandrocalamus strictus* (44.89), *Leucaena lucocephala* (35.72), *Holarrhena antidysenterica* (31.96) and *Acacia leucopholea* (25.56). Minimum IVI was recorded for *Holoptelia integrifolia, Butea monosperma, Gmelina arborea, Tectona grandis, Zizyphus nummularia* and *Anogeissus latifolia* (7.05). In any case, almost all the trees, whether they were planted, seed-based or native, were extremely important and had favourable ecological implications. It is important to note that *Acacia catechu* (*khair*) had maximum IVI in almost all plots except Plot C where *Tectona grandis* (*hagda*) had maximum IVI. *Acacia catechu* was not a native plant; it was planted through either saplings or seeds. However, *Tectona grandis* was a native that witnessed growth only in Plot C. The growth of *Acacia catechu* in all plots to a maximum extent, as indicated by its high IVI, shows that the selection of this plant was appropriate and friendly to the local ecosystem. As shown later, the survival rate of this plant was highest of all planted species. It was one of the few plants that survived and contributed to achieving higher floral diversity even though the survival rate of other planted species was less encouraging as shown later.

Acacia catechu had maximum IVI of 117.63 in Plot A also. However, it achieved less diversity here than on Plot B did. The different level of diversity achieved in each plot is shown in Figure 5.6, which also shows the species richness index across all plots. The diversity index of Plot B was estimated to be 0.93 while it was 0.85 for Plot A, which was slightly lower than that of Plot E (0.89), where the regeneration programme was initiated later. It is true that several factors, including edaphic characteristics and healthy biogeochemical cycle, contribute to achieving higher floral diversity. Needless to mention, favourable edaphic characteristics for growth of plants depend upon a balanced combination of various physico-chemical parameters. One of the most important parameters is availability of organic carbon in soil. The soil sample analysis of all five plots showed that concentration of organic carbon in Plot A was 2.24 per

Table 5.13

IVI for Trees and Shrubs

IVI	*Plot A*	*Plot B*	*Plot C*	*Plot D*	*Plot E*
1-10	Cassia auriculata Calotropis procera Acacia leucophloea Leucaena lucocephala	Holoptelia integrifolia Butea monosperma Gmelina arborea Tectona grandis Zizyphus nummularia Anogeissus latifolia Tamarindus indica	Phoenix sylvestris Ziziphus mauritiana Mitnigyana parvifolia Flacourtia ramontchi	Ipomea nil Echinochloa colona	Dandrocalamus strictus Azadirachta indica Acacia nilotica Echinochloa colona Leucaena lucocephala Ziziphus mauritiana Acacia leucophloea
10-25	Dandrocalamus strictus Ailanthus excelsa Phoenix sylvestris Acacia senegal Azadirachta indica	Azadirachta indica Jatropha curcus	Holoptelia integrifolia Leucaena lucocephala Nyctarthes arbortristis Dandrocalamus strictus	Dandrocalamus strictus Holoptelia integrifolia Jatropha curcus	Jatropha curcus Butea monosperma
25-60	Albizzia lebbeck Emblica officinalis	Acacia leucophloea Holarrhena antidysenterica Leucaena lucocephala Dandrocalamus strictus	Diospyres malanoxylon Holarrhena antidysenterica	Emblica officinalis Calotropis procera Azadirachta indica	Calotropis procera Emblica officinalis
60-100		Acacia catechu			Acacia catechu Prosopis juliflora
100-150	Acacia catechu		Tectona grandis	Acacia catechu	

Source: Field Survey

cent, which was more than that of, Plot B, 1.89 per cent. Despite having higher availability of organic carbon, floral diversity achieved by Plot A was less than that of Plot B. In this context, one must not overlook the harmful impact of uninterrupted human interference in the growth of plant species and its adverse impact on floral diversity. The latter seemed to have greatly contributed to the low level of floral diversity in Plot A.

As mentioned earlier, households adjacent to this plot were constantly interfering with the forest. They had their own reasons for entering this part of the forest—felling trees and collecting fuelwood and tree lopping. Despite the fact that their dispute with the forest department had been settled, they were reluctant to comply with the rules of management. However, no such interference took place in Plot B, which was why a favourable microenvironment was created on this plot. Consequently, it could attain a higher level of floral diversity.

Figure 5.6

Floral Diversity and Species Richness Indicies for Trees and Shrubs

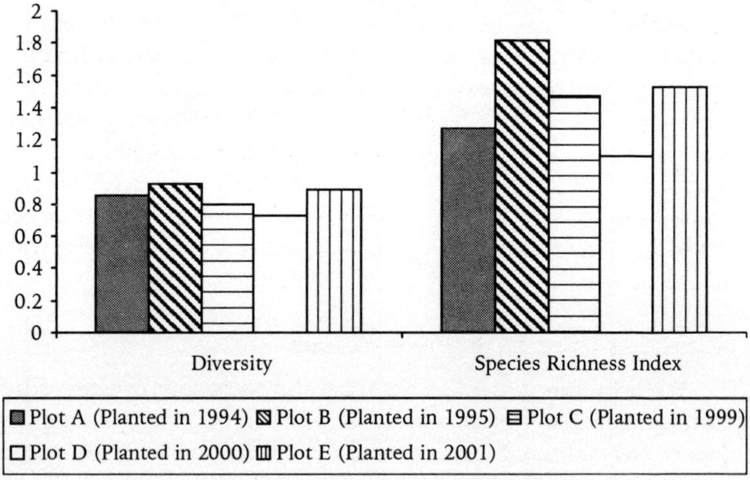

Source: Estimated based on the data collected from field survey.

What is interesting to notice that floral diversity and species richness index moved into the same direction in almost all plots (Figure 5.6). Although due ecological consideration was given to species selection, the survival rate of the planted ones, as shown latter, was comparatively less than the native ones, which were not planted. However, it was undeniable that the plants that could survive over the years enhanced the richness of the forest cover.

In any case, the extent to which floral diversity was achieved in Plot B, which was the highest, did not have much to contribute to the stability of the forest ecosystem. For the rest of the plots were yet to achieve enough floral diversity for the ecosystem to regain its resilience. With low availability of organic carbon in Plot C (1.17 per cent) and Plot D (1.17 per cent), as the soil sample analysis shows, one was not certain how these plots, for instance, could contribute to activate healthy biogeochemical cycle and boost the process of regeneration of ecosystem of the entire forest area of Dagal.

However, it did not imply that the achievement of floral diversity was too low to ensure resilience of the ecosystem. One must bear in mind that regeneration process was initiated in these plots in the recent past. But, still, it was important to consider the extent to which care had to be taken (for example, no human interference) so that the regeneration process could continue uninterruptedly. The entire forest was yet to achieve diversity beyond the present level, which did not significantly reduce its vulnerability to ecological shocks and crises. For floral diversity through regeneration was desirable, at least to the extent that it ensured safe minimum standard of the stock of renewable resources (Bromely, 1998). Even if one argues that the present level of diversity ensured safe minimum standard of the forest resources, VFPC would have to accelerate the regeneration process for the ecosystem to regain its resilience. However, one was not certain if the degraded forest could regain its resilience when the survival rate of the planted species was so incredibly low.

Surviving Species: Planted Versus Native

Based on the data collected from the secondary sources, share of each species in the total number of surviving species in each group of plants was estimated. Plant species were grouped into three categories (categories were made purely for the purpose of analysis), including sapling-based plantation, species that grew through seed sowing and species that were native. Here, share was estimated in terms of percentage of each species survived to the total number of surviving species of the same group (Table 5.14).

Table 5.14

Stock Position of Tree Species Through Plantation, Seed Sowing and under Natural Conditions as on 2004

Planted Species	*Stock*	*Seeds*	*Stock*	*Natural*	*Stock*
Dandrocalamus strictus (Bans)	38.89	Acacia catechu (Deshi Khair)	39.59	Tectona grandis (Sagwan)	65.72
Acacia catechu (Desi Khair)	29.89	Acacia leucophloea (Rheonj)	2.46	Churel	16.05
Kegar Khair	3.11	Leucaena lucocephala (Su-babool)	3.91	Diospyros melanoxylon (Tendu)	5.85
Emblica officinalis (Anwala)	10.23	Azadirachta indica (Neem)	13.97	Boswellia serrata (Salar)	0.09
Terminalia belarica (Behra)	0.27	Acacia senegal (Kumtha)	20.92	Liena coromendelica (Gurjan)	0.05
Ziziphus mauritiana (Ber Deshi)	2.89	Terminalia belarica (Behra)	2.87	Butea monosperma (Khakra)	10.28
Azadirachta indica (Neem)	1.64	Butea monosperma (Khakra)	8.65	Acacia leucophloea (Rheonj)	1.96
Albizia lebbeck (Siras)	0.97	Echinochloa colona (Sawan)	0.14		
Tectona grandis (Sagwan)	1.70	Baihan	2.88		
Churel	6.29	Kegar Khair	4.61		
Echinochloa colona (Sawan)	2.54				
Ailanthus excelsa (Ardu)	0.01				
Pithicolobium dulsi (Jungle Jalebi)	0.24				
Acacia senegal (Kumtha)	1.34				
Total	100.00	Total	100.00	Total	100.00

Source: Forest Department, Ghatol Forest Range, Banswara District, Government of Rajasthan.

Table 5.14 shows that regeneration of plants through sapling and seed sowing did not yield perceptible results. Only two-three plants such as *Dandrocalamus strictus* (bamboo) and *Acacia catechu (desi khair)* in the sapling category and *Leucaena lucocephala* (*su-babool*) and *Acacia catechu* (*desi khair*) in the seed-sowing category had the maximum share in their respective groups. However, the most encouraging aspect of the regeneration process was that the incidence of survival of the native plants was relatively higher than other two categories of plantation. One, however, did not expect such higher percentage of survival rate from the native plants since they were the ones that were adversely affected in the degeneration of the forest. *Tectona grandis* (*sagwan*) had the highest percentage share in the total surviving native plants. The relative performance of each category of plants in terms of their ability to survive in the forestland is shown in Figure 5.7.

Figure 5.7

Share of Each Group of Plants in Total Surviving Plants

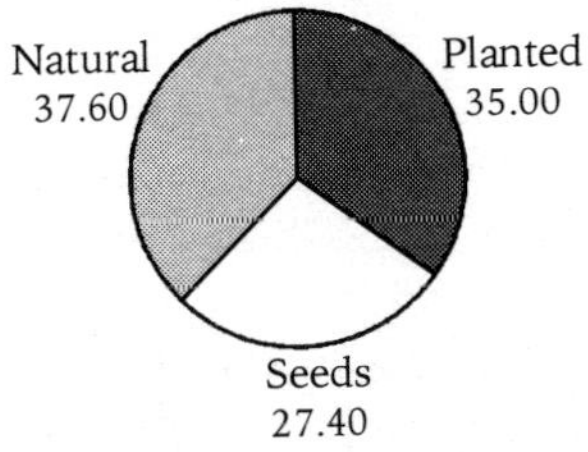

Source: Forest Department, Ghatol Forest Range, Banswara district, Government of Rajasthan.

Figure 5.7 shows that share of sapling-based plantation was 35 per cent of all surviving species taken together in all plots. However, the share of the native species was 37.60 per cent, and in respect of seed-based plants, it was 27.40 per cent.

The relatively higher growth of native plants indicated a positive change in the ecosystem, particularly, in view of the fact that planted species could not suppress growth of native species. While one must count the contribution being made by each tree species to the ecological improvement, no growth of tree species at the cost of the native ones is desirable in view of better health of the forest. For, suitable microenvironment had already been created for the survival of the microorganisms in respect of the native plants. Microorganisms are integral parts of biogeochemical cycle in that they make nutrients available for plants. From this point of view, the higher survival rate of the native plants was definitely an encouraging feature of ecological improvement. However, it does not necessarily mean that sapling-based species must have lower survival rate. What appears to be missing in the entire strategy of forest regeneration was an appropriate approach for plantation, which was why a conducive microenvironment was not created for the survival of microorganisms in all plots.

Plot-wise Analysis of Survival of Plants

In order to examine how ecological responses varied between different plots, the survival index of the plants of each group in each plot is worked out and shown in Tables 5.15, 5.16 and 5.17. It appears from Table 5.15 and 5.16 that ecological responses to the survival of the plants varied considerably in respect of most of the plants.

This happened despite the fact that all plots were part of the same forestland and had almost same edaphic and climatic characteristics. However, there were a few plants, which could survive up to 60 per cent in all plots. This was true for both sapling and seed-based species. For example, sapling-based species such as *Dandrocalamus strictus* (bamboo), *Emblica officinalis* (*aonla*), *Tectona grandis* (*sagwan*) and *Churel* could survive around 60 per cent or more in almost all plots. In some cases, of course, species as *Acacia catechu* (*desi khair*) had more than 80 per cent survival rate in all plots in the sapling-based group and little less than 80 per cent in the seed-based group.

In other words, few plants survived equally well whether they grew through plantation or seed sowing. However, there were some plants such as *Acacia senegal* (*kumta*) and *Azadirachta indica* (*neem*) that proved to be more appropriate in view of their higher survival rate since they were grown through seed. Hence, the second aspect of forest regeneration of Dagal was that there were some plants that had higher incidence of survival in case they were grown through saplings, while some others through sowing. The third dimension was that same tree species such as *Azadirachta indica* (*neem*) or *Echinochloa colona* (*sawan*) exhibited different survival rate in different plots despite being planted through the same source.

The fourth dimension was the difference that existed in the survival rates of plants grown through two different sources on the same plot. This is illustrated by examining the survival rate of two differently grown plants through sapling and seed on the same plots

Table 5.15

Incidence of Survival of Planted Species in Each Plot as on 2004

Botanical Name (Local Name)	*Plot A (1994-2004)*	*Plot B (1995-2004)*	*Plot C (1999-2004)*	*Plot D (2000-04)*	*Plot E (2001-2004)*
Dandrocalamus strictus (*Bans*)	80.42	87.65	56.53	66.22	90.71
Acacia catechu (*Desi Khair*)	90.21	90.17	87.64	81.50	96.65
Kegar Khair	45.83	28.60	Nil	Nil	Nil

contd. ...

contd. ...

Botanical Name (Local Name)	Plot A (1994-2004)	Plot B (1995-2004)	Plot C (1999-2004)	Plot D (2000-04)	Plot E (2001-2004)
Emblica officinalis (*Anwala*)	78.50	66.00	73.67	71.25	87.50
Terminalia belarica (*Behra*)	36.67	19.00	Nil	Nil	Nil
Ziziphus mauritiana (*Desi Ber*)	65.00	Nil	63.00	47.67	81.67
Annona squamosa (*Sitafal*)	7.0 (up to 98)	Nil	Nil	Nil	60 (up to 2002)
Azadirachta indica (*Neem*)	66.00	25.00	58.00	Nil	75.00
Albizia lebbeck (*Siras*)	30.50	38.75	Nil	Nil	Nil
Tectona grandis (*Sagwan*)	70.00	77.50	Nil	Nil	Nil
Churel	67.00	66.00	76.00	Nil	91.33
Echinochloa colona (*Sawan*)	45.00	37.00	46.25	38.00	80.00
Ailanthus excelsa (*Ardu*)	3.00	Nil	Nil	Nil	Nil
Pithicolobium dulsi (*Jungle jalebi*)	Nil	13.50	Nil	Nil	Nil
Bombex cieba (*Semala*)	Nil	31 (up to 97)	Nil	Nil	Nil
Acacia senegal (*Kumta*)	Nil	75.00	Nil	Nil	Nil
Maduhca indica (*Mahuva*)	Nil	Nil	7.5 (up to 2002)	Nil	Nil
Ficus religiosa (*Bada Peepal*)	Nil	Nil	30 (up to 2001)	Nil	Nil

Source: Estimated based on the data collected from Forest Ranger, Ghatol Forest Range, Banswara District.

Table 5.16

Incidence of Survival of Seed Showing Species in Each Plot as on 2004

Botanical Name (Local Name)	Plot A	Plot B	Plot C	Plot D	Plot E
Acacia catechu (Deshi Kher)	75.00	81.60	81.40	74.50	80.91
Acacia leucophloea (Rhoni)	71.67	Nil	Nil	Nil	Nil
Leucaena lucocephala (Su-babool)	41.25	68.00	Nil	35.00	40.00
Azadirachta indica (Neem)	49.29	61.60	64.17	70.25	72.66
Acacia senegal (Kumta)	86.88	86.67	90.57	50.83	72.50
Terminalia belarica (Behra)	Nil	37.50	Nil	Nil	Nil
Butea monosperma (Khakra)	Nil	73.75	Nil	84.25	45.00
Echinochloa colona (Sawan)	Nil	12.50	Nil	Nil	Nil
'Baihan'	Nil	Nil	57.50	68.00	Nil
'Kegar Khair'	Nil	Nil	Nil	77.33	85.00

Source: Same as for Table 5.15.

Figure 5.8

Survival Rate of Planted Versus Seed Sowing Species

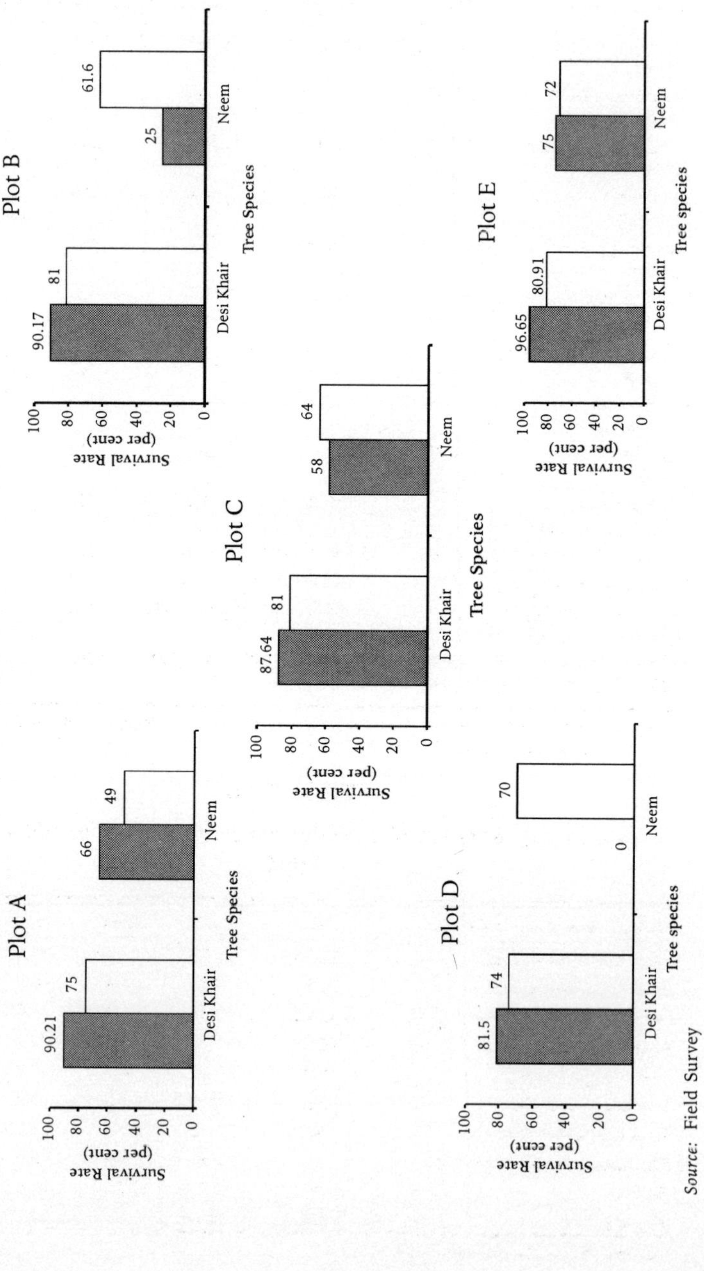

Source: Field Survey

Table 5.17

Annual Compound Growth Rate (Regeneration Rate) of Native Species

Common Name	Plot A			Plot B			Plot C			Plot D			Plot E		
	1994	2004	ACGR	1995	2004	ACGR	1999	2004	ACGR	2000	2004	ACGR	2001	2004	ACGR
Rheonj	370	965	10.06	250	650	11.20	Nil	Nil		Nil	Nil		700	735	1.64
Churel	220	850	14.47	200	1025	19.91	3700	4050	1.82	9550	11250	4.18	1900	2020	2.06
Sagwan	Nil	Nil		20	1550	62.15	30500	35500	3.08	35750	40200	2.98	1260	1350	2.33
Khakra	Nil	Nil		20	1525	61.86	3205	3550	2.07	6060	7220	4.48	Nil	Nil	
Tendu	Nil	Nil		Nil	Nil		3050	3250	0.71	3410	3750	2.40	Nil	Nil	
Salar	Nil	Nil		Nil	Nil		30	40	5.92	35	70	18.92	Nil	Nil	
Gurjan	Nil	Nil		Nil	Nil		15	25	10.76	20	30	10.67	Nil	Nil	

Note: ACGR = Annual compound growth rate.

Source: Same as for Table 5.15.

Figure 5.9

Survival Trend of Native Plants Versus Sapling and Seed Based Plants

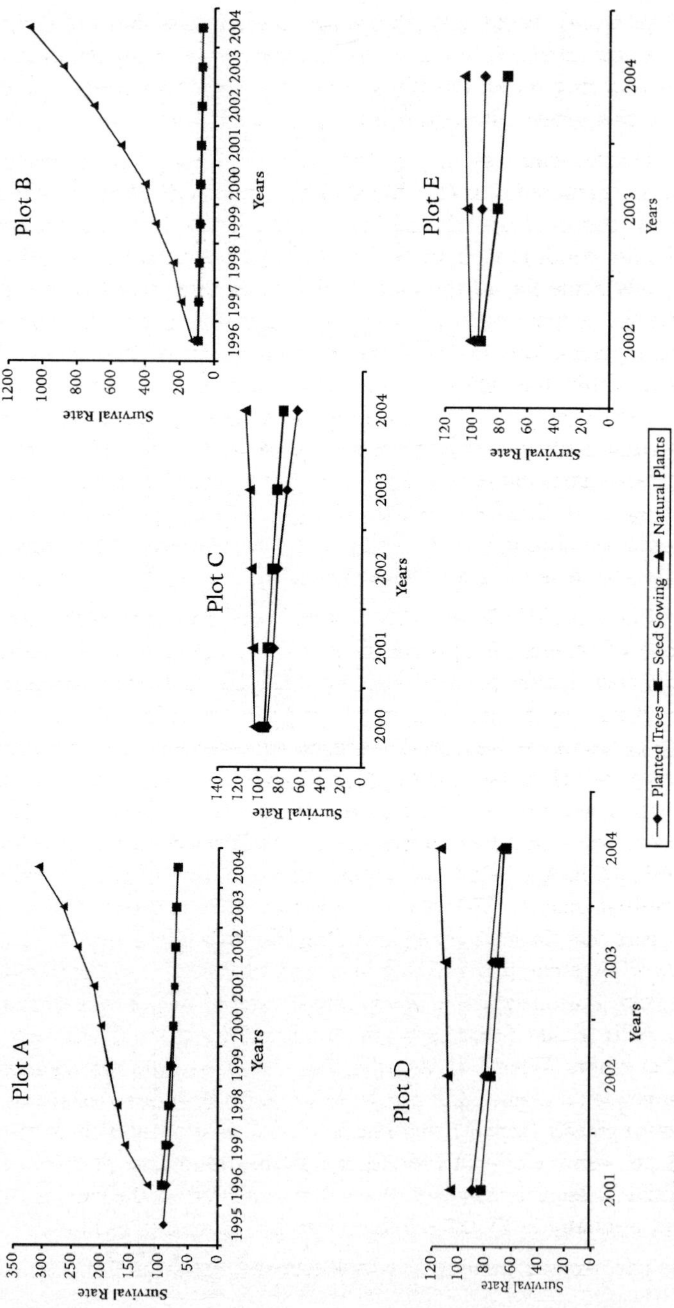

Source: Field Survey

(Figure 5.8). What appears from Figure 5.8 was that the survival rate of *Acacia catechu (desi khair)* through sapling was higher than through seed sowing on all the plots. Similarly, *neem* had better survival rate if it was grown through seed.

Hence, what assumes considerable importance in the discourse on forest regeneration is the choice of an appropriate strategy of plantation given the ecological variations in the edaphic characteristics and climatic conditions at the micro level. While climatic conditions were equally same for all the plots under study, it remains to be answered why the survival pattern varied across the plots and within the plots for many plants. Was it due to the microenvironment that seemed to vary even within the same plot and create different survival conditions for different species? If *Acacia catechu (desi khair)* could survive on all plots through sapling, why should it be grown through seed. If the edaphic characteristics did not ensure survival of sapling-based *Azadirachta indica (neem)*, it is difficult to understand why a less cost-effective method of plantation through seed sowing was not followed. What was true for *neem* was true for many other plants too.

The story of native plants was different from that of the above. The issue of regeneration of plants was more relevant to the native ones than their incidence of survival especially in a situation when moisture conditions of the soil improved due to intervention for soil and water conservation. Besides, plots taken up for regeneration were fenced. The rate at which native plants were regenerated in all plots is shown in terms of their annual compound growth rate in Table 5.17. Table 5.17 shows stock of native plants in each successive year, i.e., 1994, 1995, 1999, 2000 and 2001 and estimates their annual growth rate at the terminal year of 2004 with respect to each successive base year. It appears that *Churel, Tectona grandis (sagwan)* had higher growth rate in plot A, and B as compared to that in plots C, D and E. As far as plot C and D were concerned, the regeneration rate of only two plants namely, *Boswellia serrata* (salar) and *Liena coromendelica (gurjan)* was higher than other plants. What it meant, therefore, was that even regeneration of the native plants might differ across the plots. To be more specific, let us take *Tectona grandis (sagwan)* that regenerated in Plot B annually at the rate of 62 per cent, while in Plot C, it did so at the rate of 3.08 per cent. Similarly, *Butea monosperma (khakra)* regenerated at the rate of 61.06 per cent annually in Plot B, while it was 2.07 per cent in Plot C.

The skewed pattern of survival of trees that grew under different conditions of plantation hardly indicated a better health of the forest. This was the reason why the forest of Dagal did not achieve as much diversity as one expected in view of expenditure incurred towards its regeneration. In fact, if one strictly goes by cost-benefit analysis of regeneration of this forest, one must account more for revival of native species as compared to other forms of regeneration. The second best option could be plantation through seed sowing. The trend analysis of survival of plants under different plantation technique *vis-à-vis* native plants, as shown in Figure 5.9, might lend credible support to this hypothesis.

The survival pattern of sapling-based plants and species grown by seed shows a declining trend, while regeneration of native species shows an uninterrupted increasing trend in each successive year in all plots. The lower survival rate of sapling-based and seed sowing species showed their inappropriate selection to the edaphic characteristics and climatic conditions. However, they did not create conditions against the biogeochemical cycle of the local ecosystem that prompted regeneration of the native plants.

Figure 5.9 shows that the survival rate of the native plants in each plot over a decade or so was consistently higher than the plants that grew under different plantation techniques. It suggests that forest regeneration could depend upon native ones more, wherever they had a critical minimum stock. It does not necessarily mean other plantation technologies had no relevance to forest regeneration. As mentioned earlier, species chosen must be appropriate to the microenvironment in which plantation is to take place. It will not only ensure short-term benefits but also long-term ecological gains that the village community of Dagal seemed to have been deprived of.

VII

Institutional Prospects: Was Reconciliation Successful?

The prospect of institutional renovation remained as exigent as sustainable forest regeneration was for Dagal. While complacency about regeneration had ostensibly overtaken the collective mood of the village community, it is not certain how the approach would remain collective in future. It is not only the mode of institutional renovation being followed in Dagal, but also the absence of consensual approach in the decision making process that contributed to such apprehensions. This happened mainly due to excessive dependence of the village community on the forest department. It is true that JFM expanded wage employment opportunities for the villagers. It is also true that villagers would not perhaps have been able to bear the huge expense of forest regeneration. In a situation where they were pushed to become receivers, there was a gradual erosion of the sense of community ownership of the forest. This was likely to have an adverse impact on the institutional performance.

Hence, it is contingent to know the present level of institutional performance and the other probable scenarios and conditions of institutional performance that the Dagal village institution could have opted for. An attempt is made here to analyse probable alternative institutional scenarios based on the Backward (LR) Multiple Logistic Regression Model, following the methodology as used in other cases in the present study. It indicates the best institutional scenario that VFPC, the village institution, could have chosen to ensure its sustainability.

The objective of the model is to estimate probable alternative scenarios by means of simulating the present one. It identifies the conditions under which the best possible scenario emerges. The Multiple Logistic Model specifies functional relationships between the probable institutional success and the factors responsible for it. Here, institutional success refers to positive institutional outcome. The indicators of positive institutional outcomes of collective action were based on the perceptions of the village community, including improvement in livestock, development of the natural resource base, availability of grass and tree lopping in the forest, and decline in migration.

The responses elicited by the sample households on the variables shown in Table 5.18 have been taken into consideration for analysis. It is needless to mention that only significant variables are incorporated into the Backward (LR) Multiple Logistic Regression Model. To determine their validity and relevance, the same regression model is used on the independent variables that are responsible for the probability (Y=1) to occur.

Since the Goodness-of-Fit test in Table 5.19 shows that the model adequately fitted the data (p=0.985), all variables shown in Table 5.18 have been incorporated for estimating alternative institutional scenarios. Before estimating the results of the probability (Y=1), the values of the independent variables are estimated and shown in Table 5.20.

Table 5.18

Results of Significant Test of the Parameters

Independent Variables	*Type of Response*
Awareness about the village organisation*	Yes=1, No= 0
Member of the village organisation*	Yes=1, No= 0
Attending the meeting regularly*	Yes=1, No= 0
Participation in the activities undertaken by village organisation for natural resource management*	Yes=1, No= 0
Effective participation in decision making process*	Yes=1, No= 0
Regularity of meeting*	Yes=1, No= 0
Decision making process in the meeting*	With Majority of members=1Only a few=0
Performance of organisation"	Satisfactory=1 Not satisfactory =0
Association with village organisation"	Voluntary Labour=1 Partially Labour=2 Fully paid Labour =3
Denied to participation in the management activities of the village organisation"	Yes=1, No= 0
Access to the improved natural resource base*	Yes =1, No= 0
Freeriding (Non-compliance)*	Yes =1, No= 0
Satisfied with the distribution pattern of resources"	Yes 1, No= 0
Any conflict between villagers and the members of the village organisation*	Yes 1, No= 0
Adverse impact of forest development on Social and Economic lives*	Yes 1, No= 0

Note: *–Significant, "–Not Significant.

Table 5.19

Results of Goodness-of-fit Test of the Model

Estimation terminated at iteration number 4 because Log Likelihood decreased by less than 0.01%.

Cox & Snell –R^2	0.263
Goodness of Fit	81.776
$X_{(8)}^2$	1.854 (p=0.985)

Table 5.20

Estimated Value of the Significant Parameters

Variables	*Estimates*
Aware about the village organisation	-1.263
Member of the Village organisation	1.095
Participation in the activities undertaken by village organisation for natural resource management	1.948
Regularity of meeting	1.113
Attending the meeting regularly	0.255
Decisions making process in the meeting	5.081
Effective participation in the decision making process	-1.889
Freeriding	0.885
Conflict	-1.229
Access to the improved resource base	-0.633
Adverse impact of forest development on Social and Economic lives	-1.708

As explained earlier, the objective is to indicate the likely scenarios of institutional performance that VFPC may witness under different conditions. Different combinations of independent variables as mentioned in Table 5.18 represent different conditions. Nine scenarios are shown in Table 5.21 including the one attained by VFPC at present. This is indicated by 'F' (0.53). However, VFPC is able to achieve higher level of institutional performance or more institutional success (0.92) under the conditions indicated by scenario 'I'. The probable institutional outcomes are plotted in Figure 5.10. The major differences between these two different institutional outcomes were conflict, the degree of non-compliance of the rules and the number of members taking the decisions. Compliance plays a significant role in ensuring stability of an institution. Its importance is so significant that the probability of institutional success goes up in its presence even if villagers are not given access to the improved resource base. Table 5.21 shows that in such a situation the probability of institutional success is 0.73. This is higher than the present status indicated by the probability of 0.53—lower because incidence of non-compliance (freeriding) was so high that it outweighed the positive impact of access to the improved resource base.

Figure 5.10

Probable Institutional Scenarios of VFPC of Dagal

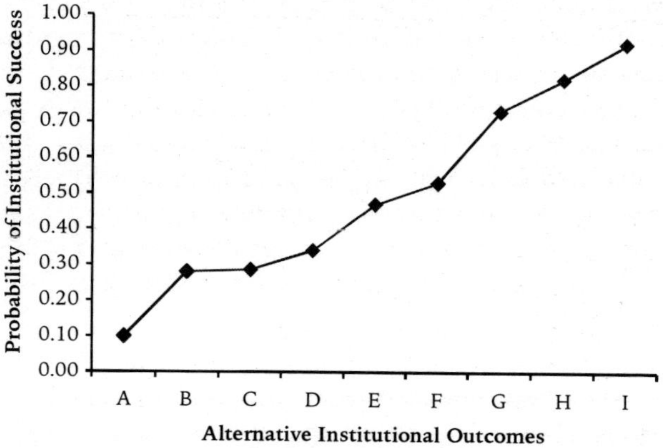

Table 5.21

Probable Scenarios of Institutional Outcome under Different Conditions in Dagal

Variables	*Probable Scenarios*								
	A	*B*	*C*	*D*	*E*	*F*	*G*	*H*	*I*
Awareness about the Institution	0	1	1	1	1	1	1	1	1
Member of the VFPC	0	0	1	1	1	1	1	1	1
Participation in the activities undertaken by VFPC for forest development	0	0	0	1	1	1	1	1	1
Effective participation in the decision making process	0	0	0	0	1	1	1	1	1
Regularity of meeting	0	0	0	0	0	1	1	1	1
Regular Attendance in the meeting	0	0	0	0	0	1	0	0	0
Decision taking process in the meeting	0	0	0	0	0	0	0	0	1
Freeriding	1	1	1	1	1	1	0	0	0
Any conflict	1	1	1	1	1	1	1	0	0
Access to the improved natural resource base	0	0	0	0	0	1	0	1	1
Adverse impact of intervention	1	0	0	0	0	0	0	0	0
Outcomes	**0.10**	**0.28**	**0.29**	**0.34**	**0.47**	**0.53**	**0.73**	**0.82**	**0.92**

As explained earlier, the incidence of freeriding, especially for fuelwood, had been a regular feature for the village community. While the village community never treated their entry into the forest for collection of fuelwood as undesirable, the forest department insisted that it must be treated so and therefore to be discouraged. This was discussed in several meetings of the VFPC. However, no action was taken against anybody who collected fuelwood and grass sometime from the protected forest without the permission of VFPC. There was also the case of the conflict between Hindula *pada* and the rest of the villagers, on the one hand, and the forest department, on the other, which had been a constant threat to collective action for improving institutional performance.

Besides, the most important factor that contributed significantly to different levels of outcome was the decision making process, which was dictated by the forest department. This went against the community wisdom and leadership that lost significance in promoting joint action. In any case, all the three factors that made significant difference to institutional outcomes were associated more with the intervention strategy followed by the forest department. It suggests that the reconciliation, sought between the earlier informal institution and the formal institution that entered the village later, was not only incomplete but also weak. Hence, it was vulnerable to shocks and crises.

VIII

Conclusion

While it is heartening to note that the regeneration of degraded forest figured prominently on the environmental agenda of the state, the design of the intervention was flawed. With only a limited understanding of rural society, especially that of tribals, Aravali, the government sponsored intervening agency, ignored local social and cultural specificities that had much to do with the behavioural pattern of the people. The agency was also oblivious that any newly introduced social practice can have the potential of being institutionalised only when it is embedded in the local culture. The uniform rules of JFM, followed by the intervening agency across the state, reduced the local cultural specificities to nothing. It overtook the functions of the village committee and left virtually no space for local initiative to grow. As a result, local initiatives taken by the village community itself to restore the forest disappeared. The villagers, in practice, became prisoners of the dependency syndrome. Even worse, they were pushed towards the receiving end. The outcome was obvious. They lost their sense of involvement with the local institution that they had developed some years ago. Observations made by two recently appointed forest guards explain it better. Asked why forest guards were changed while the *samiti* remained same, their reply was 'How do you expect change in the *samiti* when the forest department remains the same?'

The issue that concerns one is whether such an institution could sustain in a situation where rules of regeneration and conservation of the forest were imposed by the state. This is not to dismiss the role of the state—the state must assume its role as a catalytic agent to ensure growth and sustainability of local initiative—but it warrants true reconciliation between the formal (state) and informal (local) institutions for natural resource management to succeed (Dasgupta, 2001a; Dia, 1996).

6

Micro Watershed Development

Introduction

The recent three years of acute drought put the fragile, arid ecosystem of western Rajasthan in a distress situation, as so often happens in this state. However, it was not drought alone that contributed to the area's escalating vulnerability to natural disasters. The adverse impact of the mounting competing claims on the renewable natural resource base—that was constrained to grow at par with the growth of population—cannot be underrated. It was a case of scarcity growing because of the poorer rate of regeneration of natural resources. One may find production and consumption analytics more relevant to the understanding of such a crisis.

In western Rajasthan, the increasingly weak base of natural resources failed to match with the increasing consumption requirements of the rising human and livestock population. Its vulnerability to drought exacerbated with the declining rainfall during the last two decades. On the other hand, the growth and density of the human population remained almost the same and was no less than the rest of the state; population density in the arid zone grew by 28.48 per cent per sq km during 1991-2001 (Table 6.1).

Table 6.1

Zone-wise Changing Density of Population in Rajasthan (1981-2001)

(in percentage)

	1981-1991	*1991-2001*
Arid	29.86	28.48
Semi-arid	29.10	29.87
Sub-humid	22.23	24.76
Humid	29.02	27.12
Rajasthan	28.44	28.33

Source: Calculated based on data collected from Population Census, Rajasthan.

Almost low as that of semi-arid zones and more than humid and sub-humid zones, the spiralling population intensified the biotic pressure on the resources, and drove the most vulnerable to bare necessities. In the absence of migration, the only option left with the rural community, in general, and the poor, in particular, was how to help the same resource base respond to the competing claims through enhancement of environmental productivity.

Villagers of the Tibna *panchayat*, located in Jodhpur district of western Rajasthan, underwent a protracted struggle in meeting their claims on the natural resource base before they finally found the answer in a shift of their social embeddedness to new institutional practices. The new institutional practices, they perceived, would enhance productivity of their resource base and leave no room for distress to overtake their lives even under drought. The three most taxing problems that pushed them into vulnerability for many years were the loss of houses and crops with the flooding of a seasonal river called Kheton ki *Nadi*, a severe scarcity of drinking water and advancement of sand dunes leading to loss of agricultural land.

They sought government intervention only to be left with little choice. Not only were interventions limited in their scale and magnitude, but they also did not create any space for community involvement. As a result, they failed to make any significant dent in community life. All that was achieved was a lessening of the drinking water crisis to an extent, although the uncertainty remained. With the gradual erosion of survival base, the villagers realised that only collective engagement could be instrumental in addressing their problems. In this scenario, the Poorva Sainik Bahudheshiya Sahakari *Samiti* (PSBUSS), an NGO of Jodhpur, brought the villagers together and encouraged them to form village-level organisations in each village of Tibna *panchayat* while maintaining links with each other. They thus decided to develop micro watershed in their area.

The reason was simple. The residents of all four villages of Tibna *panchayat*, namely, Fateh Singh ka Bas, Bhoja ka Bas, Tibna and Timri, realised that their *panchayat* was endowed with 13 micro watersheds of different sizes covering 53-620 hectares of land, with an aggregate area of 2,916 hectares of land (Map 14). Villagers sought improvement of these micro watersheds for they thought that such a step would greatly improve their survival base. It is in this context that the present case study intends to examine how development of

Map 14

LOCATION MAP OF TIBNA PANCHAYAT

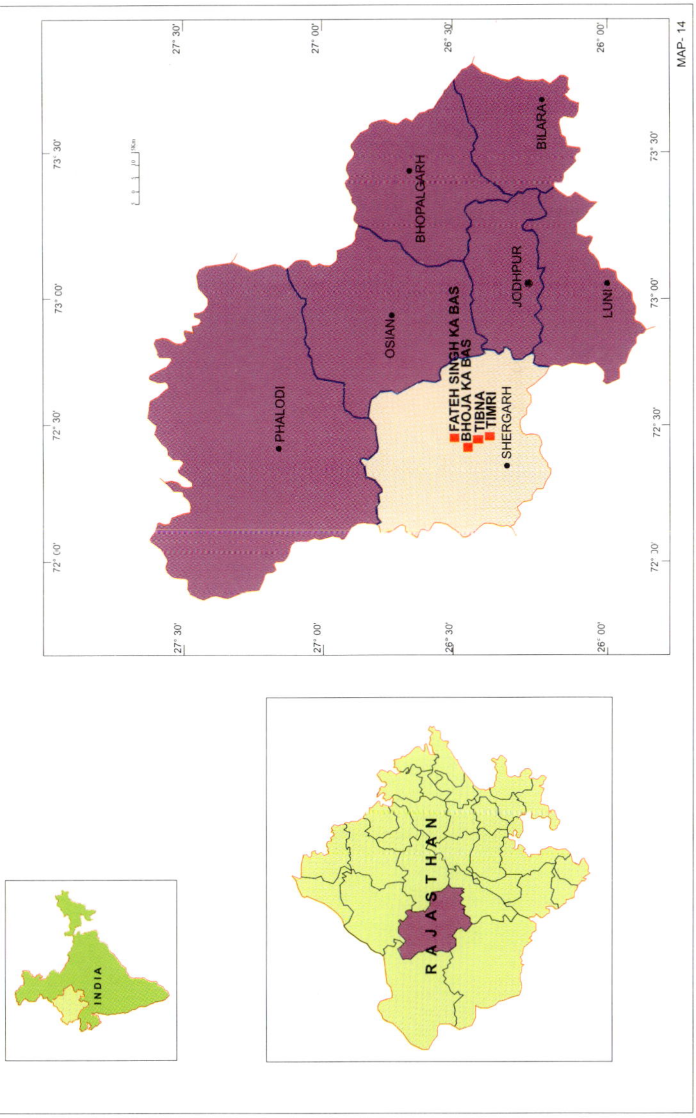

micro watershed regenerated the natural resource base of the village community under dessert conditions through institutional renovation.

Before analysing the process of change that Tibna *panchayat* witnessed during the last several years, it is vital to know the phsiographic conditions under which watershed development was sought. For the sustainability of the initiatives not only depends on the strength of the collective action, but also on the technical appropriateness of the intervention against the backdrop of the physiographic conditions. This is briefly discussed in the section II of the study. Section III overviews the interventions for watershed development through institutional renovation by several organisations, including PSBUSS. While watershed development activities are analysed in the section IV, section V examines environmental change that came with the intervention. In section VI, how household economy gains through resource development despite demographic pressure is investigated. The study is concluded in the section VII.

II

PHYSIOGRAPHY OF TIBNA *PANCHAYAT*

Unlike other river basins, no perennial river flowed by the Tibna *panchayat* area. It was only rainwater that collected in the catchment area and flowed into what was locally called *nadi*. In fact, it was an area where the most rainwater run-off flowed through three drains (*nallah* in local terms). These *nallahs* were formed by various streams that flowed in the northern and southern regions of the ridgelines in the entire watershed area. The total watershed area was 2,916 hectares, stretching from northeast to southwest of Tibna *panchayat* of Shergarh *tehsil*. As can be seen in Map 15, one major *nallah* passed through Fateh Singh ka Bas; the second flowed through Bhoja ka Bas and the third flowed through Siyandhan. These *nallahs* became active during the rainy season, and normally took the form of a fast-flowing river. Consequently, soil erosion became a constant problem during rainy season in both hilly tracts and in the sand dunes.

Kheton ki Nadi formed at the confluence of two *nallahs* that originated from Fateh Singh ka Bas and from the Siyandhan hills. These *nallahs* were surrounded by a large number of streams through which rainwater flowed to those *nallahs*. *Kheton ki Nadi* took the form of a small *nallah* from the place of its origin and passed through the hills on both sides (Map 15). Rainwater so collected in the *nallah* through these innumerable streams flowed with such high velocity that it caused considerable erosion of soil in the riverbed near Bhoja ka Bas. The riverbed, as a result, expanded. Since the same river flowed through all other villages in succession, it continued to erode soil on both sides of the river and expanded the riverbed on both sides until it discharged its water into *Bol Nadi*. *Bol Nadi* was a reservoir located at Timri.

In addition to soil erosion, which damaged agricultural prospects, the *nadi* also regularly caused collapse of houses in the low-lying areas. Then there was the added misery of the removal of sedimentation of soil after the floodwater receded. Although *Kheton ki Nadi* was a seasonal river (in fact, a water channel that activated only during the rainy season), it stood out as a threat to the livelihood of the villagers of Tibna *panchayat* for years together. The gravity of the problem was not so high when the size of the population was small. However, with the rise in human and livestock population, the habitation cover of the area of the villages expanded, as did cultivation of land. Hence, it became a challenge to train and harness the river so that it could no longer erode soil, damage agriculture and destroy houses. Besides, it was felt that the rainwater run-off must be harvested so that recharge of groundwater would increase.

The northwestern part of Tibna *Panchayat* was covered by blown sand, alluvium of the quaternary period, while the southeastern part comprised Paleozoic sandstone. The type of rocks in the watershed area was rhyolite. Another characteristic feature was that the area was infested with sandy hummocks with heights varying between 13 r to 45 r (r= relative height in metres) and rocky outcrops. The climate of the area was arid with erratic monsoon rainfall (271 mm), extreme temperatures (maximum 50^0 C and minimum 2^0 C), moderate to high wind velocity (25 kms per hour during summer) and had wind direction from southwest to northeast with high evapo-transpiration rate. The soil was sandy loam with a depth of 60 cm. The soil surface structure was sandy-to-sandy loam with 2-3 per cent clay content. Vegetation cover was poor with disparate growth of shrubs and thorny bushes.

The options for livelihood generation were limited in this harsh geo-climatic condition. On top of it, if agriculture failed due to soil erosion, on the one hand, and advancement of sand dunes, on the other, the livelihood crisis intensified. Livestock development could be an option, but ecological stress made it an uncertain proposition what with all of the above and the increasing scarcity of drinking water.

III

INTERVENTIONS AND INSTITUTIONAL DEVELOPMENT

Past Initiatives

In response to the difficulties faced by the villagers, the Public Health and Engineering Department (PHED) and other departments of government of Rajasthan along with the Hanwant Charitable Trust, an NGO, initiated some actions to improve the resource base of the Tibna *panchayat* and ensure supply of drinking water in a sustainable manner. During the early 1980s, PHED laid pipelines and installed hand pumps at several places to ensure drinking water supply. Plantation works on the riverbank by the forest department followed this. Subsequently, in 1990, the irrigation department constructed an anicut on Siyandan *nallah* for soil and water conservation. However, due to lack of maintenance, the eroded sand filled it and rendered it completely useless.

The Hanwant Charitable Trust, run by the Jodhpur royal family, raised plantation in some patches of the watershed area to stabilise sand dunes situated near the hamlets of Tibna village, where most of the households belonging to *Meghwal* and *Suthar* social groups lived. It also installed a hand pump at Tibna.

These initiatives were important in their own way and had a positive impact on the improvement of the resource base of the villages and thereby restoration of the local ecosystem. However, they were limited in their lack of comprehensive planning for water and soil conservation. There was no systematic planning for drinking water supply. The approach was *ad hoc*, imposed from above and not rooted in community management. In other words, sustainability of the initiatives was never an objective of those who intervened before Poorva Sainik Bahudheshiya Sahakari Samiti (PSBUSS) stepped in.

Entry of PSBUSS

PSBUSS, with its head office at Jodhpur, was a unique NGO, which was formed exclusively for enhancement of welfare of those retired defence personnel who lived in Jodhpur district. PSBUSS had its own reason of intervening in Tibna *panchayat*. Despite the fact that Shergarh *tehsil* was the second largest contributor of defence personnel to the country, it was one of the most backward areas of Jodhpur district. Tibna *panchayat* had the distinction of having a large concentration of ex-servicemen who were known for their dedication and sense of discipline. Notwithstanding such a large stock of human capital, no worthwhile initiative was taken by them to improve their common resource base that increasingly became too weak to contribute to the improvement of their household economy. It was only when PSBUSS showed its eagerness to intervene through collective action, that the villagers discovered their potential strength that could stop further depletion of their common resource base and bring about improvement in it.

While there is no room for complacency about the potential strength of human capital, it must not be overestimated either. What mattered significantly was the stock of social capital and its content that made objective conditions appropriate to the working of human capital. One may debate which precedes the other. It is here that the underlying forces of change that gained momentum in the village social sphere may be evaluated.

The spirit of collective action that retired defence personnel nurtured throughout their service led them to create a forum of their own in Shergarh *tehsil*. They would meet at regular intervals, share their problems, if any, and seek their redress jointly. In one such meeting of the forum, sometime during 1996, ex-servicemen hailing from Tibna *panchayat* shared their problems. They sought intervention from Col. Narendra Singh, president of PSBUSS. While Tibna *panchayat* agreed to support PSBUSS, villagers were enthusiastic about the proposed works to be undertaken by PSBUSS. At the initiative of PSBUSS, villagers of the *panchayat* gave a representation to the district rural development agency (DRDA), which asked PSBUSS to initiate development works and seek redress to the problems under the watershed development programme.

PSBUSS roped in different other agencies in its endeavour. For instance, the state remote sensing application centre located at

Jodhpur town assisted PSBUSS in planning for watershed development for one cluster of 500 hectares of land that was initially allotted for development. However, this did not appeal to most of the villagers because it was apprehended that such a piecemeal approach might inject development imbalances between villages located within the same watershed. In view of these difficulties, the initial plan was abandoned and all villages of Tibna *panchayat* were included into the planning. The entire area was then treated as micro watershed. District Rural Development Agency agreed to finance its development.

Initiative taken for micro watershed development in Tibna *panchayat* exemplified a unique case of collaboration between NGO, government, *Panchayat* and the village community. It was true that PSBUSS assigned itself a key role in shaping the nature of collaboration. However, the important aspect was whether it succeeded in bringing the community to the centre of all activities and their management to ensure sustainability of collective action. The challenge was how to make it participatory for all social groups, which were heterogeneous, and improves everybody's stake equally in the improved resource base. Hence, the immediate task of PSBUSS was to develop a village level institution to ensure sustainability of the collective action for natural resource management.

Institution Building

While examining the process, through which institution building took place in all villages, it was imperative to know how social forces evolved historically and contributed to it. The history of these villages goes back to 17th century to Jodha Khirja Singh, who was a famous warrior of the Jodhpur army. He was awarded land around the present village Khirja as *jagir* for his valour in the famous battle of Bikaner. Khirja's son, Jodha Karni Singh named the village as Khirja. His five sons who established *dhanis* where they stayed during monsoon to protect their crops inherited this. As time passed, these *dhanis* grew into small village settlements and were identified with the names of the five sons of Karani Singh. The eldest son Tej Singh stayed on at Khirja and the place came to be known as Khirja Khas; implying Khirja main. Tej Singh established the present village Khirja Tibna. Asha Singh established a *dhani* by the name Khirja Asha ka Bas. Bhoja Singh established a *dhani* named Khirja Bhoja ka Bas. The youngest son Fateh Singh was given the area presently known as Khirja Fateh

Singh ka Bas. As time passed, the descendants of five brothers continued living in their respective *dhanis*, which grew into permanent village settlements over the years. Thus, three out of the five villages inherited by sons of Karani Singh were included in Tibna *Panchayat*; Asha Ka Bas and Khirja Khas were not included. Instead, Timri was included on grounds of geographical closeness and population.

It seems only Rajputs inhabited these *dhanis* earlier. But a sedentary settlement gave rise to the need for occupational diversification. Therefore, people belonging to other castes like *Brahmins* (priests), *Meghwals* (cobblers), *Nai* (barbers), *Suthar* (carpenters), etc. were invited by the Rajputs to live in their villages. They were given land and cattle for subsistence living. This happened almost 500 years ago, according to the oral history of these villages. One of the changes that these villages witnessed over centuries was its demography. The size of population of each social group expanded uninterruptedly. However, it could never dilute the Rajput dominance. Rajputs still derived their authority from the fact that they owned the lands hereabouts and were the ones to bring people of other castes and give them employment. This laid the foundation of social relations between the Rajput community and other social groups with clearly defined power relations.

To begin with, PSBUSS organised a public meeting of all villages together. However, it was not a meeting, as told by the villagers, held for consensus building on issues related to development of their natural resource base. It was held only to inform the villagers about the role that PSBUSS was going to play. Discussions with the villagers revealed that there was hardly any scope for most of the villagers to articulate their development concerns. Nor was any attempt made to explore the areas in which the villagers could contribute. Although they were clamouring for such development initiatives, the villagers did not appreciate the manner in which PSBUSS intervened. For most of them were denied space for effective participation, barring a chosen few. Despite this underlying resentment, they were enthusiastic about what was proposed to be done by PSBUSS in their villages.

For the ease of operation, the entire watershed area was divided into three sectors and, further, into eight clusters. The latter was sought for obtaining more grants from DRDA, because, as per the rules framed by the Government of India for watershed development, they were entitled to receive funds cluster-wise. Of course, size of the cluster could not exceed 500 hectares of land. According to the rules, a sum of Rs. 25 lakh was allocated for each cluster. Table 6.2 shows the sector and cluster-wise division of the watershed area.

Table 6.2

Sector-wise Division of the Watershed Area of Four Villages

Sector A	Parts of Tibna and Timri	500 hectares
	Second part of Tibna	500 hectares
	Third part of Tibna	500 hectares
Sector B	First part of Bhoja ka Bas	500 hectares
	Second part of Bhoja ka Bas	500 hectares
Sector C	First part of Fateh Singh Ka Bas	500 hectares
	Second part of Fateh Singh Ka Bas	500 hectares
	Third part of Fateh Singh Ka Bas	457 hectares

Source: Abiyan Maru Vijay, PSBUSS publication.

Sector-wise division of the watershed area was kept in mind while forming village organisations. Accordingly, three *sanstha* came into being. It appears from Table 6.2 that Tibna and Timri formed Sector A and thus constituted the Jal Grahan Vikas Sanstha. In a similar way, a Jal Grahan Vikas Sanstha was formed for Bhoja ka Bas and Fateh Singh ka Bas, respectively. It must be remembered that no discussion was ever held with the villagers about the formation and structure of their organisations, the tasks to be taken up and a guaranteed effective participation of the villagers. Three ex-army men, namely, Roop Singh, Kisan Singh and Sultan Singh, were made president for each *sanstha*, respectively, by PSBUSS. Even members of each *sanstha* were inducted only after consultation between PSBUSS and the president of each *sanstha*. No discussion was held with the villagers before the choice was exercised.

As per the advice of PSBUSS, one committee named, Jal Grahan Vikas Samiti, was also formed in each sector. Of course, this was done on the basis of the guidelines formulated by Government of India for watershed development. An executive committee was formed in each *sanstha* comprising the president, vice president, secretary and treasurer besides the executive members. The *samiti* was constituted similarly. The functions of both the *samiti* and *sanstha* were clearly demarcated in that the *sanstha* activities were more of managerial and advisory nature, while the *samiti* was given the responsibility of performing the activities as desired by the *sanstha*.

Initially, activities were performed as per the assignment. Members were enthusiastic about organisational initiatives. However, most of them soon felt alienated from mainstream development process. For initiatives were taken by a few from the dominant *Rajput* community. Even while offering membership to the villagers, care was taken to see that the same community was represented better in the village-level organisation. The representation of *Rajputs*, in terms of their numerical strength, was the largest in sector A (Table 6.3). Though, it was not so in sector B and C, where other social groups were represented equally well. Be that as it may, the voice of the *Rajput* community gained overriding importance in the decision making process in any meeting. Needless to mention, all the *samities* and *sansthas* ceased to exist effectively in practice once PSBUSS withdrew from the scene some years ago.

Table 6.3

Caste-wise Membership of Village Organisation

Caste	*Initial Stage*			*At Present*		
	Sector A	*Sector B*	*Sector C*	*Sector A*	*Sector B*	*Sector C*
Rajput	13	8	8	13	7	10
OBC	1	5	8	3	4	4
SC	2	3	4	1	3	2
ST	1	1	1	-	1	1
Total	17	17	21	17	15	17

Note: OBC=Other Backward Caste, SC= Scheduled Caste, ST= Scheduled Tribe.
Source: Field survey.

Most of the hamlets (*dhanis*), such as Meghwalon ki Dhani, Sutharon ki Dhani and so on, were caste-based settlements. Even though the number of households of other social groups was large, if not more than that of the *Rajputs*, a serious attempt was made to gain representational characteristics of the organisation by way of giving equal weightage to other social groups.

Neema Ram Meghwal of Bhoja ka Bas was a member of the *samiti*. Initially, he was regular in attending the meeting, although he had no voice in decision making process. However, he started alienating himself from the *samiti* when the president of the *samiti*, Bhanwar Singh, took the liberty of constructing 'Gopal Ghar' on his own land.[1] It was supposed to have been constructed on the common land to which no individual could claim ownership later. According to him, it was apprehended that this Gopal Ghar, constructed with common funds, would be used by Bhanwar Singh for his own purpose. On another occasion, he brought to the notice of the members of the committee that the people living in his *dhani* were suffering from severe shortage of drinking water. He persuaded members of the *samiti* to take some concrete steps to resolve this problem. The president of the *samiti* asked him and other members to request those living in his *dhani* to contribute Rs. 20 per household so that any additional amount required for ensuring availability of the same might be met by the organisation. He went ahead with this proposal and collected money. However, when he sought organisational intervention, as decided earlier, he was told that the *samiti* had ceased to exist. Since then, he stopped attending the meeting. According to him, only a few *Rajput* families were running the show, not entertaining anyone from other social groups.

Table 6.4 shows that very few members of the village community were involved in the decision making process. Of course, regular attendance of the members in the meeting in all villages, except Tibna, was also very low. It is important to note that Timri had no representation. In almost in every respect, villagers felts that discrimination was perpetrated against them by PSBUSS and Tibna village *samiti* despite Timri being incorporated into the development plan approved by DRDA. The reason was simple. The number of ex-servicemen from the *Rajput* community living in this village was too small to become a part of the mainstream power relations and influence the decision making process. Disenchanted with the organisational arrangements made by PSBUSS, the village community of Timri sought to create a *samiti* of their own. It, however, did not materialise. The villagers registered their protest in the collector's office against such discrimination. In response to it, the collector's office ordered that all development works taken up in other villages to be extended to Timri also.

1. Gopal Ghar is a service centre for breed improvement and providing veterinary care to livestock. This was implemented under the Gopal Yojna of Government of Rajasthan. Three persons from three clusters were trained for providing services to livestock. However, at the time of the field survey, they seemed to have lost interest in it. Consequently, Gopal Ghar became defunct.

Table 6.4

Participation of the Villagers in the Meetings

	Tibna	*Bhoja Ka Bas*	*Fateh Singh Ka Bas*	*Timri*
Attendance in meeting	11	1	3	0
Regularity in participation in the decision making process	4	1	1	0

Source: Field survey.

Despite this order, Timri continued to be deprived of the benefits of river training works. Their agricultural fields continued to be inundated, while agricultural lands of other villages were saved. Although PSBUSS carried out a limited plantation work and constructed check dams in Timri, the efforts were far from adequate. Discussions with the villagers of Timri revealed that most of the development activities initiated by Tibna *panchayat* and meant for the benefit of all four villages, finally found their way to Tibna and Bhoja Ka Bas villages only with their greater concentration of influential *Rajput* families. According to them, PSBUSS reproduced the same practice, which was why Timri never figured on their intervention map in actual practice, as desired.

The principal objective of developing village-level institutions in the field of natural resource management was to evolve a regulatory framework that could change the behavioural pattern of the community towards conservation of natural resources in a sustainable manner. Equal participation and access to the resource base instantaneously help collective action to succeed and sustain. Any deviation from this basic principle may jeopardise the prospect of sustainability. Deviation is likely to occur if no change takes place in the content of social capital, which is traditionally nurtured. In the villages under study, no need was felt by the intervening agency to examine the content of social capital, its relevance to the distribution of development gains and, finally, its implications for sustainability of the collective action. The organisational structure, as it were, was planted from above since it was required as a fulfilment of the obligations. Besides, no rule of management was ever designed for conserving the improved resource base.

A sort of horizontal integration between government officials, PSBUSS and a few affluent households of the villages under study did not allow social capital to change its content. The social polarisation that stiffened further as a result, failed to crystallise the collective spirit of the villagers. This might be a reason why institutional development remained incomplete. One is, therefore, inclined to ask how development initiatives taken by PSBUSS and the *panchayat* would sustain in the absence of involvement of the village community. Before one proceeds to answer this question, it is worthwhile to examine the steps that were taken for watershed development and their impact on community living.

The watershed development programme, as mentioned earlier, included river training works and construction of water-harvesting structures; drinking water management; plantation and sand dune stabilisation and developmental activities like livestock development, construction of road and social works. All these were significant components of the programme with the main emphasis given to water and soil management in the watershed area. Activities that were taken up in all four villages are shown in Table 6.5. Table 6.5 also shows that distribution was skewed in that Timri was deprived of many of them.

Table 6.5

Activities Performed by PSBUSS in All Four Villages

S.No.	*Activity*	*TIBNA*	*BKB*	*FKB*	*TIMRI*
1	Bus stand	Yes	Yes		
2	Anicut		Yes	Yes	
3	Tube well		Yes		
4	Distribution of seeds of *sonamukhi*	Yes	Yes		
5	Plantation	Yes	Yes	Yes	Yes
6	Nursery	Yes			
7	Stadium	Yes			
8	Hand pump	Yes	Yes	Yes	Yes
9	Gopal centre	Yes	Yes	Yes	
10	Medical/animal health camp	Yes	Yes	Yes	Yes
11	Construction of *nadi*	Yes	Yes	Yes	Yes
12	LSCD	Yes	Yes	Yes	Yes
13	River training work	Yes	Yes	Yes	
14	Repair of Culvert	Yes	Yes		

contd. ...

contd. ...

S.No.	Activity	TIBNA	BKB	FKB	TIMRI
15	Tanki/Pipe line	Yes	Yes	Yes	Yes
16	Community Platform	Yes	Yes	Yes	
17	Repair of *Aganwari*	Yes	Yes	Yes	
18	School development work	Yes	Yes	Yes	Yes
19	Desilting of well	Yes	Yes		
20	*Adhyatma Kendra*/Library	Yes	Yes	Yes	
21	Repair of temple	Yes			
22	Construction of road		Yes	Yes	
23	Cremation pyres	Yes	Yes	Yes	
24	Solar light	Yes	Yes	Yes	Yes
25	Drain construction		Yes		
26	Farm pond	Yes	Yes	Yes	
	Total	22	23	17	8

Source: Jal Grahan Vikas Samiti, Tibna.

Activities were limited both in terms of their number and magnitude. They were categorised into eight parts (Table 6.6). Table 6.6 shows that around four per cent of the total development expenditure was spent in Timri. This was so even if expenditure incurred towards the construction of nursery and stadium, which were meant to be used by the villagers of all four villages together.

Table 6.6

Expenditure Incurred Towards Different Activities in Four Villages during 1998-2002

(Rs.)

Heads of Expenditure	*FKB*	*BKB*	*TIBNA*	*TIMRI*	*Total*
Water-harvesting structure	783928	254094	61006	47386	11,46,414
Livestock	326691	351662	241925	0	9,20,278
River Training	70820	702936	1599639	0	23,73,395
Drinking Water	789791	1206662	460151	107249	25,63,853
Fencing & Plantation	1382815	662505	745753	177123	29,68,196
Nursery	0	0	646888	0	6,46,888
Stadium	0	0	288704	0	2,88,704
Others	719062	685771	798262	162769	23,65,864
Total	4073107 (30.68)	3863630 (29.11)	4842328 (36.48)	494527 (3.73)	1,32,73,592 (100.00)

Source: PSBUSS, Jodhpur.

A deal was struck between *samities* and PSBUSS that individual households would have to pay 10 per cent of the total expenditure incurred for making embankment close to their land. It was the responsibility of the *samiti* to collect this money. However, contribution made by the villagers was deposited in the village fund. Even those who were working as wage labourers in any activity related to watershed development were paid 5-10 per cent less than the wage rate at which they had agreed to work. The unpaid amount was transferred to the same village fund, which was monitored by the village level *samiti*.

The purpose of creating such a fund was to meet expenses towards maintenance of the initiated activities. The accumulated amount of money in the development fund, as observed during field survey, was Rs. 5,70,000. Normally, there was no provision for spending money from the village development fund for maintenance activities unless the *samiti* agreed to do so. The practice that was followed was that each household had to contribute a stipulated amount of money as fixed by the *samiti* for carrying out maintenance works. Villagers were asked to contribute even for installation. For example, each household had to pay Rs. 500 in Fateh Singh ka Bas for laying pipelines. Similarly, each household of Timri had to contribute Rs. 200 for installing an electric motor inside the well for pumping out water.

One must bear in mind that all maintenance works taken up by the *samities* were related to the drinking water supply only. While *samities* took the initiative in managing drinking water supply, it was done without creating space for voluntary involvement of the villagers. As mentioned earlier, only a few influential households held the key to the management and maintenance works.

However, the participation of rest of the villagers in the development activities as wage labour was undoubtedly very high. Until 2000-01, as many as 49,273 man-days of work were generated and amount of money that was spent, on an average, towards wage payment was Rs. 31,53,472.[2] It was estimated to be around 24 per cent of the total expenditure incurred (Table 6.6). Creation of employment opportunities of this magnitude, especially during

2. The wage rate varied between Rs. 40-200, depending upon the deployment of skilled or unskilled wage labour. Here, the average wage rate has been taken and multiplied it by man-days of work generated.

drought years, was certainly an encouraging feature that one seeks to achieve from such intervention. However, participation as wage labour must not be construed as participation that gives momentum to collective action and eventually reinforces institutional sustainability.

IV

WATERSHED DEVELOPMENT ACTIVITIES

River Training Works

As mentioned earlier, the seasonal inundation of Kheton ki Nadi was a source of misery for almost all the villages through which it passed. The objective of the training works was to control the rainwater run-off that flowed with high velocity from upstream and inundated the downstream villages. Several water-harvesting structures, including loose stone check dams, anicuts and *bunds*, were constructed at several places for this purpose. However, the significant work was the construction of embankment on both sides of the river by piling up stones available locally. This was done to restrict the spread of the river water and, thereby, control the flood in the downstream villages, especially Tibna and Timri. This was called river training work, supposed to be taken up for 13 km stretching from Fateh Singh ka Bas to Timri (Map 15). Development grants were sanctioned by DRDA.

Actually, the training works began at Kumharon ki Dhani near Bhoja Ka Bas village, not Fateh Singh Ka Bas as sanctioned by DRDA. For the river flowed through a hilly tract up to this point and, therefore, no need was felt for embanking it since there was insubstantial soil erosion. However, after this point, it was sandy on both sides of the river, which was why soil was more vulnerable to erosion.

As can be seen on Map 15, the river training work was not taken up to Timri as planned initially. The work was stopped abruptly at Tibna, which was away from the Bol Nadi reservoir, despite the fact that the flat, low-lying land lying between Tibna and Timri was comparatively more vulnerable to seasonal floods. It is curious why training works were not taken up to Bol Nadi (Timri) for which grant was received from the government. While some members of the *samiti* accused the villagers for being unprepared to contribute as stipulated,

Map 15

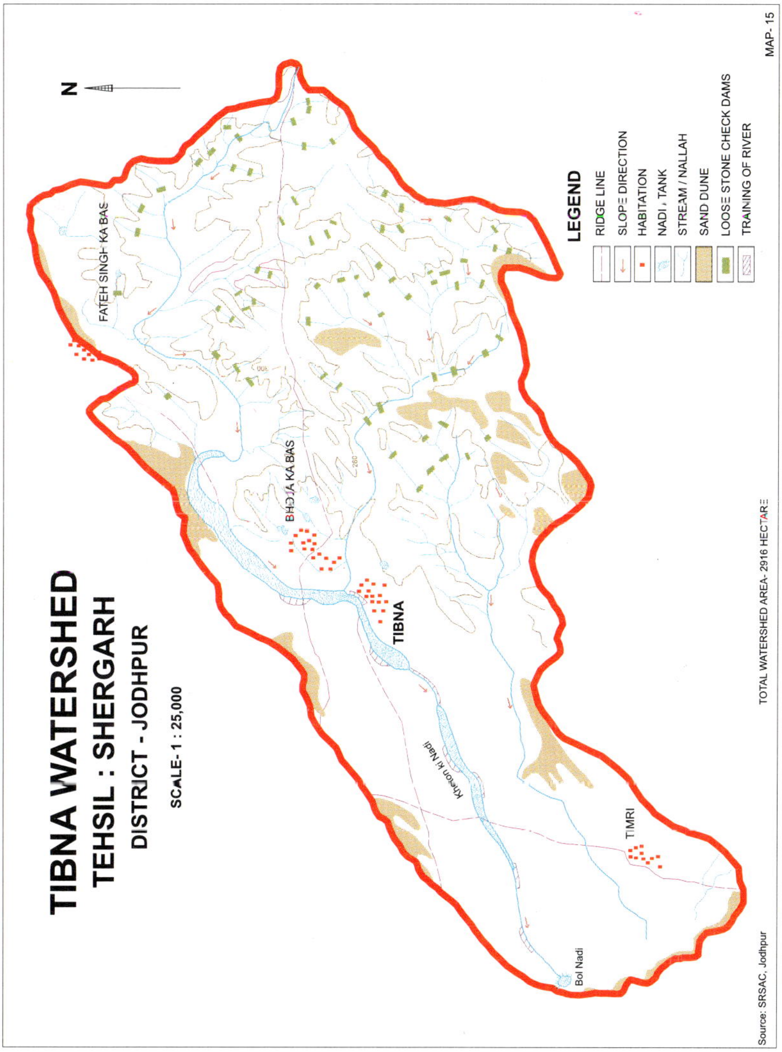
TIBNA WATERSHED
TEHSIL : SHERGARH
DISTRICT - JODHPUR
SCALE- 1 : 25,000
N
FATEH SINGH KA BAS
BHOJA KA BAS
TIBNA
TIMRI
Bol Nadi
LEGEND
RIDGE LINE
SLOPE DIRECTION
HABITATION
NADI , TANK
STREAM / NALLAH
SAND DUNE
LOOSE STONE CHECK DAMS
TRAINING OF RIVER
TOTAL WATERSHED AREA- 2916 HECTARE
Source: SRSAC, Jodhpur
MAP- 15

some other members accused one of the villagers whose cooperation was vital but missing.

All these arguments, however, were meaningless since both technical and physical aspects were considered at the planning stage. Enquiry revealed that *Suthars*, a subordinate social group that had practically no voice in any village affair, owned agricultural plots adjacent to the channel through which river water passed. Their subordinate position did not allow them to gain access to the development inputs provided by the government. Had they been members of the *Rajput* community, in general, and ex-servicemen, in particular, the river training works would have been completed. This was a direct outcome of the social discrimination that traditional power relations perpetrated through the renovated village institution. Another insight learnt from this was that the failure or success of the state in providing equal access to development gains was largely dependent on how institutional development corresponded to social capital and its content. In other words, given the same power relations, no institutional renovation could be of any help for the disadvantaged to gain equal access to development benefits. The *Suthars* in Timri exemplified this glaringly.

Besides, there were several reasons to be critical about the training work in terms of its technical soundness and, hence, sustainability. Why were other methods not tried when the long-term capacity of the present work in preventing soil erosion was not certain? For, it must be borne in mind that the entire activity was carried out by only lining up loose stones. It was quite likely that the water would start seeping through the gaps found between stones and might gradually intensify inundation and destabilise riverbanks. In such a situation, plantation of grass and trees might be of great help to protect the riverbank from erosion. It was done, but in a limited way, in small patches of the embankment near Bhoja ka Bas village and Tibna only. Additionally, the height of the embankment was too low to prevent water from spilling over, once it flowed with high velocity in case of heavy rains.[3]

3. Discussions with engineers of the department of soil and water conservation, Government of Rajasthan, revealed that there were several cost-effective and technically sound methods that could have been followed to ensure sustainability. It was suggested that the construction of earthen bund, made up of compact and heavy soil and covered with stone, would have been a better alternative in terms of technical soundness and sustainability of the structure. Another suggestion was construction of bund with sand to be covered by stone, which could have been bound with wire mesh to make them fast and to thus maintain the height of the embankment.

Besides, why was no spur constructed where the river meandered and entered Bhoja ka Bas? The river channel was prone to sedimentation that accumulated in increasing order unless steps were taken regularly to desilt it. Since no system of desilting was in place, it was quite likely that the entire exercise taken for river training works could prove futile in future. However, despite the lack of technical soundness and several other limitations, benefits that accrued to the villagers cannot be overlooked although one might be apprehensive about the sustainability of these works.

Construction of anicuts and loose stone check dams (LSCD) was taken up in the watershed area as part of the river training works. While there was only one anicut constructed earlier near Bhoja ka Bas, three more were added at the initiative of PSBUSS—two in Bhoja ka Bas and one in Fateh Singh ka Bas. The Fateh Singh ka Bas anicut had a special significance, as reported by the members of the village *samiti*. It would not only restrict rainwater run-off that flowed with high velocity, but also recharge groundwater in the neighbourhood. The other two anicuts were constructed on private lands, as mentioned earlier, in Bhoja ka bas, upstream of Tibna and Timri. Here, again, 10 per cent of the cost of construction was borne by the private farmer. The decision of constructing the same on private lands was of great help to the villagers, in general, and private farmers, in particular. While agriculture on private land was boosted by rainwater harvesting and recharging groundwater, the initiatives arrested soil erosion and, finally, reduced the intensity with which sedimentation was taking place through this channel.

Fateh Singh ka Bas and Bhoja ka Bas villages were chosen for construction of a large number of LSCD for both these villages were upstream. As many as 960 and 890 LSCDS were constructed in Fateh Singh ka Bas and Bhoja ka Bas, respectively (Map 15). Only 129 of theses were constructed in Tibna and Timri. The reason why there was maximum concentration of LSCDs in the first two villages was that the terrain was hilly and rocky. However, this was not the case with Tibna and Timri where the terrain was sandy and flat. LSCD is a unique method followed for preventing soil erosion at a low cost. It is an embankment of small size (both in terms of height and area covered) constructed on the slopes of hillocks. The purpose is to arrest rainwater run-off to the extent possible as determined by the capacity of LSCD, before the same flows down to the river. As it flows downwards, it crosses many such LSCDs, which is why its velocity declines gradually, and so does soil erosion. LSCD also contributes to the expansion of vegetation cover on its terrace due to accumulation of rainwater. This, in turn, improves biodiversity of the area.

While one may not dispute the strategically appropriate choice for construction of LSCD in a terrain like the one under study, one must know whether they were sustainable. As discussed later, there was substantial improvement of vegetation cover wherever LSCD was constructed. It lent incredible support to the sustenance of livestock even in perpetual drought. Besides, its impact on improving groundwater level cannot be underrated. However, the question is how and whether they could continue to do so in the absence of collective effort of the villagers for their maintenance. During field survey, it was observed that many LSCDs were in poor condition, which was why their terrace lacked vegetation cover. Most of the villagers felt that it was the responsibility of PSBUSS to repair them. They never thought that village *samiti* must perform the job. They failed to take initiative on their own to do the job for they had lost the involvement that they had exhibited at the initial stage.

Management of Drinking Water Supply

The shortage of drinking water supply was a perennial problem in these villages. It gradually became severe in the last two decades with the increasing frequency of drought and population pressure. In order to ease the severity, Public Health Engineering Department (PHED), Government of Rajasthan intervened sometimes during early 1980. Although an improvement was witnessed in the availability of the drinking water with this intervention, it continued to be a threat to sustainable livelihood.

Before the entry of PHED in Tibna *panchayat*, the village community was dependent on the traditional practice of collection of drinking water. They used camel carts fitted with iron tanks with capacities of 500-600 litres of water and filled them up with water from a tank or tube well located at a village called Ketu, 5-6 km away from Tibna. This was stored in what was locally called *tankli* (small-sized tank) inside the house. Those who did not have camel cart would buy it at Rs. 10-15 per cartload of water. Besides, villagers had ponds on their agricultural lands for harvesting and storing rainwater to meet both human and livestock needs. This traditional practice was not only time-consuming but also full of drudgery.

Laying a pipeline from Ketu to Fateh Singh Ka Bas was the first job that PHED undertook. A ground level reservoir (GLR) was then constructed in Fateh Singh ka Bas to store water. Now the same camel carts, instead of travelling up to Ketu for carrying water, began to travel to the point where GLR was constructed at Fateh Singh ka Bas. GLR was a small-sized storage tank, which was designed appropriately and constructed at seven different locations including Bhoja Ka Bas, Tibna and Timri. PHED installed machines in two open wells located at Tibna and Timri to pump out water and supply to all GLRs. In addition to this arrangement, it installed three hand pumps in Tibna and Bhoja Ka Bas during that time. By now, it had also installed 29 hand pumps in all four villages of Tibna *panchayat*.

The objective of PSBUSS was to ensure the availability of potable water within a radius of one km for consumption of both human and livestock population under Janta Jal Yojana. With this end in view, it:

- Desilted the common well at Tibna, fitted a booster pump in it so that water was pumped out of it at a greater speed, and reached even those hamlets, which were located far away from the main village;
- Extended the supply system of piped water by laying more pipelines and joining them to the main pipelines carrying PHED supply of drinking water;
- Desilted a community well at Timri and fitted a booster pump in it; and
- Installed hand pumps.

While these steps were extremely important to the villagers facing critical availability of drinking water, no involvement of the villagers, except a few, was solicited by PSBUSS.

It may be noted that each household was charged an initial amount before the pipelines were installed or tanks were brought to use in order to avail of drinking water. Charges differed between the villages (Table 6.7). Charges were fixed mainly depending upon the altitude of the village. In fact, altitude made a significant difference to the cost of supply of water. For instance, villagers of Fateh Singh ka Bas, as Table 6.7 shows, were charged Rs. 500 per household because this village was located at a relatively higher altitude and drinking water was supplied from Bhoja ka Bas, located at a lower altitude. Supply of water from lower to higher altitude required superior quality of pipe and other supporting infrastructure facilities that required higher investment cost.

Table 6.7

Charges Paid by the Households for Obtaining Supply of Drinking Water

Village	*Initial Payment Per Household*	*Charges Paid by Each Household After Installation/ Construction*
Tibna/Timri	Rs. 300.0	Nil
Bhoja ka Bas	Rs. 250.0	Rs. 20.0 per month
Fateh Singh ka Bas	Rs. 500.0	Nil

Note: Those who could not afford to pay the initial amount as decided, worked as wage labour equivalent to the amount of money fixed.

Source: Field survey.

The case of regular monthly payment by the villagers of Bhoja ka Bas, as shown in Table 6.7, was an exceptional one. It was a sort of a special case in that Jal Grahan Vikas Samiti charged this amount from the users for meeting maintenance cost of tube well installed by PSBUSS. The reason why regular payment was made by the villagers of Bhoja ka Bas was that the arrangement of drinking water supply through tube well in this village was looked after by the same village organisation. One person was appointed by the *samiti* on salary. However, management of all sources of drinking water, including the tube well installed by PSBUSS in other villages, was in the hands of PHED. Be that as it may, a brief illustration of a case like the one as shown above shows that villagers had no inhibition in paying for the services rendered for supplying drinking water. The lesson that one learns from this is that, if infrastructure is made available to the villagers and remains useful for them, they are willing to pay the price for the services.

In the absence of adequate supply of drinking water from the well from where water was pumped out by PHED, initiative taken by PSBUSS brought considerable relief to the villagers. Field survey revealed that villagers were happy to pay users' charges if drinking water was supplied regularly. Needless to mention, both sources including PSBUSS and PHED were equally responsible for ensuring supply of drinking water. While villagers made payment for the management of supply of drinking water initiated by PSBUSS, no such system existed in the case of PHED-initiated activities.

One of the noticeable changes that villagers witnessed was installation of a large number of tube wells after PSBUSS intervened. Table 6.8 shows that as many as 44 community-owned hand pumps were installed after intervention up from seven. Similarly, private households installed nine tube wells, mainly for providing irrigation to agriculture.

Table 6.8

Sources of Drinking Water before and after Intervention

	Before Intervention			*After Intervention*		
	Private	*Common*	*Total*	*Private*	*Common*	*Total*
Well	6	4	10	8	4	12
Tube well	3	0	3	12	1	13
Hand pump	0	7	7	0	44	44
Total	9	11	20	20	49	69

Source: Field survey.

However, not all sources could remain active. After a few years of installation, many hand pumps, for example, turned non-functional due to a variety of reasons including increasing pressure on groundwater. One of major reasons was, of course, lack of care and maintenance for hand pumps installed by PSBUSS. This was, perhaps, why five out of 12 hand pumps installed by PSBUSS, as shown in Table 6.9, were found to be out of use for the last few years.

Table 6.9

Sources of Supply of Drinking Water in Use

	PHED			*PSBUSS*		
	In Use	*Useless*	*Total*	*In Use*	*Useless*	*Total*
Hand pump	27	3	30	7	5	12
GLR	8	2	10	0	1	1
Tank (*Hod*)	0	0	0	27	9	36
Total	35 (87.5)	5 (12.5)	40 (100.0)	34 (69.4)	15 (31.6)	49 (100.0)

Source: Field survey.

Table 6.9 shows that 87.5 per cent of all sources of drinking water installed by PHED were active, while it was 69.4 per cent in the case of PSBUSS. It may be noted that 27 tanks (*Hod*) constructed at the initiative of PSBUSS were active, while 9 of them were not so. Of the 27 tanks, around 16 were partially active in the sense that whenever water was available sufficiently enough to reach these tanks, they became useful. Otherwise, they remained out of use.

However, supply of water to the tanks in some cases was a serious problem due to bad management on the part of the *samities*. There were several instances, as field survey revealed, where the relatively more powerful households forcibly got supply connection to their own houses from the main line without community approval before the same reached the tanks. It was a case of deprivation in that weaker social groups could not afford to register their resentment against such practices. For example, around 50 households were found to be engaged in such a practice in Tibna village. The irony was that nobody was penalized by the village organisation for this offence because the supply was managed by PHED. However, those who took such unapproved connections (around 20 households) at Bhoja Ka Bas were charged Rs. 10 extra (over and above Rs. 20 which was paid by those who collected water from tanks) per month by Jal Grahan Vikas Samiti of the same village. The fact that these villagers paid the amount regularly without any resentment shows that community management could ensure equitable enforcement of the rules across all sections of population.

The incidence of non-functional tanks was concentrated more in those hamlets with socially weaker groups like *Suthars*, *Nais* and *Meghwals*. They suffered due to two reasons. First, whenever water was supplied from the common well, pressure was less. Therefore, the chances of water reaching them were less. Second, powerful households forcibly got unauthorised connections, as seen above. These groups of households managed to meet their requirements through hand pumps. Such an incidence gained unduly greater proportion in Tibna, in particular, and other villages, in general (Map 16). Map 16 also shows that the partially active tanks were concentrated nearer to the sources of origin such as the common well and tube well. Mainly *Rajputs* lived in these areas. In this situation, the partial availability of water supply had much to do with inadequate supply of water in that those *Rajput* households who lived nearer to the sources gained more access to it.

Discrimination against the disadvantaged was perpetrated mainly in the case of supply of piped water. However, it did not aggravate the scarcity for hand pumps were installed at several places where they could access them. This was really a boon for them during three years of drought. Table 6.10 shows that around 86 per cent of the sample households were brought within one km reach of the water supply. This was made possible through the combined initiatives taken by PHED/Tibna *panchayat* and PSBUSS.

Table 6.10

Distance Covered by the Sample Households for Accessing Drinking Water

(No.)

	Before Intervention	*After Intervention*
Up to 100 metres	14	42
100 – 1000 metres	35	60
1000 – 2000 metres	25	8
Above 2000 metres	45	9
Total	119	119

Source: Field survey.

Loss of sustainability was yet another aspect of development, which seemed to have been ignored. For, no attention was paid to the drafting of groundwater at an increasing rate against its declining status. The fact that several tanks turned useless and partially useless (69 per cent of the total number of tanks constructed) gave enough indication of this. Besides, the village community continued to be dependent on PHED. Despite development of a village institution, management and distribution of drinking water continued to remain in the hands of PHED. One may not find any difficulty in it if enough space is created for involvement of the villagers in its management. In such a situation, chances of deprivation of the poor, as it happened in Tibna, would be much less. Lastly, no room should be left for the powerful to influence norms of distribution and perpetrate discrimination against the poor.

Construction of farm ponds was also taken up for meeting drinking water needs, especially of those who were working in the agricultural fields. Construction of these ponds was felt necessary for people had to carry drinking water for both humans and livestock from the nearby sources, which, in many cases, were located in distant places. In such a situation, as reported by the villagers, one person had to be engaged for this job. As many as 16 farm ponds were constructed in all villages, except Timri—nine such ponds were constructed in Fateh Singh ka Bas alone. These were constructed on private agricultural lands in that households belonging to *Rajput* community owned 14 of them.

At the time of construction, as per the understanding between PSBUSS and the village community, 10 per cent of the cost of construction was to be borne by those farming households who were likely to benefit from these ponds. However, nobody made any contribution except the owners of the agricultural land where ponds were constructed. No rule was framed to enforce others to make the payment. This also illustrates community institutional failure in that private households gained although community gains were the central objective of the intervention. It was true that the owner of the land prevented nobody from collecting drinking water. However, it was also true that he could do so and claim absolute stake in it if he so desired.

Plantation and Sand Dune Stabilisation

The advancement of sand dunes was a serious problem that grew over the years. It was a source of vulnerability since it rendered agricultural lands useless. Even in those lands where cultivation was possible, their productivity declined rendering agricultural production cost ineffective. Besides, it caused loss of vegetation cover in several places and forced the villagers to look for other places away from their own villages for their animals to graze. In other words, advancement of sand dune reduced the prospect of livestock economy that villagers had always fallen back upon in the event of uncertain agricultural production. These two together created a pressing demand for stabilisation of sand dunes. As many as 12 sand dunes spread over the watershed area were brought under stabilisation programme. Initially at the initiative of the government and later PSBUSS, the village community implemented the stabilisation programme through plantation of tree and grasses.

Table 6.11 shows that altogether 759.5 *bighas* of land were brought under plantation programme by different agencies including the Forest

Department, Government of Rajasthan, Hanwant Trust and PSBUSS. Several observations can be marked from Table 6.11. First, sand dunes were preferred to plain areas for plantation—540.5 out of 759.5 *bighas* of land were brought under plantation. There were several benefits from plantation of sand dunes. Apart from stabilising, it could turn them into dependable sources of grass/ fodder and tree lopping. Second, sand dunes were preferred for plantation on private lands (410 out of 432 *bighas*) more than common lands (130.5 out of 327.5 *bighas*).

Table 6.11

Areas (sand dune and plain) Brought under Plantation in Tibna Watershed

(in Bigha)

Villages	*Private*		*Common*		*Total*		*Total*
	Sand Dune	*Plain*	*Sand Dune*	*Plain*	*Sand Dune*	*Plain*	
Tibna	80	6	15	22	95	28	123
Timri	0	0	35	0	35	0	35
Bhoja ka Bas	30	6	17.5	165	47.5	171	218.5
Fateh Singh ka Bas	300	10	63	10	363	20	383
Total	410	22	130.5	197	540.5	219	759.5

Source: Field survey.

It was difficult to ascertain why strategy for stabilisation programme was designed to focus more on private sand dunes than common ones. Perhaps, it was thought that private owners might take better care of the plants. This would effectively reduce the incidence of shifting of sand dunes. It, in turn, would lead to expand the scope for more agricultural operation. Besides, private owners contributed 10 per cent of the cost incurred. Third, private households did not desire to go in for plantation in the plain areas since they preferred to keep them for agriculture. It is important to note that PSBUSS did not give much importance to initiating plantation programme on common lands.

Table 6.11 further shows that maximum private sand dunes were brought under plantation programme in Fateh Singh ka Bas, which was an upstream village. It is shown later that sand dune stabilisation in this village through plantation had significant

Map 16

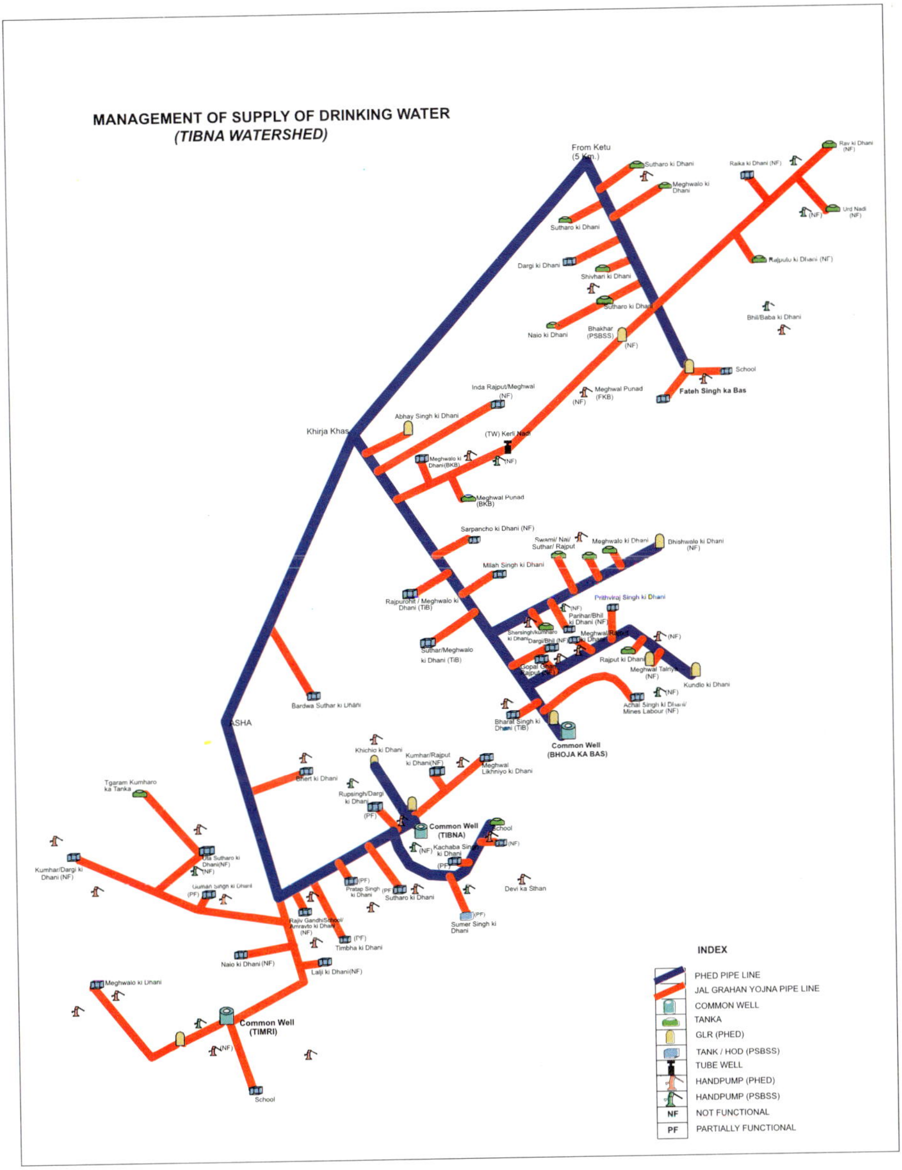
MANAGEMENT OF SUPPLY OF DRINKING WATER
(TIBNA WATERSHED)
From Ketu (5 Km.)
Sutharo ki Dhani
Meghwalo ki Dhani
Raika ki Dhani (NF)
Rav ki Dhani (NF)
Urd Nadi (NF)
(NF)
Sutharo ki Dhani
Dargi ki Dhani
Shivhari ki Dhani
Rajpulu ki Dhani (NF)
Sutharo ki Dhani
Naio ki Dhani
Bhakhar (PSBSS)
(NF)
Bhil/Baba ki Dhani
School
Fateh Singh ka Bas
Meghwal Punad (FKB)
(NF)
Inda Rajput/Meghwal (NF)
Abhay Singh ki Dhani
Khirja Khas
(TW) Kerli Nadi
Meghwalo ki Dhani(BKB)
(NF)
Meghwal Punad (BKB)
Sarpancho ki Dhani (NF)
Swami/ Nai/ Suthar/ Rajput
Meghwalo ki Dhani
Dhishwalo ki Dhani (NF)
Milah Singh ki Dhani
Rajpurohit / Meghwalo ki Dhani (TiB)
Prithviraj Singh ki Dhani
(NF)
Parihar/Bhil ki Dhani (NF)
Shersingh/kumharo ki Dhani
Dargi/Bhil (NF)
Meghwal/Rajput ki Dhani
(NF)
Suthar/Meghwalo ki Dhani (TiB)
Gopal Ghar Rajput (NF)
Rajput ki Dhani
Meghwal Talriya (NF)
Kundlo ki Dhani
(NF)
Achal Singh ki Dhani/ Mines Labour (NF)
Bardwa Suthar ki Dhani
ASHA
Bharat Singh ki Dhani (TiB)
Common Well (BHOJA KA BAS)
Khichio ki Dhani
Kumhar/Rajput ki Dhani(NF)
Meghwal Likhniyo ki Dhani
Bhert ki Dhani
Tgaram Kumharo ka Tanka
Rupsingh/Dargi ki Dhani
(PF)
Common Well (TIBNA)
School
(NF)
Kachaba Singh ki Dhani
(NF)
(PF)
Kumhar/Dargi ki Dhani (NF)
Ula Sutharo ki Dhani(NF)
(NF)
Guman Singh ki Dhani
(PF)
(PF)
Pratap Singh ki Dhani
(PF)
Sutharo ki Dhani
Devi ka Sthan
(PF)
Sumer Singh ki Dhani
Rajiv Gandhi School/ Amravto ki Dhani (NF)
(PF)
Timbha ki Dhani
Naio ki Dhani (NF)
Lalji ki Dhani(NF)
Meghwalo ki Dhani
Common Well (TIMRI)
(NF)
School
INDEX
PHED PIPE LINE
JAL GRAHAN YOJNA PIPE LINE
COMMON WELL
TANKA
GLR (PHED)
TANK / HOD (PSBSS)
TUBE WELL
HANDPUMP (PHED)
HANDPUMP (PSBSS)
NF NOT FUNCTIONAL
PF PARTIALLY FUNCTIONAL

positive impact on livestock development. In terms of the contribution made by the agency for plantation programme, it can be seen that PSBUSS stabilised 692 of 759.5 *bighas* (91 per cent) of land under plantation (Table 6.12).

Table 6.12

Areas Brought under Plantation by Different Agencies

(in Bighas)

Agencies	*Private*	*Common*	*Total*
PSBUSS	402	290	692
Forest Department	nil	37.5	37.5
Hanumant Trust	30	nil	30
Total	432	327.5	759.5

Source: Field survey.

It may be noted that all social groups owned private sand dunes that were brought under the plantation programme. It means that benefits of stabilisation was never allowed to be concentrated in a few individual households as it happened after intervention in some other cases in these villages. No owner was allowed to use his/her sand dune for animals to graze or collecting tree lopping for two years after plantation was done. They were under complete supervision and control of PSBUSS during this period. However, after this period was over, to whatever extent benefits were derived from plantation in the private sand dunes, they went to the owners only. Owners did not allow fellow-villagers to gain access to their sand dunes either for collecting tree lopping or grass or grazing their animals. Fellow villagers could benefit from the stabilisation programme to the extent that the incidence of shifting of sand dunes declined. While such positive externalities of the stabilisation programme cannot be denied, the fact that private owners of the sand dunes gained substantially from it cannot also be disputed. Hence, initiative taken for sand dune stabilisation was geared to serve the interests of a few individual farming households more.

In order to examine whether sand dunes were stabilised and could contribute to the well-being of the villagers or not, enquiry was made with 10 households chosen from all social groups in each village on the basis of a few parameters shown in Table 6.13. Although the stability of sand dunes was primarily determined by the extent to which it achieved floral diversity, an attempt is made here to capture villagers' perception about it (Dhabriya, 1988). Floral diversity is examined later.

Table 6.13

Response of the Villagers on Sand Dune Stabilisation

Parameters	*Tibna*	*Bhoja ka Bas*	*Fateh Singh ka Bas*	*Timri*
Shifting of sand dune (stabilised)	9	9	8	7
Increase in water level	5	4	6	2
Increase in soil fertility	4	5	5	3
Change in river course	0	0	1	1
Increase in vegetation cover	8	7	9	5
Increased cultivated area	3	2	1	0

Source: Field survey.

In the absence of any scientific tools for measuring the extent to which sand dunes were stabilised, the interview method was adopted. Table 6.13 shows that the majority of the households interviewed in all villages observed that there had been a decline in the shifting of sand dunes. However, response of the households of Timri was relatively less encouraging in this regard. In any case, the decline was not steep enough to yield positive impact on the rise of the cultivated land area to a conceivable extent. Since performance of sand dune stabilisation in Timri was relatively low, its farming area did not witness any rise. As far as increase in water level is concerned, except Fateh Singh ka Bas where 6 out of 10 respondents observed that it had increased, response in rest of the villages was not all that encouraging. Response of the respondents of Timri indicated lowest achievement in this regard also.

It may be argued that groundwater level increases as sand dunes are stabilised by means of deepening vegetation cover (Singh, 1998). With no information available about the status of groundwater in the sand dunes, villagers living at the foothills of the dunes were asked if they had observed any significant change in groundwater level after the plantation programme. Their responses reveal that Fateh Singh ka Bas gained comparatively more than other villages (Table 6.13). However, it was difficult to say whether plantation in sand dunes contributed to the rise in groundwater in this village. For many other villagers observed that the construction of the anicut in the village contributed to the increase in groundwater level.

The increase in vegetation cover, as reported by the villagers, appeared to be comparatively less in Timri than that in other villages. Table 6.11 shows that plantation was done only in 35 *bighas* of common land in Timri. In the absence of care and maintenance, the progress of stabilisation of the sand dunes in this village was much lower. A disproportionate implementation of the stabilisation programme did not allow Timri to become an equal development partner of other villages. This again exemplifies how discrimination was built into the design of intervention. Further, it had to be discovered how successful the stabilisation programme was wherever the scale of its implementation was large. For, success largely depends upon the extent to which floral diversity is achieved over the years.

V

ENVIRONMENTAL CHANGES

While floral diversity is linked to improvement in soil characteristics, together the two aspects contribute immensely to the elevation of local ecosystem. This was especially so on the sites where the plantations programme was implemented.

Floral Diversity

In order to examine whether the sites achieved floral diversity at a higher level or not, an attempt is made here to compare them with the status of floral diversity of those sites that were not brought under the plantation programme. It may be reasonable to expect a higher level of floral diversity on the former. Data was collected from both sites including the sites located around the anicuts, sites on the sand dunes and on the riverside. In all, 10 sites were chosen for sampling of vegetation cover in the watershed area including those under plantation programme and those that were not. The objective was to see how the areas brought under plantation fared in expanding floral diversity as compared to the areas without plantation. The same standard method as mentioned in Annexure 2 was followed for sampling that was done in the pre- and post-monsoon periods of 2005. Indices including importance value index (IVI), diversity index and species richness index were estimated for both intervention and non-intervention sites.

Importance Value Index (IVI)

The importance value index shows the hierarchical order of species that has much to do with the stability of the ecosystem. For the order of the species must be friendly not only to the ecosystem in terms of suitability, but also should be able to maintain the biogeochemical cycle in a sustainable manner. The latter is linked more to biodiversity in that a diverse biological system encourages more biotic interactions. Higher level of biotic interaction, in turn, ensures a dynamically stable biogeochemical cycle. Table 6.14 shows IVI for trees and shrubs of both intervention and non-intervention sites.

Table 6.14

IVI for Trees and Shrubs in both Intervention and Non-intervention Sites

IVI	*Intervention Site*	*Non-intervention Site*
1-10	*Lycium barbarum* *Tecomella undulata* *Maytenus emarginata* *Ziziphus mauritiana* *Tephrosia purpurea* *Aerva tomentosa* *Dalbergia sissoo* *Cocculus pendulus*	*Salvadora oleoides* *Azadirachta indica* *Grewia tenax* *Aerva tomentosa*
10-25	*Prosopis juliflora* *Prosopis cinereria* *Zizyphus nummularia* *Capparis decidua* *Acacia senegal* *Leptadaenia pyrotechnica* *Aerva pseudotomentosa* *Calotropis procera*	*Lycium barbarum* *Zizyphus nummularia* *Capparis decidua* *Leptadaenia pyrotechnica* *Aerva pseudotomentosa* *Calotropis procera*
25-60	*Maytenus emarginata*	*Acacia tortilis* *Prosopis cinereri* *Tecomella undulata* *Acacia senegal*
60-100	*Acacia tortilis*	

Source: Field survey.

While *Acacia tortilis,* followed by *Maytenus emarginata,* had higher IVI in the intervention sites. *Acacia senegal, Tecomella undulata, Prosopis cineraria and Acacia tortilis* had higher IVI within the range of 25-60 in the non-intervention sites. If the analysis is confined to the range of IVI above 25, Table 6.14 shows that the two sites had almost different species of trees. While *Acacia tortilis,* followed by *Maytenus emarginata,* occupied the highest level of dominance in the hierarchy of plantation on the intervention sites, three different species, in addition to *Acacia tortilis,* were largely found in the non-intervention sites.

Apart from being found suitable for the arid climate, *Acacia tortilis* had various uses. It was used as green fodder. Besides, its fruits, bark, gums were extremely useful. The growth of *Prosopis cineraria* had other serious and positive implications for environmental conservation because of its extensive root system that stabilises shifting sand dunes. It was also useful as windbreak shelterbelt and in afforestation of dry areas. Besides, it had many other features that contributed to the improvement of the organic matter of soil.[4] However, an increased density of this tree eliminates normal ground vegetation. While it was present on both sites, the non-intervention sites had two other species of *Maytenus emarginata* and *Tecomella undulata,* which were equally palatable and eco-friendly and seemed to have contributed greatly to the floral diversity of these sites. For example, *Tecomella undulata* acted as a soil-binding tree by spreading a network of lateral roots on the top surface of the soil, thus helping in stabilising shifting sand dunes and acting as a shelterbelt.[5]

Table 6.15 shows the importance value index for grass and herb species. The analysis shows that the pattern of growth of grasses and herbs on the two sites was distinctly different in terms of hierarchical positions. While *Cenchrus biflorus* occupied the dominant position on the intervention sites, within a range 60-100 IVI, *Tribulus alatus* grew maximum on the non-intervention sites—a cause of concern because this grass species was not eco-friendly. It was, however, palatable. Grass species falling in the range of 25-60 IVI were eco-friendly and palatable. The problem with the intervention sites was that the non-palatable, non-eco-friendly *Heliotropium striogosum* had tremendous

4. "The role of *prosopis cineraria* in an agro-pastoral system in western Rajasthan" by H.S. Mann and K.A. Shankarnarayan, Central Arid Zone Research Institute, Jodhpur, *www. Ilri.cgiar.org/InfoServ/webpub/fulldocs/Browse_in_Africa/Chapter49*

5. *http: //en.wikipedia.org/wiki/Tecomella_undulata.*

growth here with the same IVI. Interestingly, it had much less growth and the lowest IVI on the non-intervention sites (Table 6.15). It has to be remembered that grass species were never planted on the intervention sites. Despite this, their growth pattern on both sites in terms of their hierarchical order was different. It may be argued that had there been no intervention for plantation of tree species the growth pattern of grass species would have been similar. In any case, no noticeable difference was found to exist in the palatability and eco-friendliness of the grass species on both the sites.

Table 6.15

IVI for Grass and Herb Species in both Intervention and Non-intervention Sites

IVI	*Intervention site*	*Non-intervention site*
1-10	*Corchorus tridense* *Euphorbia granulata* *Indigofera cordifolia* *Convolunlus microphyllus*	*Heliotropium strigosum* *Corchorus tridense* *Lasiurus sindicus* *Aristida funiculata*
10-25	*Boerhavia diffusa* *Brecharia ramosa* *Dactyloctenium aegypfium* *Aristida funiculata*	*Eragrostis pilosa* *Dactyloctenium aegypfium* *Aristida adscendens* *Crotalaria burhia* *Boerhavia diffusa* *Indigofera cordifolia*
25-60	*Eragrostis pilosa* *Tribulus alatus* *Heliotropium strigosum*	*Brecharia ramosa* *Cyterus bulbosus* *Cenchrus biflorus*
60-100	*Cenchrus biflorus*	*Tribulus alatus*

Source: Field survey.

Other indices have a direct bearing on IVI. The analysis of IVI indicated that the non-intervention sites had no significant difference in floral diversity with the intervention sites in the case of the tree species. Table 6.16 shows that non-intervention sites had floral diversity (1.01 accompanied by higher species richness index (1.33) in respect of trees. However, in the case of intervention sites, it was 1.06 and 1.29, respectively. While one had reasons to expect that the intervention sites must achieve relatively more diversity and species richness, it never happened so.

Table 6.16

Diversity and Species Richness Index of both Intervention and Non-intervention Sites

Indices	*Intervention Site*		*Non-intervention Site*	
	Tree	*Grass*	*Tree*	*Grass*
Diversity index	1.06	0.94	1.01	1.00
Species Richness Index	1.29	0.58	1.33	0.57

Source: Field survey.

Table 6.16 further shows that no significant difference was found to exist in respect of diversity of grasses and herbs between the two types of sites. Although, intervention sites achieved comparatively little more diversity in respect of trees, the difference was not significantly higher. In other words, one might argue that intervention for plantation did not make much difference to the floral diversity of the area. It shows that the plantation programme was not designed appropriately in that it had a limited choice of tree species or there was an inappropriate selection of trees, which failed to survive.

Besides, implementation was handicapped with lack of maintenance and care that were not given adequate importance. As noted earlier, most of the sites brought under plantation were privately owned in which the village institution had hardly any role to play. In such a situation, the responsibility was naturally fixed with the private owners of the sites and intervening agency such as PSBUSS. However, efforts made to meet this responsibility seemed inadequate. It is not certain whether the current level of vegetation cover would sustain in future in the absence of improved floral diversity.

Soil Characteristics

No difference was found between the soil characteristics of both sites either. Needless to say, arid zone soil is poor in terms of fertility, for physical, chemical and biological processes and their interactions take place at a slow pace due to the scarcity of the required moisture. The plant cover is sparse, which is why enrichment of soil through plant residues is poor. Inadequate moisture availability reduces the microbial activity leading to poor recycling of plant residues. There are reasons to expect reversal of this process in those places like the one under study where intervention is made for improving the vegetation cover. While enrichment of soil and expansion of floral diversity are linked to each other, floral diversity through plantation is expected to bring about an improvement in soil characteristics in a place like Tibna which was barren and infertile.

It is in this context that it has to be examined whether floral diversity, to whatever extent it was achieved on the intervention sites, made any significant difference to their soil characteristics in comparison to the non-intervention sites. In view of this, soil samples were collected from as many as 15 sites while following the same selection criteria as adopted for investigating floral diversity. Physico-chemical analysis that was conducted included pH, Electrical conductivity (EC), Organic carbon (OC), Organic matter (OM), Nitrogen (N), Phosphorus (P) and Potassium (K).

The results of pH, EC, OC and OM are shown in Table 6.17. Both the intervention and non-intervention sites had almost equal status in terms of soil characteristics. pH was recorded at 8.29±0.14 and 8.21±0.20 for the intervention and non-intervention sites, respectively. While it did not differ significantly between the sites, it was within the normal range of 7.0-8.5. pH indicates the acidity and alkalinity in the soil, while electrical conductivity describes the presence of ions in soil through which salinity is determined. Since pH was found to be normal on both sites, they were never prone to acidity or alkalinity. No significant difference was found in respect of EC on both the sites as well. It was recorded at 0.46±0.11 mmhos/cm and 0.44± 0.19 mmhos/cm on intervention and non-intervention sites, respectively. Since EC had values <0.75 mmhos/cm, the soil was normal and salt free (Ashman and Puri, 2002).

Table 6.17

Soil Characteristics of Intervention and Non-Intervention Site

Parameters	*Intervention Site*	*Non-Intervention Site*
pH	8.29±0.14	8.21±0.20
Electrical conductivity (mmhos/cm)	0.46±0.11	0.44±0.19
Organic carbon (per cent)	0.17±0.05	0.16±0.03
Organic matter (per cent)	0.29±0.09	0.27±0.06

Source: Field survey.

The fertility of soil depends upon the presence of organic carbon, organic matter, nitrogen, phosphorus and potassium. Table 6.17 shows that accumulation of organic carbon on intervention sites was 0.17 per cent, while it was 0.16 per cent in non-intervention sites. This was contrary to what was expected; it was expected to be higher, especially on the intervention sites. Growth of trees exerts ameliorative effect on the soil by loosening the subsoil and improving the permeability through the action of root system (Agarwal *et al.*, 1982). However, the low level of floral diversity seemed to have resulted in lower levels of organic carbon, especially on the intervention sites. Similarly, the organic matter did not significantly differ between the two sites at 0.29 per cent and 0.27 per cent, respectively (Table 6.17).

Organic matter is a source of nutrients such as nitrogen, which is released continuously by microbial activity. It prevents compaction, increases filtration, aeration and the moisture-holding capacity of the soil. The amount of organic matter is an index of soil fertility. However, such a process was yet to mature on the intervention sites, not to talk about the area where no plantation took place. In other words, floral diversity, to whatever extent it was achieved on the intervention sites, failed to improve soil fertility. The analysis of variance shows that the difference between the two sites was not statistically significant ($p>0.05$).

Figure 6.1

Availability of N P K in Intervention and Non-intervention Sites

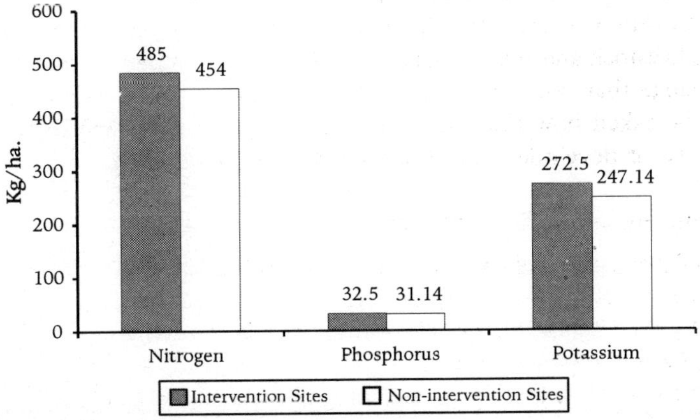

Source: Field Survey

However, nitrogen and potassium were slightly more on the intervention sites (Figure 6.1). Nitrogen was available to the extent of 485±151.85 kg/ha on the intervention sites and 454±104.38 kg/ha on non-intervention sites. Similarly, phosphorus was available in little more quantities on the intervention sites. The status of nitrogen content in the soil shows that nitrogen availability was well within the medium range (the standard range of available nitrogen in soil is 280-560 kg/ha) on both sites.

It cannot be disputed that the nitrification process on the intervention sites was activated due to litter fall, which was why there was a little more availability of nitrogen. However, it was too low to raise its availability beyond the medium range. The availability of nitrogen on the non-intervention sites was also within the medium range. Hence, the fertility status was not noticeably different between the two sites, especially if these results were understood against the poor availability of organic carbon. ANOVA shows that difference between two sites in respect of all these nutrients was not statistically significant ($p>0.05$).

VI

RESOURCE DEVELOPMENT AND HOUSEHOLD ECONOMY

The villagers witnessed a discernible improvement over the years in respect of their land and livestock resources because of watershed development. These were accompanied by enrichment of the stock of human resources through educational progress. While development of land, livestock and human resources were integral parts of the process of change that was sought through micro watershed development, it may be asked how the improved resource base could sustain in the absence of development of community institution.

Changing Land Use Pattern

As discussed earlier, the claims on available stock of renewable resources intensified in the arid part of the state because of the rise in both human and livestock population. The villages under study were no exception to this. In such a situation, one of the most dependable options left to the villagers was to bring more unused cultivable land under cultivation. In an area like this, cultivation was carried out only when it rained. Hence, the objective was to bring more lands under cultivation, if they were available, so that the volume of crop production could be raised in the same season. Of course, improvement in the stock of livestock resources always provided a route of escape from such vulnerable conditions. But then, greater availability of fodder was required to match the increased stock of livestock resources.

As far as the area under cultivation is concerned, villagers mainly depended upon rain-fed crops and brought large percentages of cultivable land under cultivation. Analyses of the land use pattern from 1988-89 until recently show this (not all years are shown in Table 6.18). Three different years were chosen randomly and shown in Table 6.18 only to indicate how land use pattern changed over the years after microwatershed development was sought. Only 53 hectares of land (2 per cent of the cultivable [C+D] land) had been brought under cultivation (Table 6.18). The rest was left uncultivated during 2002-03, one of the worst hit years by drought that these villages had witnessed in the recent past.

However, this was not true of the rest of the drought years. For example, 84.19 per cent of the area was cultivated during 2001-02, another drought year (not shown in Table 6.18). Villagers could cultivate rain-fed crop during this year. The only difference was that cultivation of some area was possible in 2002-03 due to the availability of irrigation. Similarly, only 69.33 per cent of the cultivable land was cultivated in 1996-97 when the villages received normal rainfall. In any case, it was noticeable that 53 hectares of land was brought under irrigation even under severe drought conditions during 2002-03. This happened only in two downstream villages namely, Tibna and Bhoja ka Bas. The field survey revealed that these two villages had never witnessed irrigated cultivation in the past. Once micro watershed development programme was implemented, groundwater was available even for cultivation.

Tibna and Bhoja Ka Bas witnessed this remarkable achievement during a drought year. The irony, however, was that such a change was confined to two villages only. Table 6.19 shows that with the availability of irrigation, there was crop diversification in these villages from traditional rain-fed crops such as *bajra, moong* and *moth* to wheat, mustard, chilli, onion, etc. Even *bajra* was brought under irrigated cultivation in both the villages. Although the scale of the operation was low, it exemplified how agricultural performance could be improved in the arid zone even under drought conditions.

Table 6.18

Land Use Pattern of All Four Villages Together during 1988-89, 1996-97 and 2002-03

(Land in hectare)

	1988-89	1996-97	2002-03
A. Total area	3512[6] (100.0)	3511 (100.0)	3511 (100.0)
B. Land not available for cultivation	855 (24.35)	854 (24.32)	854 (24.32)
C. Available for cultivation but not cultivated	535 (15.23)	815 (23.21)	2604 (74.17)
D. Cultivated land	2122 (60.42)	1842 (42.46)	53 (1.51)
Irrigated	0	0	53
Unirrigated	2122	1842	0

Source: Tehsil office, Shergarh *Tehsil*, Jodhpur.

Table 6.19

Crop Diversification During Drought Years in Tibna and Bhoja ka Bas Villages

(Land in hectare)

Crops	2001-02		2002-03	
	Tibna	Bhoja ka Bas	Tibna	Bhoja ka Bas
Bajra			17.48	1.62
Wheat	4.1	0.3	8.1	1.6
Methi	1.8			
Mustard	0.3		12.9	1.6
Eshabgole	5.2	1.9	8.1	
Zeera	16.7	3.2		
Onion	0.6	0.3	1.0	
Chilli	1.2	0.3	0.8	0.3
Total	29.9	6.0	48.38	5.12

Source: Tehsil office, Shergarh, Jodhpur district.

6. During this year, the total geographical area of all four villagers increased by one hectare as compared to other years under consideration.

However, agricultural gains were largely concentrated, as the field survey revealed, in the affluent farmers in these villages, for they could afford to install tube wells for irrigating their lands and take risks. The largest number of affluent farmers, including medium and large farmers, belonged to the *Rajput* community (Table 6.20).

Table 6.20

Households of All Villages in Terms of their Social Background and Landholdings

Social Background (Caste)	*Landless*	*Marginal*	*Small*	*Medium*	*Large*	*Total*
SC	6	93	49	19	Nil	167
ST	0	13	4	6	Nil	23
OBC	20	108	52	98	3	281
Rajput	3	66	84	182	10	345
Other general categories (*Brahmin + Bania*)	2	3	Nil	Nil	Nil	5
Total	31	283	189	305	13	821

Source: Field survey.

Although the population of other social groups increased over the years, the dominance of the *Rajputs* continued unchallenged in terms of numbers, and ownership and control over land resources in all the villages. It was further facilitated by the fact that a large number of members of this community were ex-servicemen. Ten out of 13 large farmers (77 per cent) and 182 out of 305 medium farmers (60 per cent) belonged to the *Rajput* community alone. Hence, its relative economic affluence as well as numerical strength helped this community to occupy the supreme position in the social and economic hierarchy of the village community. This had a large bearing on social relations.

Changing Educational Background

The rich stock of human capital along with a steady improvement in formal education over the years did not bring about any change in the traditional power relations. Just about 20 years ago, the entire *panchayat* was almost non-existent on the literacy map. Only 12 per cent of the population was considered literate (Table 6.21). It, however, increased to 45 per cent in 2001. While this achievement can hardly be dismissed, it is difficult to understand why the educational progress was not as encouraging as desired in the presence of the large numbers of ex-servicemen and good educational infrastructure. Whatever be the stock of human capital and the scale of its improvement, it failed to dilute the traditional power relations that continued to hold the key to institutional renovation or any development intervention that was sought in these villages.

Table 6.21

Changing Stock of Literate Population in All Four Villages

	1981	*2001*
Total population	3146	4806
Literate population	388(12.33)	2169(45.13)

Source: Census of India (Population, 1981 and 2001), Government of India.

Table 6.22 shows that educational progress was more pronounced in the age group of 5-14 years. Women's education was the highest in this age group. However, as the age group increased, the educational level declined. In any case, the good stock of literate population of the sample households, around 62 per cent, was a positive feature of development. More importantly, around 20 per cent of the literate population of the sample households had received formal education above middle school. However, the performance of women was dismal at less than one per cent of the stock of literate population.

Table 6.22

Educational Progress of the Sample Households in All Four Villages

Age Group	*Illiterate*		*Literate*										
			Primary (1-5)		*Middle (6-8)*		*Sr. Secondary (9-12)*		*Above 12*		*Total*		
	Male	*Female*	*Male*	*Female*	*Male*	*Female*	*Male*	*Female*	*Male*	*Female*	*Male*	*Female*	*Total*
5-14	7	21	86	78	23	3	0	0	0	0	109	81	190
14-30	23	88	31	25	50	11	37	1	18	0	136	37	173
30-60	33	84	21	7	26	1	26	0	9	0	82	8	90
>60	19	17	11	0	4	0	4	0	1	0	20	0	20
Total	82	210	149	110	103	15	67	1	28	0	347	126	473

Source: Field survey.

Improvement in the stock of human capital had no correspondence whatsoever with sustainable collective action and management of watershed. One may argue that the design of intervention by PSBUSS was not appropriate enough to create scope for utilising the improved human resource base of the villagers for institutional renovation. It also failed to nurture collective leadership. One might attribute it to the design of intervention made by PSBUSS in that neither was scope created to give rise to development of local leadership nor was there any attempt to seek institutional development devoid of traditional power relations.

Changing Status of Livestock Resources

Table 6.23 shows that there had been a consistent rise in livestock population of all varieties, except sheep. The latter witnessed a decline in 2003. The noticeable feature of livestock resources was that the population of large animals such as cows, buffaloes and camels witnessed a rise even in drought years. The status of livestock can better be understood in terms of cattle units. Table 6.23 shows that livestock population in terms of cattle unit increased from 2,890 in 1992 to 4,095 in 1997. Although the population declined marginally in 2003, it was higher than what it was during 1992.

Table 6.23

Changing Stock of Livestock Resources

Animals	*1992*		*1997*		*2003*	
	No	*Cattle Unit*	*No*	*Cattle Unit*	*No*	*Cattle Unit*
Cow	748	748	1247	1247	1277	1277
Buffalo	103	206	231	462	320	640
Camel	87	174	122	244	141	282
Goat	5041	1260	4733	1183	4956	1239
Sheep	2009	502	3836	959	1821	455
Total	7988	2890	10169	4095	8515	3893

Note: Cattle unit is expressed in terms of (1) 1 cow = 1 cattle, (2) 1 buffalo = 2 cattle, (3)1 camel = 2 cattle (4) 1 goat = 0.25 cattle and (5) 1 sheep = 0.25 cattle.

Source: Field survey.

No noticeable change took place in livestock holding per household over the years. In other words, while the number of households increased with the growing population, no noticeable change was witnessed in the average livestock holding per household. However, it is important to underline that the stock of animals, except sheep, per household witnessed only a marginal decline (Table 6.24).

Table 6.24

Stock of Animals per Household

Animals	*1997*	*2000*	*2003*
Cow	1.68	1.66	1.50
Buffalo	0.31	0.40	0.39
Goat	6.37	7.31	6.12
Sheep	5.16	5.63	2.30

Source: Field survey.

Individual perception of the households regarding livestock progress in the village after the intervention supports these observations. According to 72 per cent of the sample respondents, no improvement was witnessed in the stock of animals per household after intervention. (Table 6.25).

Table 6.25

Perception of the Sample Respondents Regarding Improvement of Stock of Animals after Intervention

Category of Farming Households	*Improvement*	*No Improvement*	*Total*
Landless	0	1	1
Marginal	7	26	33
Small	4	14	18
Medium	20	42	62
Large	2	3	5
Total	33 (27.73)	86 (72.27)	119 (100.0)

Source: Field survey.

It is true that no improvement took place at the household level. It is also true that individual stock did not fall significantly despite persistent drought for three consecutive years. One might treat this as an achievement and, therefore, account for it as positive impact of watershed development. The availability of grass, tree lopping and fodder (through cultivation) even in drought years definitely reduced the uncertainty regarding fodder and encouraged villagers to hold back at least the minimum stock of their animals rather than allowing it to decline.

The fact that the average size of livestock holding per household remained almost the same between normal rainfall and drought years indicated a positive impact of watershed development, even more discernible when hardly any animals were taken on migration for grazing. In other words, the increased pressure on the local resource base with the rise in aggregate stock of animals was balanced by the corresponding rise in availability of fodder even during drought years. Nevertheless, rich farmers could retain more number of animals as compared to the poor farmers during drought years. Of course, livestock per household among the richer households was much more comparatively; they had their own support system for maintaining it.

It must be remembered that livestock resources were equally important to the rich farmers as a means of survival. Since they had relatively larger stock of animals, the major benefits of watershed development accrued to them. The average holding of livestock per rich farming household (medium and large farmers) was higher than that of the poor farming households (landless, marginal and small farmers) in respect of each type of animal during both the drought years under consideration (Table 6.26).

Table 6.26

Size of Livestock Holding Per Rich and Poor Farming Household

	2000					*2003*				
	Cow	*Buffalo*	*Sheep*	*Goat*	*Bullock*	*Cow*	*Buffalo*	*Sheep*	*Goat*	*Bullock*
Poor farmers	1.38	0.23	2.08	5.79	0.10	1.08	0.21	0.69	4.79	0.08
Rich farmers	1.88	0.54	8.39	8.49	0.12	1.84	0.52	3.55	7.15	0.06
Average per household	1.66	0.40	5.63	7.31	0.11	1.50	0.39	2.30	6.12	0.07

Source: Field survey.

One significant factor that did not allow change in the average stock, especially of milch animals, per household was the absence of commercial dairy activity. Though a milk dairy was set up at Tibna some years ago, it was closed because milk procurement was much less than the optimum quantity required to keep the unit functional. Whatever quantity of milk was produced was consumed at the household level, even poor households. The average household consumption of milk was around 13 quintals in 2003 just two quintals down from 15 quintals in 2000 (Table 6.27) providing nutritional security to a conceivable extent. Although sale of *ghee* was not a dominant feature of the milk economy in these villages, the field survey revealed that each houschold, sold *ghee* worth Rs. 450 on an average in 2003. In Fateh Singh ka Bas, small farmers could gain considerably more from the sale of *ghee*.

Table 6.27

Availability of Milk Per Household during Two Drought Years

(in quintal)

Farming Households	*2003*	*2000*
Landless	1.50	1.50
Marginal	6.94	10.76
Small	8.33	9.69
Medium	16.68	18.69
Large	20.61	26.79
Average per household	12.75	15.33

Source: Field survey.

In fact, Fateh singh ka Bas gained relatively more from livestock by producing more milk. Several favourable conditions made it possible for this village to perform better in livestock as compared to other villages. First, members of the village institution were more active and excelled in managing their resource base, including livestock. The village person, who was trained as *Gopal*, turned out to be equally active and able to implement a breed improvement programme for both large and small animals in a significant manner. Second, this village was endowed with a large hilly tract that facilitated growth of grass. The majority of private sand dunes were brought under the plantation programme. Third, almost half of the LSCDs (49 per cent of LSCDs constructed by PSBUSS) were concentrated in this village; LSCD in hilly areas created better conditions for expanding grass cover making an important contribution to the development of livestock, especially small ruminants. If the average production of milk per household were to be taken as one of the important indicators of livestock development, intra-village analysis shows that it was the Fateh Singh ka Bas that appeared to have scored highest of all villages under study (Table 6.28).

Table 6.28

Village-wise Average Production of Milk Per Household after Intervention

(in Quintal)

Villages	*2003*	*2000*
Bhoja ka Bas	7.66	13.14
Fateh Sigh ka Bas	19.33	20.88
Tibna	13.10	14.36
Timri	11.80	15.56
Average	12.75	15.33

Source: Field survey.

While drought was severe in 2003, it had relatively lower intensity in 2000. As it appears from Table 6.28, the household milk production was highest in Fateh Singh Ka Bas in both years at 19.33 quintals and 20.88 quintals, respectively. Although it declined in all villages in 2003, it was much less in this village as compared to others. The construction of anicuts, the largest presence of LSCD coupled with better implementation of the breed improvement programme and relatively successful plantation of sand dunes together contributed to livestock development in this village. Hence, the intervention made by PSBUSS had differential impact on livestock development too.

The analysis of land use pattern in the preceding section shows how Tibna and Bhoja ka Bas were benefited comparatively more from watershed development in respect of agricultural performance. The present analysis shows how Fateh Singh ka Bas could gain more in livestock development in terms of increased milk production. Apart from having higher average holding of cows per household, the average holding of goats per household with greater incidence of breed improvement was also much higher in this village at 2.05 and 10.67, respectively (Table 6.29). The number of both of these animals was much higher than the average of all four villages.

Table 6.29

Livestock Holding per Household in Each Village

Village	*Cow*	*Buffalo*	*Sheep*	*Goat*	*Bullock*
Bhoja ka Bas	1.21	0.38	2.1	5.62	0.17
Fateh Singh ka Bas	2.05	0.24	7.29	10.67	0
Tibna	1.43	0.45	0.45	4.4	0.02
Timri	1.64	0.36	3.09	7.82	0.18
Average	1.5	0.39	2.3	6.12	0.07

Source: Field survey.

As mentioned above, this village had better grazing facilities encouraging villagers to rear more small ruminants as compared to other villages, thus acquiring an additional advantage in generating an income through sale of small ruminants. On an average, the annual income from sale of animals here was estimated to be Rs. 3,185.7 per household. However, it was only Rs. 626 in Bhoja ka Bas, Rs. 944 in Tibna and Rs. 155 in Timri.

Greater emphasis here on livestock development was essentially an expression of the villagers' coping strategy that evolved in response to the increasing vulnerability to impoverishment. The scope for agricultural prospect was limited, as mentioned earlier, due to the undulating and rocky topography of the land and water scarcity. Therefore, livestock gained overriding importance in this village as a stable source of household income despite less availability of feed and fodder except grass. Table 6.30 shows the average feeding pattern of animals followed by each sample household in all four villages.

Fodder was mainly procured from two sources, namely the household's own source, which was free, and the village informal market with its tradition of facilitating buying and selling of fodder between the villagers through the age-old form of exchange. Data was collected on the prices at which feed and fodder were bought and sold within the same villages. The quantity of feed and fodder are expressed here in monetary terms in order to show a comparative picture based on the same denominator. Each household's average requirement of each item per annum is shown in terms of its percentage share in the total requirement of the same (Table 6.30). Table 6.30 shows that Fateh Singh ka Bas had to depend on agricultural by-products and fodder from the same informal market that existed in other villages since they were not adequately available in their own village. However, there was greater availability of tree lopping in this village, which was why they depended less on purchase from the informal market. Each household on an average purchased tree lopping worth Rs. 2,093 annually, while they could obtain it worth Rs. 4314.0 from their own source.

Table 6.30

Relative Dependence of Each Household on Informal Village Market for Feed and Fodder in All Four Villages

(in percentage)

	Bhoja ka Bas		*Fateh Singh ka Bas*		*Tibna*		*Timri*	
	Own Source	*Market Source*	*Own Source*	*Market Source*	*Own Source*	*Market Source*	*Own Source*	*Market Source*
Grass (Green + Dry)	100.0	0.0	100.0	0.0	100.0	0.0	100.0	0.0
Fodder	53.0	47.0	34.6	65.4	46.6	53.4	53.7	46.3
Food grain	32.5	67.5	19.9	80.1	47.8	52.2	4.2	95.8
Concentrate	0.0	100.0	0.0	100.0	0.0	100.0	0.0	100.0
Tree lopping	90.3	9.7	67.3	32.7	86.5	13.5	96.3	3.7
Average per household	59.2	40.8	39.2	60.8	57.2	42.8	60.0	40.0

Source: Field survey.

Table 6.30 shows that Fateh Singh ka Bas was dependent on the informal village market largely for fodder and food grain, purchasing 65 and 80 per cent, respectively. It was more than purchased by households in other villages, on an average. The watershed development intervention seemed to have engaged Fateh singh ka Bas into a form of exchange that motivated it to harness its development potential for livestock. It had never happened earlier. It was partly intended, as it were, and partly a logical outcome of interaction that was sought between the local ecosystem and the forces of change. While on the one hand, Bhoja ka Bas and Tibna could achieve better agricultural performance than witnessed earlier, on the other, Fateh Singh ka Bas, the upstream village, increasingly saw its survival in the prosperity of its livestock sector.

Despite having assumed so much importance, livestock sector, however, was not remunerative, if costs of all varieties of feed and fodder were accounted for. Table 6.31 shows the average annual gross income/value generated by livestock and expenditure incurred in respect of each village under study. It shows that the value of livestock products annually on an average per household, taking all villages together, was Rs. 17,997.1, while cost of maintenance of livestock was Rs. 22,370.0. As mentioned earlier, the value of feed and fodder was estimated on the local prices that did not adequately represent scarcity prices or competitive prices. If they were taken into the consideration, the average maintenance cost of the same stock of animals per household would have been much higher than the average income from livestock per annum.

In such a situation, one might have reason to doubt the sustainability of the livestock economy. Even if valued at local prices, the cost of maintenance exceeded the gross value generated in the livestock economy turning it unsustainable. However, the free availability of natural resources did not allow this to happen (Table 6.31). One must bear in mind that grasses, grazed upon by animals, were never brought into the system of monetary valuation. It is needless to say that they served the great purpose of rearing small ruminants especially. If they were accounted for and incorporated into the valuation, the maintenance cost of the same stock of animal would be much higher than shown.

Table 6.31 shows that the role of non-priced component (no payment was made; however, pricing was done to find out the magnitude of those that were available free of cost) was mainly responsible for the livestock economy to sustain. Villagers could see their gain in livestock economy so long as the non-priced component of the expenditure was higher than the priced one. Here, in the present case, the average non-priced component expenditure was Rs. 12,083.4 and the priced component was Rs. 10,286.6. The higher the magnitude of the non-priced component of the expenditure, the better the prospect of livestock sector.

Agricultural by-products, grass and tree lopping available free of cost constituted the non-priced component of the maintenance cost of livestock.[7] Now, if one concentrates only on the latter alone (grass and tree lopping), it is worked out to be equivalent to Rs. 5,894.12 in monetary terms, or 26.3 per cent of the average household

7. Although value of agricultural by-products is considered unpaid, actually it is not so. Investment is made in agricultural production. This reflects local prices at which valuation is done.

maintenance cost of livestock (of Rs. 22,370) per annum (Table 6.31). In other words, if the household had to pay this amount, the livestock economy would have ceased to be a sustainable source of livelihood. One can well imagine how much importance the non-priced component of expenditure gains in the absence of the availability of agricultural by-products (in case no agricultural operation). In other words, the higher the availability of free grass and tree lopping the higher the prospect of livestock. This happens if the natural resource base improves further. The present analysis shows that the livestock sector would have been unsustainable if fodder, grass and tree lopping ceased to be available free of cost.

Table 6.31

Average Annual Gross Value Generated from Livestock and Expenditure Incurred Towards its Maintenance per Sample Household

(Rs.)

Villages	Income	Expenditure			
		Not Purchased			
		Agriculture	Non-agriculture (COL+PL)	Purchased	Total
Bhoja ka bas	11,375.1	5,240.22 (30.8)	4,827.27 (28.4)	6,924.2 (40.8)	16,991.7 (100.0)
Fateh Singh ka Bas	28,341.9	5,662.72 (18.8)	6,155.87 (20.4)	18,342.9 (60.8)	30,161 (100.0)
Tibna	17,991.4	6,599.78 (29.6)	6,138.11 (27.6)	9,540.00 (42.80)	22,277.9 (100.0)
Timri	15,736.4	6,186.00 (32.1)	5,353.29 (27.8)	7,707.00 (40.00)	19,246.3 (100.0)
Average per household	17,997.1	6,189.28 (27.7)	5,894.12 (26.3)	10,286.6 (46.0)	22,370.0 (100.0)

Note: COL= Common Open Land; PL= Private Land.
Source: Field survey.

In the villages under study, it appears that these were mostly available from private lands rather than common ones. Table 6.32 shows the annual average collection of grass, tree lopping and fuelwood from both common and private lands per household. The practice of collecting tree lopping in these villages was unique in that villagers never chopped branches off trees. In other cases under study, villagers did so, collected the leaves and finally used the same branches as fuelwood. However, in this part of the state, only leaves were collected to allow them to grow again shortly for reuse.

The practice of following this traditional norm exemplified the informal institution's influence on natural resource management.

The major source of gains from the common land in these villages was grass, grazed mainly by small ruminants. However, villagers were more dependent on their private lands for fuelwood and tree lopping (see Table 6.32). This was true even for grass.

Table 6.32

Average Annual Collection of Grass, Fuelwood, and Tree Loppings per Household from Common and Private Lands

(Qtls.)

Category of Farming Households	Grass (Dry+Green)			Fuelwood			Tree Lopping		
	Common Land	Private Land	Total	Common Land	Private Land	Total	Common Land	Private Land	Total
Marginal	0.75	14.25	15.0	0.93	6.73	7.66	0.65	8.99	9.64
Small	1.34	13.32	14.66	0.87	8.29	9.16	0.07	10.00	10.07
Medium	0.93	21.56	22.49	1.06	10.76	11.82	1.42	16.59	18.01
Large	0.0	17.00	17.0	0.0	16.52	16.52	0.0	29.60	29.6
Average per household	0.89 (4.73)	17.92 (95.27)	18.81 (100.0)	0.94 (9.06)	9.43 (90.93)	10.37 (100.0)	0.93 (6.28)	13.89 (93.72)	14.82 (100.0)

Source: Field survey.

Table 6.32 shows that the common land was never a source of collection of fuelwood, grasses and tree lopping for large farmers. Even the poor farmers collected very little from it. The average collection of fuelwood per household per year was 10.37 quintals of which 9.06 per cent was collected from the common lands. Similarly, 4.73 per cent of the total requirement of grass and 6.28 per cent of tree lopping was collected from the common lands. Common land could have ensured adequate availability of fodder if it had been protected and allowed to grow by following a systematic conservation method as done in some other places that are under the present study.

As observed earlier, plantation on common lands failed to enhance floral diversity and consequently, an insignificant vegetation cover despite renovation of the village institution. No rules of management were put in place to regulate the behaviour of households for making best use of the common land. In the absence of such rules, scope for livestock development that, as seen earlier, witnessed no prospect after the intervention was made. Plantation of trees, whatever is the extent, could meet the requirements of the villagers from private lands and sand dunes. It was important to note that farmers, irrespective of their land holdings, followed the practice of planting trees on their own private land since they could raise only one crop in a year.

Although the gross annual household income from livestock varied significantly, it was not so in agriculture. Table 6.33 shows that the gross value generated by agriculture per household in Fateh Singh ka Bas was Rs. 8,548.27, which was slightly lower than the rest of the villages. It would have been still lower, had the farmers of this village not gone in for *sonamukhi* cultivation.[8] Earnings through the non-farm sector were made mainly through pensions since each village had ex-servicemen. Table 6.33 shows that the average gross value generated by livestock per household were more than that of agriculture in each village. On an average, it was Rs. 17,997 in the year under study, while that of agricultural produces was Rs. 10,025 per household. It shows how significant the contribution of livestock was to the household economy. One must see the relevance of microwatershed development to the improvement of livestock economy in this light.

Table 6.33

Average Gross Income per Sample Household from Different Sources

Village	*Agriculture*	*Livestock*	*Other Sources Including Non-farm*	*Total*
Bhoja ka Bas	9,147.32	11,375.07	38,597.93	59,120.32
Fateh Singh ka Bas	8,548.27	28,341.90	33,685.71	70,575.88
Tibna	10,946.14	17,991.41	50,802.41	79,739.96
Timri	10,301.43	15,736.36	37,807.27	63,845.06
Average per-household	10,025.03	17,997.08	43,606.39	71,628.50

Note: Gross income of Agricultural and livestock products of the households was estimated based on their respective local prices.

Source: Field survey.

8. In this village, around 40-50 households were engaged in *sonamukhi* cultivation in their own lands. At the initiative of PSBUSS, seeds of *sonamukhi* was spread on the common lands of this village and other two villages, namely Bhoja ka Bas and Tibna. It was a medicinal plant and had two varieties—palatable, growing naturally and economically insignificant; non-palatable for humans but a good source of livelihood generation. Its leaves were sold at the rate of Rs. 20 per kg, while seeds were sold at the rate of Rs. 80 per kg.

VII

Conclusion

One of the most insightful lessons that can be learnt is that no amount of increasing claims on the natural resource base of the village community in arid Rajasthan, even under drought conditions, can reduce the scope for livelihood generation if the intervention strategy is designed appropriately. If the design were taken seriously by the intervening agency, the technical and human aspects of intervention would never be at variance with each other. The combination of the two, if conceived properly, is sure to ensure sustainability of the improved resource base that is sought through external intervention. This was precisely where PSBUSS, as an intervening agency, failed to blend the two in an appropriate manner. While the intervention for river training and plantation works was inadequate, one had reason to question its technical soundness.

One might, then, doubt the sustainability of these works and their continued positive impact on the village economy. The absence of community institution in the real sense that initiated collective action in the first place might aggravate the situation further. In this case, the lack of space for the village community at large to participate in the intervention went against the human aspect of the initiative. This gave rise to a sectarian leadership that allowed development initiatives to originate outside rather than within the community. Discrimination perpetrated by the powerful against the disadvantaged and the uneven distribution of the development gains between the two was the direct outcome of the sectarian politics. This continued to happen despite improvement in the educational background of the village community.

These limitations, however, were not strong enough to overshadow the positive changes, at least for now, that the village economy gained from watershed development. The notable achievement was the availability of drinking water. Besides, the retention of the minimum stock of livestock and assured nutritional security for the household members even during three years of drought are not small achievements either. Not to talk about the prospect of irrigated farming that the villagers could not even dream of earlier. These were some triumphs that may not be of large magnitude, but are definitely sources of hope of change. They drive home the point that the material conditions of even drought prone, arid Rajasthan can be changed if the human (community institution) and technical aspects of intervention are blended in a manner that the disadvantaged were equally benefited from it.

7

Women's Gain

Introduction

The direct involvement and participation of women in the development process has always been sought in both policy and practice. In the case of natural resource management (NRM), it has even greater legitimacy because of the close association between women and nature. While eco-feminists have tried to trace their association in some mystical commonality between women and nature, both of which are subject to exploitation at the hands of patriarchy (Ortner, 1974; Shiva, 1989), environmental feminists have tried to do the same in material reality instead of resting on an ideological illusion (Agarwal, 1992). A mixed bag of oft-assumed answers is offered to questions like 'participation of women in natural resource management for what.'

Many consider it necessary to incorporate women's welfare oriented approaches to further women's interests and find solutions to the problems women face while carrying out activities associated with natural resources. However, others consider it essential to increase programme efficiency and sustainability. According to some others, issues of equity and empowerment will never be addressed if women fail to directly participate in NRM initiatives. All these and many other perspectives offer answers, but only in parts. However, several researches have contradicted these assertions and suggest that a deeper analysis is needed to answer these questions.

The present analysis is mainly confined to empirically verify how women have gained from the local institutional management of an improved natural resource base in all the case studies under consideration. The gains may be varied, ranging from participation in the decision making process to the achievement of gender equity and self-esteem. This chapter is divided into four parts. The first part of the chapter briefly overviews the existing theoretical approaches on women and NRM. This lends critical support to the broad framework within which the present analysis is carried out. Section II empirically verifies the extent of participation of women in the management of natural resources in all the case studies. While section III examines the various tangible and non-tangible gains that accrued to women, section IV concludes.

Approaches

Before looking at specific approaches that explore the intrinsic connection between women and NRM, a brief discussion on the general approaches to women and development seems to fall within the sphere of relevance of the present study. One of the basic assumptions on which the welfare approach (1950-1970s) rested was that women were passive recipients of development; they were never active participants in the development process (Moser, 1989; 1993). As a reaction to this approach, several alternative approaches, including that of equity, anti-poverty, efficiency and empowerment, were suggested in 1970s, clubbed together under the broad category of Women in Development (WID). Of these, the empowerment approach was considered central to women as most beneficial to them. Women's participation in the development process would facilitate incorporation of their needs, concerns, aspirations and perspectives in all sectors, at all levels of society and in all programme implementation processes. These were vital for development to sustain (Claridge *et al.*, 1999). During 1970s, a shift towards viewing empowerment of women within the Gender and Development (GAD) perspective became evident. It suggested that the problems of women must be addressed against the backdrop of social relations of power that existed between women and men. One must not look into their peripheral status, be it in the field of natural resource management, in isolation.

Environmental concerns and sustainability issues became central to the development policy during 1990s, following the publication of the *Brundlandt Report* in 1987. Emphasis was laid on local control, participation and management of natural resources with an equal emphasis on direct involvement of women in NRM. This strategic approach gained momentum with the opening up of eco-feminist discourses, which argued that there was a bond between women and nature with spiritual origins—both were victims of patriarchy and capitalism. In the words of Shiva, "Women and nature are intimately related, and their domination and liberation are similarly linked. The women's and ecology movements are, therefore, one, and are primarily counter-trends to a patriarchal maldevelopment" (Shiva, 1989).

However, the eco-feminist discourses were indifferent to the gender analysis in which men and women were relational, socially constructed, culturally specific and negotiated categories (Jackson, 1993). The mystic assumption of eco-feminism has led to many simplistic assumptions about women's role as 'natural managers'. The advocacy of women's participation in NRM initiatives was based on these assumptions only. However, as several studies have shown, it led to increased workload for women. While being critical of eco-feminism, environmental feminism provided a new perspective in that it traced the roots of the close association that existed between women and nature in material reality. Eco-feminism in the West, and especially in the United States, conceptualised the link between gender and environment primarily in ideological terms. However, in developing countries, the women's struggle for survival highlighted the material basis for this link and set the background for an alternative formulation to eco-feminism (Agarwal, 1992).

The material basis was a gender-based division of labour and distribution of property and power. In this formulation, the link between women and the environment was seen as structured by a given gender and organisation of production, reproduction and distribution. Ideological construction of gender and nature and their relationships were then seen as (interactive) part of this structure (Agarwal, 1992). Feminist environmentalism challenges both the ideological construction of gender and the notion of people's relationship with the environment. It suggests transformation of the actual division of work and distribution of resources between the genders (Sara, 1999). While dealing with the question of why women's participation in NRM initiatives was necessary, environmental feminists raised the issues of entitlement, efficiency and empowerment (Agarwal, 1997). All these had no place in eco-feminism.

There is no denying the fact that women have been burdened with tasks like collection or gathering of fuelwood, cutting of grass or fetching of water in most developing societies across the globe. While their engagement in these activities forms the basis of the assumption of close association between women and nature, it must, however, be seen as an outcome of gendered household division of labour. The increased participation of women in NRM initiatives without a shift in the household division of labour led to increase workloads for women. This, in turn, pushed gender equity further down. While this is not to negate the importance of women's participation, one may wonder how women could promote their participation without the shift, in the household division of labour. While household division of labour directly corresponds to gender inequity, one may argue that no breakthrough could be achieved in inequitable gender relations of power without a change in the household division of labour. The issue is, therefore, whether new institutional practices as followed for management of natural resources could bring about any such shift in the household division of labour and lead women to gain from it. If so, how does it accrue to them and whether it lends credible support to them to empower themselves, are some relevant questions that have been responded to in the following sections.

Women's Participation

In view of its root in material reality, the rationale sought to justify women's participation is that women's participation in NRM initiatives creates space for them to derive larger benefits. It is also argued that their active involvement is necessary for effective management of natural resources and ensuring sustainability of the initiatives taken. The third, and perhaps most powerful, argument is that participation empowers women. There are several ways empowerment issue can be examined. Participation in the decision making process is considered one of the significant routes through which women feel empowered.

However, this appeared to have been missing in all the areas under the present study. It may be recalled that the main decision making bodies were the executive committees that were formed in all village development organisations. Active participation of women in the decision making process opens up opportunities for them to articulate their own perceptions about resource management, in particular, and village development, in general. In other words, intervention must be designed in a manner that eventually widens the scope for households and women to benefit more than they could have done in the absence of their participation.

Table 7.1 shows that the number of women members in each village organisation was too low to be accounted for. In Chhota Saradhana and Singhariya, where pastureland was regenerated, there was no women's participation in planning and decision making activities. The field survey revealed that women were integrated as members of the executive committee only because it was a prerequisite for registering the village organisation under the Societies Act. It remained a formality since the women played no active role in the executive process. They did not attend the meetings regularly, and even when they did they always took a backseat and hesitated in articulating their concerns and wishes. Enquiries to the women members revealed that they did not even know when the meetings took place. However, interestingly, they were all—irrespective of their social background, especially in Chhota Saradhana—concerned about the sustained development of their village pastureland; concerned enough to keep a watch on violations of management rules and to inform the president about it.

Table 7.1

Women Members in the Executive Committee of the Village Institutions

Village	*Male*	*Female*	*Total*
1. Chhota Saradhana	10	1	11
2. Singhariya	9	2	11
3. Bhanwata-Koliyala	9	2	11
4. Samra	24	1	25
5. Bijanwas and Bikarni	8	1	9
6. Tibna/Timri	15	2	17
7. Bhoja ka Bas	17	0	17
8. Fateh Singh ka Bas	19	2	21
9. Dagal	10	3	13

Source: Field survey.

In Dagal, three women were members of the executive committee of the forest protection committee (FPC) as shown in Table 7.1. Along with this, a separate women's advisory committee was also formed. These women members were initially active and regularly participated in the meetings, volunteering to work for forest development in several ways. However, they could never become an integral part of the decision making process, partly also because of old social attitudes that conditioned their responses. They considered other members to be their elders and, hence, were hesitant to speak before them. It may be mentioned that the women's advisory council never became functional, it existed only on paper. Interestingly, the so-called president of the committee was herself found unaware of its existence.

Here again, women were found equally concerned about the growth and protection of the forest. However, there was no evidence to suggest that any effort was made to incorporate women's perceptions in the management of the improved resource base in any of the organisations mentioned above. What was true of these villages was true of Bhanwata-Kolyala, Samra, Bijanwas and villages of Tibna *panchayat*. Institutional renovation failed to create space in the social sphere for women to assert themselves by means of active participation in the administrative process. In other words, they were deprived of potential gains of empowerment through this route.

However, women's active participation was visible as wage labour in the various activities taken up since the beginning of the intervention for regeneration of the resource base. They were deployed by the village organisation in the same way as the male members of the community. But, then, this was largely a wage-earning activity with little or no community interest. This kind of involvement must not be construed as women's participation in management. Participation of women as wage labourers was sure to result in women working for the intervention, rather than vice versa as was indicated earlier by many studies (Blumberg, 1995; Elson, 1991; Kabeer, 1994; Moser, 1993).

Gains from Intervention

Women, however, gained from the intervention in several ways, both tangibly and intangibly. These gains began to accrue to them once their natural resource base witnessed an improvement with the institutional intervention. First, there was a marked reduction in the long distances being covered by women for collecting fuelwood, grass and tree loppings in almost all study areas; there was a noticeable difference in the case of Chhota Saradhana and Singhariya. The intervention here of reviving the pasturelands reduced the physical toil and drudgery which women were subjected to earlier apart from their vulnerability to other unforeseeable problems that endangered their lives. They collected fuelwood and grazed their animals in constant fear of being caught by the forest officials and fined. This used to demoralise them, but they had no recourse than to silently reconcile to it for the sake of survival. Post-intervention, they were relieved from this anxiety.

Almost similar trends were found to have set in other villages. It was, however, not necessarily true that the village community in each study area was able to meet its requirements fully from the resource base that it had improved. Much depended on the rate of regeneration and its compatibility with the growing requirement of the resources. Wherever there was a mismatch, women had to revert to the old practice of walking long distances for fodder and fuelwood. Of course, there were a few cases, as shown in the preceding chapters, which exemplified how discrimination was perpetrated against the powerless. In such cases too, women had to suffer.

Otherwise, the decline in the distance amounted to reduction of time consumed for collection. Besides, women now did not have to go as frequently with these resources now available in greater quantity in one place. More importantly, timings were fixed for their collection in that they could forage only in the stipulated number of days in a year. Such an institutional arrangement brought in a certain degree of change in the household division of labour in almost all villages under study except the Tibna *panchayat* villages. On the other hand, in some households, male and female members shared the burden of collection, and, further, they transported it on camel/bullock carts. These practices were absent earlier in that the engagement of male members was far less than what happened now.

Table 7.2 shows the extent of engagement of both male and female members of the sample households in collection of fuelwood, grass and tree lopping in each village. Engagement is shown here in terms of percentage of male and female members to the sample households in each village. In other words, it shows percentage of households to the total sample households that encouraged male members to join female members in collection, etc. For example, 26 per cent of the total sample households in Chotta Saradhana engaged male members in collection of fuelwood, 36 per cent in grass and 30 per cent in tree lopping. However, in the same village, the figures were 76 per cent, 66 per cent and 72 per cent, respectively, of the sample households that continued to engage female members to perform the same activity. While the rise in male engagement was a positive development, one might attribute it largely to introduction of new institutional practices. This was true especially in those villages where the institutions remained alive.

Table 7.2

Collection of Fuelwood, Grass and Tree Lopping by Male and Female Members of Sample Households in Each Village

(sample households in percentages)

	Fuelwood		*Grass*		*Tree lopping*	
Village	*Male*	*Female*	*Male*	*Female*	*Male*	*Female*
Chhota Saradhana	26.00	76.00	36.00	66.00	30.00	72.00
Singhariya	21.19	92.37	18.64	83.05	15.25	87.29
Bhanwata-Kolyala	22.64	81.13	5.66	83.02	9.43	62.26
Samra	9.59	87.67	13.70	73.97	21.92	50.68
Bijanwas	32.71	61.68	32.71	63.55	20.56	30.84
Bikarni	31.43	82.86	22.86	80.00	22.86	37.14
Dagal	14.10	85.90	17.95	83.33	11.54	60.26
Tibna	57.14	47.06	50.42	55.46	57.14	46.22
Total	28.75	74.57	26.86	72.04	25.43	56.40

Source: Field survey.

On the other hand, there were the villages of Tibna *panchayat*, presenting an altogether different picture. More than 50 per cent of sample households here engaged male members in foraging for fuelwood and fodder; the number of households engaging women was much less than other villages. But this was so because these villages had the largest number of Rajput households whose social norms discouraged women members from venturing out.

The evolving new practices put women in a more comfortable situation. They could now derive intangible benefits in several forms that contributed immensely to the enhancement of their well-being. First, the ensured availability of fuelwood and fodder reduced uncertainty and anxiety. Second, it spared some time for women for some productive work inside the house; they also had some time for rest that enhanced their reproductive efficiency. Third, they could now better look after their children, which they could not afford to do earlier. Fourth, the time earned allowed them to engage more in agricultural activities. Fifth, they could work as wage labour and earn subsistence living for the household as in the case of Singhariya. All these intangible benefits not only induced a shift, however small, in the gender division of labour, but also brought in a certain degree of freedom.

Conclusion

One may argue that the benefits that accrued to the womenfolk in different forms as discussed above confer higher degree of freedom to them. This is, of course, another form of empowerment of women in that they now acquired the ability to make choices, which they were denied earlier. These choices had a bearing on shaping their livelihood (Kabeer, 1999). It is true that participation in a public decision making forum, such as the village development committee, would have helped reverse their traditional exclusion. It might have increased their self-confidence in asserting their rights in relation to public bodies in general (Agarwal, 1997).

However, their limited and weak representation in the village organisation failed to open up the route to their empowerment. An improvement in the household entitlement to improved natural resources caused due to institutional renovation filled this gap to a considerable extent. The fact that collection of grass, fuelwood, etc. was no longer an activity of women alone, shows clearly how institutional renovation can bring about a change in the household division of labour in favour of women. It is in this context that one must not overlook how household entitlement of the resources, not individual women member of the household, in the institutional context causes breakthrough in broadening space for women to empower them.

Hence, within the framework of household entitlement, the expanded ability of women to make choices denied earlier on the one hand, and a shift in the household division of labour (or gender division of labour) on the other definitely improved their well-being. One must recognise the fact that the change in the household division of labour, consequent to institutional renovation, created favourable conditions for women to free their productive forces and realise their potential at least in those cases where institutional practices were followed for managing natural resources.

8

Summing Up

Central to the present enquiry is the exploration of the relevance and effectiveness of a communitarian institutional approach for sustainable management of natural resources in a feudal society like that of Rajasthan. While institutional development for resource management is closely verified within the power theoretic framework, the study encompasses several aspects of change as an offshoot of intervention. In its journey from case to case, the study searches for a conceivable strategy for equitable management of renewable natural resources in the public domain. It has followed proven methodologies for empirical testing. And, it makes several findings that have serious implication for livelihood crisis that rural Rajasthan is passing through. These findings are summed up as follows.

Power Relations, Social Processes and Institutional Sustainability

The study reveals that community institutional approach for the revival of common resource base gained social legitimacy and momentum when the latter had degraded to the point of no return. Although the history of the formation of social capital and its content varied across the case studies, the involvement of the community, despite social and economic differentiation, displayed social processes that exemplified a high degree of mutual trust, reciprocity and social networking. All these together created conditions for the emergence of new institutional practices for restoration of the degraded ecosystem and its development in a sustainable manner, be they pastureland regeneration, revival of degraded forests or development of farm forestry.

However, such favourable social processes around new institutional practices failed to sustain and evolve on their own in some other cases—the revival of traditional practices of rainwater harvesting and micro watershed development, for example. The new institutional practices began to collapse in these cases when the degraded natural resource base started gaining back its resilience. This happened, as the study revealed, due to strong influence of the traditional power structure. The traditional power structure stemmed from numerically strong caste background coupled with relative economic affluence. This was responsible for the reversal of social processes that initially lent credible support to the institutional approach for management of the natural resource base. Ultimately, the institutional prospect was damaged in such cases.

On the other hand, in the case of pastureland development, degraded forest and farm forestry, although the social processes varied from each other, they created favourable objective conditions for new institutional practices to sustain. For instance, the case of pastureland development in Chhota Saradhana shows that the alternative credit delivery system developed by the villagers themselves gave enough freedom even to the poor to assert and participate actively in the organisational matters of the village body. They no longer had to act according to the dictates of the more affluent. Perhaps, this was the fallout of the participatory strategy for all followed by the external agency from the initial stage of restoration programme.

The case of revival of traditional forest by the tribals in Dagal village shows how the socially and economically homogeneous base of the villagers crystallised their social capital for revival of their dying forest. But there is also the story of Bijanwas and Bikarni. These villages lacked such a homogeneous base. However, that did not deter them from developing alternative sources of grass production in the common revenue land by encroaching. It did not allow conflict to sharpen between the powerful and the powerless over sharing of grass in the protected land where the forest was raised. Conflict over control of the improved farm forest between the two villages does have the potential to jeopardise institutional prospects in the future. But, as the study reveals, such a conflict is located within the ambit of power relations of a different kind—the powerful of one village have never given scope to the other village in managing the improved resource base. It may not be always true that new institutional arrangement has to break down in the event of traditional power relations rearing up and wielding influence on the enforcement characteristics of the arrangement. The Singhariya pastureland restoration case shows otherwise. The powerful members of the village community never allowed the renovated institution to collapse, for they gained comparatively more from the restoration of the pastureland and would, therefore, have lost more if the new institutional arrangements had collapsed.

Hence, the study reveals that it was possible for the renovated institutions to sustain in the study areas distinctly under two conditions. The first condition exemplifies democratic approach. It ensured democratic participation of all, non-discriminatory enforcement characteristics of village institution and equitable distribution of collective benefits. And, the second one exemplifies an authoritarian approach that manifested in several forms in practice. It succeeded to ensure all that are mentioned above in some cases, while it failed in some other cases. But, institutions sustained in both cases. For instance, villagers of Dagal and Bijanwas-Bikarni continued to hold that the outside authority namely government and the 'company' (IFFDC), respectively owned the forest. It was a form of an authoritarian approach that was imposed from outside. It worked well in enforcing institutional rules and making distribution of the collective benefits equitable. In any case, the village institutions were shaped and made to sustain by the outside agencies, which was why active participation of all sections of the villagers was never sought. Singhariya presents another form of authoritarian approach that made the village institution to sustain but never allowed collective benefits to distribute equally. In this case, the traditionally powerful people within the village, not outside of it exercised authoritarian approach.

Can institutional sustainability be achieved through development of human capital? It is argued that educational development contributes to enrichment of social capital (Mayer, 2004). The latter, in turn, may ensure institutional sustainability. However, no such relationship was found in the villages under study. The improvement in social networking, mutual trust and sense of reciprocity was very much a part of the localised cultural context. The variation in the social capital seen over the years, the study shows, had no correspondence with the improvement in education. It is true that Putnam neglected the educational tradition that had fostered the civic community and contributed to better institutional performance in northern Italy (Putnam *et al.*, 1993). However, as the present study shows, the Rajasthan cases that experienced better institutional performance did not witness improvement of social capital because of human capital enrichment.

Entitlement to Improved Resource Base

The powerful largely determined access to the improved resource base as the scarcity conditions of resource availability stiffened. This is the learning from the case studies on pastureland development in Singhariya and micro-watershed development in the villages of Tibna *panchayat*. However, this did not happen in Chhota Saradhana and Dagal where new institutional practices were followed under more democratic social processes as mentioned earlier. In these cases, the poor and the disadvantaged gained equal access to the improved resource base.

The study on revival of traditional rainwater harvesting practices in upstream and downstream villages in the same river basin shows another dimension of unequal access to the recharge of groundwater that needs special attention. The lack of comprehensive and appropriate planning for rainwater harvesting led to unequal gains in recharge of groundwater for upstream at the expense of downstream. The study shows that the approach followed for management of land and water harvesting was essentially village-centric. It was never basin-centric. As a result, agricultural prospects were better upstream at the cost of downstream.

The study also shows that, despite increased recharge of groundwater, upstream gradually entered into the unsustainable zone because of the mismatch between the production and consumption of groundwater. The potential social hazards were invincible in such a situation, especially when the *gram sabha*, the local institution was practically found to be inoperative and riddled with a host of conflicts. One is not certain whether such a village institution can ever be revived when private initiative for rainwater harvesting purely for private benefit has gained overriding importance under the sponsorship of the external agency. In such circumstances, one can hardly rule out unequal access to the improved groundwater resources even upstream.

Ecological and Institutional Resilience

The study shows that achievements made in restoration of the degraded ecosystem through new institutional practices were incredible whether it was revival of pastureland, forest, desert or traditional rainwater harvesting practices. More so, because this was despite the persistent drought that the state witnessed consecutively for three years. The case study on pastureland revealed that there had been a steady improvement of floral diversity in its ecosystem accompanied by improvement in the physico-chemical characteristics of soil. The higher nitrogen content and organic carbon in soil as found in the protected lands can be cited as some examples in this regard. No significant difference was found between upstream and downstream in terms of diversity index. However, in terms of species richness index, upstream achieved more floral diversity as compared to downstream, which witnessed considerable decline in the number of individual species due to deforestation.

It was found that in respect of development of farm forestry and revival of degraded forest, contribution of the native species to the improvement of the local ecosystem, by means of expanding floral diversity, was comparatively more than that of the planted ones. For the survival rate of the planted species, the study shows, was lower than the native ones. Inappropriate selection of the planted species against the backdrop of edaphic characteristics and climatic conditions led to this. The villages of Tibna *panchayat* did not witness significant ecological improvement either due to inappropriate or limited selection of planted species. However, there was little improvement of soil fertility in the intervention site. Such an achievement indicated signs of restoration of the ecosystem. However, it has to be remembered that the restored ecosystem sustains as long as the institutional practices sustain.

For instance, sand dune stabilisation in Tibna *panchayat* was successful in a certain sense. It helped agriculture perform better by arresting shifting of sand dunes. Plantation of tree and grass species in dunes increased the level of groundwater that, in turn, led to a rise in the increase of irrigation and cultivation of commercial crops, although they were not of higher magnitudes. However, it was not certain whether it would sustain in the event of breakdown of their village institutions. Similarly, it is not clear the extent to which the revival of traditional rainwater harvesting practices could continue to retain improvement in floral diversity in the upstream village with the collapse of its *gram sabha*. Uninterrupted deforestation and negative impact of uneven distribution of groundwater have reduced the scope for ecological improvement through expansion of floral diversity in Samra, the downstream village. The decline of ecological resilience in this village is associated more with the loss of resilience of the village institution.

Gains for the Village Economy

There were several ways, direct and indirect, through which the village economy gained with the revival of the degraded local ecosystem. Although access to the improved resource base for some subordinate classes was an issue in some cases, the overall gains made by the village economy with ecological restoration must not be discounted. It largely reduced uncertainty of fodder availability. Livestock was one of the important areas in which the village economy gained in a variety of ways. The villagers could not only retain minimum stock of their animals, but also did not have to migrate for grazing their animals despite three years of consecutive drought. What was more important, it ensured nutritional security even to the poor households to the extent possible through production and consumption of milk. In Chhota Saradhana, which revived its village pasture, the villagers opened a dairy that remained active even during drought years.

It was noted, in most cases, that institutional intervention led to an increase in the stock of large animals and decline in the stock of small ruminants. It is important to know what kind of institutional intervention must be sought to encourage growth of both large and small ruminants in a drought prone state like Rajasthan. The economics of livestock production reveal that it is unsustainable in almost all study areas (except Bhanwata-Koliyala), if gross earnings from livestock are strictly compared with the costs of feed and fodder only. If cost of grazing is incorporated into the calculation, livestock economy will be proved further unsustainable.

Despite being unsustainable, purely from the point of economic calculus, livestock economy survives and continues to be an integral part of the household economy. It happens because grass and fodder are available to the village community almost 'free of monetary cost'. Had the villagers paid for these resources, one does not know whether livestock economy would have collapsed. Agriculture was also boosted up in some villages in terms of bringing more area under irrigation and enhancing crop productivity (Bhanwata-Koliyala, Tibna and Bhoja-ka Bas). However, one must be cognisant of the fact that agriculture in one part of the watershed may lose while the other part gains in case soil and water conservation management is not taken comprehensively in the entire water basin. Samra, the downstream village, shows this.

The institutional management of natural resources created employment opportunities for the villagers in the study areas. Villagers benefited through obtaining wage employment in various activities undertaken including land preparation, plantation, construction of rainwater-harvesting structure, maintenance, etc. While all these steps together generated employment opportunities on the one hand, the common resource base of the villages was regenerated. One must recognise these two inter-linked events while addressing the issue on employment generation.

Women's Gain

Institutional intervention brought about some differences in the household division of labour. The collection of fuelwood and fodder ceased to be women's chore alone. Since availability of both was ensured in one place and access was equitable in villages such as Chhota Saradhana, Dagal, Bijanwas and Bikarni, the women were in a more comfortable situation in that they could derive intangible benefits in several forms that contributed immensely to the enhancement of their well-being.

First, easy availability of fuelwood and grass reduced uncertainty and anxiety to a certain extent. Second, women could now utilise the time that they used to spend for collection for fuelwood and fodder in carrying out some productive activities inside the house. Or, they had some time for physical rest that enhanced their productive efficiency. Third, they could now better look after their children, which they could not afford to do earlier. Fourth, they could utilise their time in carrying out agricultural activity. Fifth, they could earn subsistence living for the household by engaging in wage labour. All these benefits together not only inducted a shift, however small it was, in the gender division of labour, but also brought in a certain degree of freedom. Within the framework of household entitlements, they enhanced the ability of women to make choices denied earlier, induced a shift in the household division of labour (or gender division of labour), and improved women's well-being. However, the single most important limitation of the institutional intervention was that women were yet to overcome the cultural hurdle that alienated them from greater participation in the village level organisation.

Institutional Performance and Capacity Building

Village institutions continued to remain active in the case of pastureland development (Chhota Saradhana and Singhariya), farm forestry development (Bijanwas and Bikarni) and restoration of the degraded forest (Dagal). However, their performance in terms of achieving higher level of outcome varied. The higher level of institutional outcomes was measured on a set of parameters. These were, for example, regularity in attendance of meetings, effective participation in the decision making process, equity in enforcement of rules of conservation, non-compliance (freeriding), etc.

The study revealed that Chhota Saradhana achieved the highest level of institutional performance as compared to others. The important aspect of institutional renovation that took place in Dagal for reviving the dying forest was the lowest institutional achievement. The study further revealed that it was possible for Chhota Saradhana to achieve highest level of institutional performance because all parameters, as mentioned above, played an equally important role. Most important of all was that the entire village community, irrespective of social, economic and gender differentiation, was equally concerned about the protected village pastures. Members from each household felt that they had a stake in its development and, therefore, had the right to assert their participation in the decision making process. No room was left for discrimination against any household member. However, in other cases, either there was the problem of unequal access due to discriminatory enforcement of the rules of management or problem of non-compliance or absence of effective participation of the villagers in the decision making process.

An inappropriate and inadequate intervention strategy followed by the external agency also contributed to poor institutional performance or breakdown of institution. For instance, the process through which the village level organisation came up under the auspices of joint forest management in Dagal was antithetical to greater involvement of the village community. The gradual alienation of the villagers from the organisation led to increase in the incidence of non-compliance of rules. A lack of involvement and participation of the villagers was also responsible for poor institutional performance in Bijanwas and Bikarni. The other problem with the intervention strategy was that it failed to take note of traditional power relations in the village and how they influenced development activities of the village community, which was economically and socially heterogeneous. This was also a reason why appropriate strategies for building up institutional capacity failed to evolve in some cases and resulted in collapse of new institutional arrangements (Tibna, Bhanwata-Koliyala).

Market and Community Institution

The successful cases of institutional development for pastureland, farm forestry and degraded forest reveal that the community institution is capable of creating a market mechanism on its own. In all these cases, the village institutions decided the rate at which grass, fuelwood and tree lopping were to be sold to the villagers. Whatever be the extent of benefits, such an institutional arrangement definitely provided cost advantage to the villagers. Besides, the availability of the resources in their own village commons had a tendency to depress the price of these resources in the open market. In other words, the open market prices would have been higher than what existed at present, had there been no improvement of the village commons and availability of the resources. The case of Bijanwas grass market shows clearly that the price of grass in the private market was marginally higher than administered price of the society. The availability of fuelwood and dry fodder in Chhota Saradhana and Dagal provide equally important insights into the interaction between the market and the community institution.

9

In Search of an Alternative Perspective

Ecology, Poverty and Local Institutions

Perhaps, it is not wrong to say that growing scarcity of renewable natural resources besieges a state like Rajasthan. It is an obvious outcome of cognisable mismatch between the increasing rate of consumption requirement and the regeneration of resources. It is not true that no automatic adjustment between the two is taking place. It takes place without making any noise, but not without the cost of human suffering. Political activism in the state is yet to recognise the adverse future implications for human survival of this adjustment. It continues to remain peripheral to the development paradigm being followed by successive political regimes. However, civil society cannot remain quite. How can it do so when its survival is at stake? The people collectively take action wherever they can, to revive their pastureland, forest, groundwater aquifers, etc. They endeavour to grow and manage their resources in as equitable and sustainable manner as they can by switching over to new institutional practices.

The present study, an interdisciplinary one, is an enquiry into these initiatives taken by the village community in five selected areas spread over all the agro-climatic zones. Apart from tracing the process through which local institutions were renovated, it goes into detailed investigation of how traditional power relations shape institutional moorings. The crux of the problem is how equitable institutional outcomes are or who gains and who loses. It all depends on whether the new institutional practices are followed outside the sphere of traditional power relations or not. If they are, outcomes will be equitable as long as enforcement of institutional practices is not discriminatory against the disadvantaged. It is here that one must underline the essence of equitable institutional outcomc. It may not be an oversimplification to say that equitable institutional outcome is central to the sustainability of the variety of institutions that the present study addresses. What is, therefore, needed is to accord it an explicit recognition in any discourse on institutions as an important condition to achieve institutional sustainability.

A serious limitation of interventionists is that they tend to overlook it at the time they engage themselves in designing their interventions for institutional renovation. Consequently, they end up with a frustrating experience. While the appropriateness of communitarian approach to natural resource management may not have parallel in states like Rajasthan, one must look for space for operationalising it with the given social and economic heterogeneity. It is certainly not difficult for the interventionists to discern it because the state is increasingly becoming vulnerable to natural calamities like droughts. This creates a homogeneous base for a socially and economically heterogeneous population (Jodha, 2001).

This may be the reason why institutional interventions in some cases could succeed in restoring a degraded ecosystem by expanding biodiversity. And, by doing so, they could ensure their livelihood in a drought that prevailed for more than three years. Besides preventing livestock migration, the intervention did not allow their stock of animals to fall below the level required for ensuring nutritional security and subsistence living. At the same time, without trivialising these achievements, one must learn lessons from them that bring the missing links between poverty reduction and ecological restoration to the centre of the development discourse. In rural Rajasthan, incidence of impoverishment has much to do with the degree of vulnerability of the people to the growing crisis of renewable natural resources. The latter has paralytic impact on the local economy that can hardly be explained by the standard tools of economic analysis.

What seems to be increasingly becoming relevant is to ask who or which village has more command over a rich stock of renewable resources, be it drinking water, irrigation water, land with good forest cover or pastureland with better yield of grass, etc. Hence, within the political economy framework, there is now room for the debate to shift on factors leading to conflict for sharing renewable natural resources. It revolves around skewed distribution of command over rich stock of renewable resources, not necessarily on command over resources only. Conflict is unavoidable due to the increasing gap between the low regeneration of resources and the demand (mainly due to a burgeoning population and per capita consumption). While it is imperative to resolve such conflicts, it is worthwhile to do so within the emerging perspective of the political economy.

Conflicts are already in sight within different layers of the rural community. Potential conflicts are there for the asking between two groups of population spread over several villages within the same river basin or canal (upstream and downstream villages), socially powerful and powerless within the same village or between the rural and urban populations. Resolution of these conflicts must be sought in the public domain, definitely not in the private domain. Sen's argument on value formation and public discussion, as emphasised by Buchanan in the role of development preferences, finds relevance to this context as an approach to seek conflict resolution. For they might influence individual behaviour and help in bringing about policy changes through political processes (Sen, 1995).

One might look at value formation as an expression of capacity-building of the community. However, one must not misconstrue capacity-building to mean that the community is now capable of responding to market signals. In the public domain, where the entire community values natural resources and their management uniformly, it must use market forces to its advantage by forming rules for the market as it makes its own rules for managing natural resources in a sustainable manner. The market practices that the villagers evolved themselves for cultivation of green fodder and its sale in Chhota Saradhana, as shown in chapter II, exemplifies it. Hence, if pricing is to be done for achieving equitable distribution of resources, it has to be an integral part of the process of value formation at the community level.

It is here that one may have to emphasise on the direct participation of the poor, without which resource conservation remains far from being sustainable (Vyas, 1991). Hence, value formation that must provide strong legitimacy to institutional sustainability may be based not necessarily on mutual vulnerability or dependencies like 'tail Enders and head Enders' as shown by Ostrom and Gardner in the context of irrigation (Ostrom and Gardner, 1993; Bardhan, 1993). For, the logic of mutual dependencies, if extended in a context like the present one, may hide power relations that yield social and economic inequality. Hence, one must take cognisance of asymmetric participants against the backdrop of the power relations and, accordingly, strategy for cooperation must evolve. Ensuring the participation of the poor, as mentioned above, and assuring them of some tangible benefits could be some underlying values (Vyas, 1991), based on which strategy may be formulated.

There is no discernible alternative to value formation at the community level. It is estimated that presently rural Rajasthan draws sustenance from its natural resource base to a phenomenal extent. The value of the quantity of fuelwood and grass that rural households collect from their resource base 'free of monetary cost' is estimated to be equivalent to be around Rs. 7,900 crore annually (Annexure 4). This is almost equal to the annual budget of the state. If tree lopping and grass being regularly grazed by animals are also calculated, the estimated value will be much more. One must bear in mind that only a small part of the state's resource base has so far been brought under a restoration programme by either government, or civil society organisations, or both.

It should not be overlooked, however, that the intensification of the water resource crisis in the state is gradually leading villagers to voice their concerns. It is now transformed into agitational politics as surfaced recently in the state. The underlying factor in the massive protests staged by villagers at several places in the state, including the capital, has direct and indirect bearing on the scarcity conditions of natural resources Rajasthan is passing through. State cannot shy away from taking initiative in the management of the natural resource base of the state. It must look for strengthening its vertical integration with the civil society organisations, if necessary, and local institutions of the village community. It is true that self-governing common-pool resource communities have ability to perform at high level of equity and efficiency (Ostrom, 2000). However, it is also true that they failed to demonstrate the same in many other cases. A diverse experience of community management of natural resources such as these do not necessarily suggest that state must come forward wherever community fails. What is important for one to seek is an appropriate management strategy of common-pool natural resources whose property rights are yet to be defined clearly. State must have a role to assure that the benefits from the local resource base are not expropriated by the powerful (Dasgupta, 2001a). On the other hand, market must be looked upon as a force to complement the process through which village community seeks to achieve its sustainable management, rather than direct it.

References

Agrawal, A. (2001). "Common Property Institutions and Sustainable Governance of Resources", *World Development*, Vol. 29(10).

Agarwal, Bina (1992). "The Gender and Environment Debate: Lessons from India", *Feminist Studies* Vol. 1(18).

———. (1997). "Environmental Action, Gender Equity and Women's Participation", *Development and Change* Vol. 28. Institute of Social Studies. Oxford, UK: Blackwell Publishers.

Agarwal, R.R., J.P.S. Yadav and R.N. Gupta (1982). *Saline and Alkali Soil of India*. New Delhi: Indian Council of Agricultural Research.

Ahluwalia, M. (1997). "Representing Communities: The Case of a Community-based Watershed Management Project in Rajasthan, India", in *Community Based Sustainable Development*: Consensus or Conflict, *IDS (Sussex) Bulletin*, Vol. 28(4): 23-24, October.

Ambasht, R.S. and A.N. Maurya (1970a). "Reproductive Capacity of Dichanthium annulatum in Relation to Biotic Factors", *Tropical Ecology*, Vol. 10: 186-93.

———. (1970b). "Effect of Certain Biotic Factors on Vegetation Reproductive Capacity of Dichanthium annulatum", *Science and Culture*, Vol. 36: 354-56.

Appadurai, Arjun (2004). "The Capacity to Aspire: Culture and the Terms of Recognition", in V. Rao and M. Walton (eds.), *Culture and Public Action: A Cross Disciplinary Dialogue on Development Policy* – Palo Alto. California: Stanford University Press. (from Mausuri and Rao, 2004.)

Ashman, M.R. and G. Puri (2002). *Essential Soil Science*. Blackwell Publishing Company. India: Replika Press Pvt. Ltd.

Bebbington, A., Scott Guggenheim, E. Olson and M. Woolcock (2004). "Exploring Social Capital Debates and the World Bank", *The Journal of Development Studies*, Vol. 40(5): 33-64.

Baland, J.M. and J.P. Platteau (1998). "Wealth Inequality and Efficiency in the Commons, Part II: The Regulated Case", *Oxford Economic Papers* 50, pp.1-22.

———. (1997). "Wealth Inequality and Efficiency in the Commons, Part I: The Unregulated Case", *Oxford Economic Papers* 49, pp.451-82.

———. (1996). *Halting Degradation of Natural Resources: Is there a Role for Rural Community*. Oxford: FAO and Clarendon Press.

Ballabh, Vishwa and Balooni Kulbhushan (2002). "Why Local Resources Management Institution Decline? A Comparative Analysis of *Van* (Forest) *Panchayats* and Forest Protection Committees in India", *World Development*, Vol. 30(12).

Bardhan, P. (1999). "Rational Fools and Cooperation in a Poor Hydraulic Economy" in K. Basu, P. Patnaik and K. Suzumum (eds.), *Choice, Welfare and Development–A Festschrift in Honour of A.K. Sen*. Oxford.

———. (1993). "Symposium of Management of Local Commons", *Journal of Economic Perspective*, Vol. 7(4): 87-92.

Barun, J.V. and T. Feldbrugge (1988). "Institutional Aspects of the Handling of Crisis and Disaster in Developing Countries", *Economics*, Vol. 57.

Bhattacharya, Dwaipayan, N.G. Joyal, Bishnu N. Mahapatra and Sudha Pai (ed.) (2004). *Interrogating Social Capital, The Indian Experience*. New Delhi: Sage Publications.

Bhimaya, C.P., R.N. Kaul and L.D. Ahuja (1967). "Forage Production and Utilization in Arid and Semi Arid Rangelands of Rajasthan", in Proceedings of Joint Symposium of National Institute of Science, India and ICAR on *India's Food Problem*. New Delhi. pp.215-223.

Bhise, S.N. (ed.) (2004). *Decolonising the Commons*. New Delhi: National Foundation for India and Udaipur: Seva Mandir.

Black, C.A. (1965). *Methods of Soil Analysis*. Vol. 2. Madison: American Society of Agronomy.

———. (1968). *Soil Plant Relationships*. INC London: John Willy & Sons.

Blumberg, Rae Lesser (1995). "Introduction: Engendering Wealth and Well-being in an Era of Economic Transformation", in Rae Lesser Bhimblrg, Cathy Rakonoski, Irene Tinker and Michael Monteon (eds.), *Engendering Wealth and Well-being in Empowerment for Global Change*. Westview: Bonlder.

Bourdieu, Pierre (1983). "The Forms of Capital" in J.G. Richardson (ed.), *Handbook of Theory and Research for the Sociology of Education*. New York: Greenwood Press. (from M.C. Carroll and J.R. Stanfield (2003). "Social Capital, Karl Polanyi, and American Social and Institutional Economics", *Journal of Economic Issues*, Vol. 37(2).)

Bradbury, M., S. Fisher and C. Lane (1995). *Working with Pastoralist NGOs and Land Conflict in Tanzania*. London: IIED.

Bromley, D.W. (ed.) (1998). *The Handbook of Environmental Economics*. Oxford: Blackwell Publishers Ltd.

Cameron, John (with Ramtarak Hans and Cole. Ken) (ed.) (1995). *Poverty and Power: The Role of Institutions and the Market in Development*. New Delhi: Oxford University Press.

Casson, M. and N. Wadeson (1998). "Communication Costs and the Boundaries of the Firm", *International Journal of the Economics of Business*, Vol. 5(1). (from Dolfsma and Dannreuther, 2003.)

Chakraborty, Rabindra Nath (2001). "Stability and Outcomes of Common Property Institutions in Forestry: Evidence From the Terai Region of Nepal", *Ecological Economics*, Vol. 36(2): 341-53, February.

Chapman, H.D. and P.F. Pratt (1961). *Methods of Analysis for Soils, Plants and Waters*. California: University of California.

Chapman, S.B. (1976). *Methods in Plant Ecology*. Oxford: Blackwell Publishers Ltd.

Chopra, K. and G. Kadekodi (1999). *Operationalising Sustainable Development Economic-Ecological Modelling for Developing Countries*. New Delhi: Sage Publication.

Chopra, Kanchan (2002). "Social Capital and Development Process, Role of Formal and Informal Institutions", *Economic and Political Weekly*, July 13.

———. (2005). "Economics, Ecology and the Environment, A Developing Country Perspective" in Rao, Hanumatha *et al.* (eds.) *Indian Economy and Society in the Era of Globalisation and Liberalisation*. New Delhi: Academic Foundation.

Claridge, C.L., B.R. Frank and J.J. Mott (1999). "The Environment, Development and Women: What is the Connection?" *Journal of Rural Development*. Hyderabad: NIRD.

Coleman, James S. (1988). "Social Capital and the Creation of Human Capital", *American Journal of Sociology*, Vol. 94.

Costanza, R., B.S. Low , E. Ostrum and J. Wilson (2001). "Eco-system and Human Systems: A Framework for Exploring the Linkage" in R. Costanza, B.S. Low, E. Ostrom, J. Wilson (eds.), *Institutions, Eco-systems and Sustainability*. USA: Lewis Publishers.

Curtis, J.T. (1959). *Elements of Ecology*. New York: John Wiley and Sons.

Damodaran, A. (1991). "Tragedy of the Commons and Comedy of Common Property Resources", *Economic and Political Weekly*, Vol. 38, September 21.

Dasgupta, M., H. Groundvoinnet and M. Romani (2004). "State Community Synergies in Community-Driven Development", *The Journal of Development Studies*, Vol. 40(3): 27-58.

Dasgupta, Partha (2001a). *Human well being and the Natural Environment*. Oxford University Press.

———. (2001b). "Poverty and Environmental Resource Base" in Sankar Ulaganathan (ed.), *Environmental Economics*. New Delhi: Oxford University Press.

Dayton-Johnson, Jeff and Pranab Bardhan (2002). "Inequality and Conservation on the Local Commons—A Theoretical Exercise", *Economic Journal*, Vol. 112: 481.

Dhabriya, S.S. (1988). *Desert Spread and Desertfication, Environmentalist*. Jaipur: Modern Printex.

Dia, Mamadou (1996). *Africa's Management in the 1990s and Beyond, Reconciling Indigenous and Transplanted Institutions*, The World Bank, Washington, D.C.

Dixit, M. *et al.*, (2001). *Reappraisal of Groundwater Resource of Alwar District*. Ground Water Department, Government of Rajasthan.

Dolfsma, Wilfred and Charlie Dannreuther (2003). "Subjects and Boundaries: Contesting Social Capital-Based Policies", *Journal of Economic Issues*, Vol. 37(2), June.

Draper, N.R., H. Smith (1966). *Applied Regression Analysis*. John Wiley & Sons. Inc.

Durston, John (2001). "Social Capital-part of the problem, part of the solution—its role in the persistence and overcoming of poverty in Latin America and the Caribbean", Conference on *Social Capital and poverty reduction in Latin America and the Caribbean: Towards a new paradigm*, Santiago, Chile, 24—26, September, 2001.

Easter, K.W. and K. Palanisami (1986). "Tank Irrigation in India and Thailand: An Example of Common Property Resources Management", University of Minnesota, Department of Agricultural and Applied *Economics Staff Papers*, August (from Bardhan, 1999).

Ekins, P., S. Simon, L. Deutsch, Carl Folka and R.D. Groot (2003). "A Framework for the Practical Application of the Concepts of Critical Natural Capital and Strong Sustainability", *Ecological Economics*, Vol. 44.

Elias, N. and J.L. Scotson (1965). *The Established and the Outsiders: A Sociological Inquiry into Community Problems*, London, and Lass (from Dolfsma and Dannreuther, 2003.)

Elson, D. (1991). "Structural Adjustment: Its Effects on Women", in Tina Wallace with Candida March (eds.), *Changing Perceptions: Writings on Gender and Development*. Oxford: Oxfam.

English, Nature (1994). *Sustainability in Practice Issue 1: Planning for Environmental Sustainability*. Peterborough, UK. (from Groot *et al.*, 2003.)

Espeland, W.N. (2001). "Value Matters", *Economic and Political Weekly*, Vol. 36(21), May 26.

Falk, P. (1994). *The Consuming Body*, Sage, London (from Dolfsma and Dannreuther, 2003.)

Fine, Ben (2001). *Social Capital Versus Social Theory: Political Economy and Social Science at the Turn of the Millennium*. London: Routledge.

Government of India (1991). *Population Census*.

Government of Rajasthan (2000). *Statistical Abstract*. Jaipur: Directorate of Economics and Statistics.

Groot, R.D., Johan Perk, A. Chiesura, A. Vliet (2003). "Importance and Threat as Determining Factors for Criticality of Nature Capital", *Ecological Economics*, Vol. 44.

Galbraith, J.K. (1983). *The Anatomy of Power*. Boston, USA: Houghton Mifflin Company.

Hanna, S. and M. Munasinghe (1995). "Property Rights in Social and Ecological Context". Stockholm: Beijer Institute and Washington DC: World Bank (from Munasinghe and Reid, 2005).

Harada J. and Nario Yasuda (2004). "Conservation and Improvement of the Environment in Dam Reservoirs", *Water Resource Development*, Vol. 20(1): 77-96.

Hardin, Russell (1982). *Collective Action*. Baltimare and Md: Johns Hopkins University Press.

Harriss, John (2001). *Depoliticizing Development: The World Bank and Social Capital.* p.6. New Delhi: Leftward Books.

Hayek, Friedrich (1973). *Law, Legislation and Liberty.* Chicago: University of Chicago Press. (from Redmond, 2004.)

Hazra, C.R. (1995). "Soil and Water Conservation Aspects of Range Grasses and Legumes", *All India Co-ordinated Research Project on Forage Crops.* Jhansi: Indian Grassland and Fodder Research Institute.

Hazra, C.R. and D.P. Singh (1992). "A Silvi-pastoral Approach for Revegetation of Barren Hills and Waste Lands—An Experience at Gaharawa Watershed in Bundelkhand", J. Soil Wat. Conserv. Vol. 36(3&4) India, pp.105-144.

Hazra, C.R., D.P. Singh and R.N. Kaul (1994). *Greening of Common Lands in Jhansi through Village Resource Development—A Case Study.* New Delhi: National Afforestation and Eco-Development Board.

Heller, P. (1996). "Social Capital as a Product of Class Mobilisation and State Intervention: Industrial Workers in Kerala, India", *World Development,* Vol. 24(6): 1055-72.

Hesse, P.R. (1970). *A Textbook of Soil Chemical Analysis.* London and Colchester: William Clowes and Sons Limited.

Holling, C.S. and G.K. Meffe (1999). "Command and Control and the Pathology of Natural Resource Management", in Stanley, I. Dodson *et al.* (eds.), *Reading in Ecology.* Oxford University Press.

Holling, C.S. (1973). "Resilience and Stability of Ecological Systems", *Annual Review of Ecological Systems* Vol. 4: 1-24.

Ikeda, Keiko (1995). "Gender Differences in Human Loss and Vulnerability in Natural Disasters: A Case Study from Bangladesh", *Indian Journal of Gender Studies,* Vol. 2(2).

India Farm Forestry Development Co-operative Ltd. (IFFDC) (2001-02). *Annual Report.* Udaipur.

Jackson, Cecile (1993). "Environmentalism's and Gender Interests in the Third World", *Development and Change,* Vol. 24: 649-77.

Jackson, M. (1958). *Soil Chemical Analysis.* Oxford: Prentice Hall Inc. and Co..

Jairath, Jasveen (1999). "Participatory Irrigation Management: Experiments in Andhra Pradesh", *Economic and Political Weekly,* October 2 (from Menon, Ajit *et al., op. cit*).

Jayaraman, T.K. (1981). "Farmers' Organisations in Surface Irrigation Projects: Two Empirical Studies from Gujrat", *Economic and Political Weekly,* September 26.

Jodha, N.S. (1992). "Rural Common Property Resource: A Missing Dimension of Development Strategies", *World Bank Discussion Paper* No. 169. Washington DC: World Bank.

———. (2001). *Life on the Edge, Sustaining Agriculture and Community Resources in Fragile Environments.* New Delhi: Oxford University Press.

Kabeer, N. (1994). *Reversed Realities: Gender Hierarchies in Development Thought.* London: Verso.

———. (1999). "Resources, Agency, Achievements: Reflections on the Measurement of Women's Empowerment", *Development and Change,* Vol. 30(3): 435-64.

Kelman, R. and J. Kelman (2002). "Water Allocation for Economic Production in a Semi-arid Region", *Water Resource Development,* Vol. 18(3): 391-407.

Khan, Ahmed (1999). "A Far Cry? Revisiting the Prospects of Community Co-operation in Rural South Asia", *The Indian Journal of Social Work,* Vol. 60(2): pp.233-49. April.

Krishna, Anirudh (2003). *Active Social Capital Tracing the Roots of Development and Democracy.* New Delhi: Oxford University Press.

Leach Melissa, Robin Mearns, Robin Scoones (1997a). "Institutions, Consensus and Conflict, Implications for Policy and Practice", *IDS Bulletin,* Vol. 28(4), October.

Leach, Melissa, Robin Mearns and Ian Scoons (1997b). "Challenges to Community Based Sustainable Development—Dynamics, Entitlement and Institutions", *Institute of Development Studies Bulletin,* Vol. 28(4). Sussex, U.K.

Leach, Melissa, Merns Robins and Scoones Ian (1997c). "Community-based Sustainable Development: Consensus or Conflict?", *IDS Bulletin,* Vol. 28(4), October.

Leff, Enrique (1993). "Marxism and the Environmental Question: From the Critical Theory of Production to an Environmental Rationality for Sustainable Development", in *Capitalism, Nature and Socialism, A Journal of Socialist Ecology,* Vol. 4(1): 44-66. USA.

Mahmud, Simeen (2003). "Actually How Empowering is Microcredit?", *Development and Change* Vol. 34(4). Institute of Social Studies. Oxford, UK: Blackwell Publishing. pp.577-605.

Mansuri, Ghazala, and Vijayendra Rao (2004). "Community-based and –driven Development: A Critical Review", *The World Bank Research Observer,* Vol. 19(1): 1-39.

Mayer, P. (2004). "Making Democracy Perform: Human Development and Civic Community in India", in Dwaipayan Bhattacharya *et al.* (eds.), *Interrogating Social Capital.* Sage India.

Menard, S. (1995). *Applied Logistic Regression Analysis (Quantitative Applications in the Social Science).* Sage Publications Series 7 (106).

Menon, A, P. Singh, Esta Shah, S. Lele, S. Paranjape and K. Joy (2005). *Understanding Community-Based Natural Resource Management in South Asia.* Bangalore: CISED, ISEC Campus.

Ministry of Construction (1999). Wetlands and Nature at the Gosho Dam (Morioka: Kitakami River Integrated Dam Management Office, Tohoky Regional Bureau) in Japanese.

Mohapatra, B.N. (2001). "Social Connectedness and Fragility of Social Capital: View from an Orissa Village", *Economic and Political Weekly,* Vol. 36(8): 665-72, February 24.

Mohnot, S.M. (2003). *The Maru Gochar Yojana, Proposal.* Jodhpur: The School of Desert Sciences.

Moser, C. (1989). "Gender planning in the Third World: Meeting Practical and Strategic Gender Needs", *World Development,* Vol. 17(1): 1799-1825.

———. (1993). *Gender planning and Development: Theory, Practice and Training.* London: Routledge.

Mosse, D. (1997). "The Symbolic Making of a Common Property Resource: History, Ecology and Locality in a Tank-Irrigated landscape in South India", *Development and Change,* Vol. 28: 467-504.

Mukherjee, Neela (2002). "Measuring Social capital, Forest Protection Committees in West Bengal", *Economic and Political Weekly,* Vol. 37(29), July 20,.

Munasinghe, M. and W. Reid (2005). "The Role of Eco-systems in Sustainable Development" in Sengupta, N. *et al.* (ed.), *Bio-diversity and Quality of Life.* New Delhi: Macmillan India Ltd and Indian Society for Ecological Economics.

Murthy, H.N., Rambabu P. Joshi and B. Singh (1990). "Effect of Clear Filling and Reforestation on Run off and Peak Rates in Small Watersheds", *Indian Forester* Vol. 102(4): 219-26.

Murthy, M.S. and R. Ravindra (1975). "Allelopathic Effects of Aristida Adscensionis", *Oecologia,* Vol. 18: 243-49.

Myers, Norman and Julian. L. Simon (1994). *Scarcity or Abundance? A Debate on the Environment.* W.W. Norton and the Economy, New York (from Mc Neely, Jeffrey A. "Conservation and the Future: Trends and Options Towards the Year 2025", in Sengupta, N. *et al. op. cit.*)

Nadkarni, M.V. (2002). "Commons, Development and the Poor", in D.K. Marothia (ed.), *Institutionalising Common Pool Resources.* New Delhi: Concept Publishing Company.

Norgaard, R.B. (1984). "Co-evolutionary Agricultural Development", *Economic Development and Cultural Change,* Vol. 32(3): 525-46. (from Perrings, 2005.)

Nozick, Robert (1993). *The Nature of Rationality.* Princeton University Press (from Redmond, 2004.)

Odum, E.P. (1971). *Fundamentals of Ecology.* 3rd Edition. Philadelphia: W.B. Saunders and Co.

Oakerson, R.J. (1986). "A Model for the Analysis of Common Property Problems" in NRS (From Singh, Katar, 1994a).

Ortner, S. (1974). "Is Female to Male as Nature is to Culture?", in Z. Rasaldo and L. Lamphere (eds.) *Women, Culture and Society.* California: Stanford University Press. pp.67-88.

Ostrom, E. (1990). *Governing the Commons: The Evolution of Institutions for Collective Actions.* New York: Cambridge University Press.

———. (2000). "Collective Action and the Evolution of Social Norms," *Journal of Economic Perspective,* Vol. 14(13).

Ostrom, E. and Gardner, Roy (1993). "Coping with Asymmetries in the Commons: Self-Governing Irrigation Systems can Work", *Journal of Economic Perspective,* Vol. 7(4): 93-112.

Padre, S. (2002). *Rainwater Harvesting.* Thrissur: Altermedia.

Pangare, Ganesh, T. Jamal and R. Hooja (2002). *Community Based Rainwater Harvesting-process Documentation of the Activities and Approach of TBS.* Indian Network in Participatory Irrigation Management. (Unpublished.)

Parpart, Jane, L. (1993). "Who is the Other?: A Postmodern Feminist Critique of Women and Development Theory and Practice", *Development and Change,* Vol. 24(4). London, Newbury Park and New Delhi: Sage.

Patel, Jashbhai (1997). *Story of a Rivulet Arvari: From Death to Rebirth.* Alwar, Rajasthan: Tarun Bharat Sangh.

Perrings, Charles (2005). "The Economics of the Environment: Lessons from Ecology", in Sengupta, N. *et al.* (eds.), *Bio-diversity and Quality of Life.* pp.439-464. New Delhi: Macmillan India Ltd. and Indian Society for Ecological Economics.

Platteau, J.P. and Anita Abraham (2002). "Participatory Development in the Presence of Endogenous Community Imperfechious", *The Journal of Development Studies,* Vol. 39(2): 104-36, December.

Putnam, Robert D., Leonard Robert and Y. Nanetti Raffaella (1993). *Making Democracy Work: Civic Traditions in Modern Italy.* Princeton: Princeton University Press. (from Krishna, 2003.)

Qiang, Zhu (2003). "Rainwater Harvesting and Poverty Alleviation: A Case Study in Gausu, China", in *Water Resource Development,* Vol. 19(4), December.

Rathore, Jai Singh (2004). "Drought and Household Coping Strategies: A Case of Rajasthan", *Indian Journal of Agricultural Economics,* Vol. 59(4).

Ray, Sunil (1997). *Natural Resources, Organization and Technology Linkages.* Jaipur: Rawat Publications.

———. (1999). "Declining Production Conditions of Raw W001, Analysis of Emerging Conflicts in Sheep Husbandry in Rajasthan" *Economic and Political Weekly,* Vol XXXIV, May 15.

———. (2003a). "Institutionalism and Development of Capabilities of the Poor: Some Issues" in Surjeet Singh and Varsha Joshi (eds.), *Institutions and Social Change.* Jaipur: Rawat Publications.

———. (2003b). "Self-goal Choice *Vs* Institutional Choice, A Case of Institutional Renovation for Pastureland Development in a Rajasthan's Village", *IDSJ Working Paper* 130. Jaipur: Institute of Development Studies.

———. (2004). "Dairy Industry in Rajasthan, Problems and Prospects", *Monograph.* Jaipur: Institute of Development Studies.

———. (2007). "Sustainable Livestock Development in Rajasthan–Some Issues", in Vyas, V.S. *et al., Rajasthan, the Quest for Sustainable Development.* New Delhi: Academic Foundation.

Ray, Sunil and Jagdish Sharma (2004). "Social Capital, Local Institutions and Entitlement: A Case of Pastureland Development in Singhariya Village", *IDSJ Working Paper* 141. Jaipur: Institute of Development Studies.

Ray, Sunil and Alka Upadhyay (2004). "Restoration of Degraded Pastureland Ecosystem in Semi-Arid Rajasthan—A Study of Two Villages", *IDSJ Working Paper* No. 137. Jaipur: Institute of Development Studies.

Redmond, William H. (2004). "On Institutional Rationality", *Journal of Economic Issues*, Vol. 38(1), March.

Samantray, R. (1998). "Johad: Watershed in Alwar District, Rajasthan", *UN Inter Agency Working Group on Water and Environment Sanitation*. New Delhi. (From Shresth and Devidas, 2001.)

Samuels, Warren J. (1990). "Institutional Economics and the Theory of Cognition", *Cambridge Journal of Economics*. (Review of Economics of Institutions: A Manifesto for a Modern Institutionalist Economics by Geoffrey Hodgnon).

Sara, A. (1999). *Workshop report on Mainstreaming gender in Natural Resource Management*. Anand: Institute of Rural Management.

Saxena, N.C., Madhu Sarin, R.V. Shah and Tushaar Shah (1997). "Independent Study of Implementation Experience in Kanara Cirde", *Review Committee Report*. Bangalore: Karnataka Forest Department. (From Menon *et al.*, 2005.)

Sen, Amartya (1987). *Ethics and Economics*. Oxford: Basil Blackwell.

———. (1995). "Rationality and Social Choice", *The American Review*, Vol. 85(1), March

Sengupta, Nirmal (2001). *A New Institutional Theory of Production, An Application*. New Delhi: Sage Publications.

Sengupta, Ramprasad (2002). "Human Well-being and Sustainable Development", Review article of the book on Human well-being and natural environment by Parth Dasgupta (2001), *Economic and Political Weekly*, Vol. 37(42): 4289-94, October 19.

Shannon, C.E. and W. Weaver (1963). *The Mathematical Theory of Communication*. Urbana: University Illinosis Press.

Shiva, V. (1989). *Staying Alive: Women, Ecology and Development*. London: Zed Books.

Shresth, S. and S. Devidas (2001). *Forest Revival and Water Harvesting Community Based Conservation at Bhanwata-Koliyala, Rajasthan*. Pune: Kalpavriksh.

Simon, J.L. and H. Khan (1984). "The Resourceful Earth: A Response to Global 2000", in Mcneely *op. cit.* New York: Basil Blackwell.

Simpson, E.H. (1949). "Measurement of Diversity", *Nature*, Vol. 163: 688.

Singh V.P. and S.K. Jain (2002). "Watershed Management: New Approaches and Paradigms", *Annals of Arid Zone*, Vol. 41(3-4): 233-262, September–December.

Singh, J.S. (1967). *Seasonal Variation in Composition, Plant Biomass and Net Community Productivity in the Grasslands at Varanasi*, Ph.D. Thesis. Varanasi, India: Banaras Hindu University.

Singh, Katar (1994a). *Managing Common Pool Resources, Principles and Case Studies*. New Delhi: Oxford University Press.

———. (1994b). "Role of Tree Growers' Co-operatives in Promoting Farm Forestry in India", *Working Paper* No. 58. Anand: Institute of Rural Management.

Singh, Savindra (1998). *Geomorphology*. Gorkhpur: Vasundhara Publication.

Staveren, V.I. (2003). "Beyond Social Capital in Poverty Research", *Journal of Economic Issues* Vol. 37(2).

Tang, S.Y. (1991). "Institutional Arrangements and the Management of Common Pool Resources", *Public Administration Review*, Vol. 51(1): 42-52, (from Bardhan, 1999).

Threshow, M. (1970). *Environment and Plant Response*. New York: Mc Graw Hill.

Veblen, T. (1919). "Why is Economics not an Evolutionary Science", in T. Veblen (ed)., *The Place of Science in Modern Civilisation and Other Essays*. New York: Russel and Russel (from Samuel, 1990.)

Vyas, V.S. (1991). "Agrarian Structure, Environmental Concerns and Rural Poverty", *Elmhirst Memorial Lecture*. Tokyo: International Association of Agricultural Economists, August 22.

Wade, Robert (1988). *Village Republics: Economic Conditions for Collective Action in South India*. Cambridge: Cambridge University Press.

———. (1994). *Village Republics: Economic Conditions for Collective Action in South India*. San Francisco: Institute for Contemporary Studies.

Warren, Mark R, P.J. Thompson and Susan Saegert (2001), "The Role of Social Capital in Combating Poverty", in Susan Saegert, J.P. Thompson and Mark R. Warren (ed.), *Social Capital and Poor Communities*. New York: Sage Foundation.

White, T. Anderson and C. Ford Runge (1994). "Common Property and Collective Action: Lessons from Co-operative Watershed Management in Haiti", *Economic Development and Cultural Change*, Vol. 43(1): 1-41, October.

Yadav, J. P. S. and K. Singh (1970). "Tolerance of Certain Forest Species to the Varying Degree of Salinity and Alkalinity", *Indian Forest*, Vol. 96: 587-99.

Yadav, M.S. and M.P. Rajora (1995). "Pasture Development Research in Western Arid Region", in C.R.Hazra and Bimal Misri (ed.), *New Vistas in Forage Production*. All India Co-ordinated Projects for Research on Forage Crops. Jhansi: Indian Grassland and Fodder Research Institute.

Annexure 1

Guidelines for Estimation of Groundwater Recharge

Title	*Explanation*
• Year	
• Potential Zone Area (sq km)	
• Water level fluctuation (m)	
• Specific Yield (per cent)	
• Gross *kharif* draft (Agriculture + Domestic) mcm	25 per cent of gross agriculture draft of particular year plus domestic draft of 120 days of that year
• Monsoon recharge from groundwater irrigation Rgw (mcm)	25 per cent of gross agriculture draft of particular year multiplied by factor depending on range of water level (25 per cent for <10m, 15 per cent for 10-25m and 5 per cent for> 25m water level)
• Monsoon recharge from seepage from canal Rc (mcm)	
• Monsoon recharge from surface water irrigation Rsw (mcm)	
• Monsoon recharge from Tank & Ponds Rt (mcm)	
• Rainfall Recharge Ri (mcm)	Col. No. 2x 3x 4 + 5 – 6 –7– 8–9
• Gross Agriculture draft (mcm)	No. of wells x operational days x yield per day (ltr)
• Gross Domestic Draft (mcm)	No. of wells x 365 x yield per day (ltr)
• Normal monsoon Rainfall r (normal) (m)	Normal monsoon rainfall for the particular year
• Monsoon Rainfall ri (m)	Actual monsoon rainfall for the particular year
• (ri X ri) (Rainfall X Rainfall monsoon) (m)	Square of col. No. 14 (Required for the constant (a) (b) for regression analysis
• (Ri X ri) (Monsoon Rainfall ri X Recharge Ri)	Multiplication of Col. 10 & 14
• Block	
• Type of Area	
• Potential zone	
• S1 (Sum of Rainfall in monsoon) (m)	Sum of Col. No. 14
• S2 (Sum of Rainfall recharge in monsoon) (mcm)	Sum of Col. No. 10

Contd...

...contd...

Title	*Explanation*
• S3 (Sum of Rainfall X Rainfall monsoon) (m)	Sum of Col. No. 15
• S4 (Sum of monsoon Rainfall X Recharge)	Sum of Col. No. 16
• (a) (Constant)	{(No. of Assessment years x Col. No. 23) – (Col. No. 20 x 21)}÷{(No. of Assessment year x Col. No. 22) – Square of Col. No. 20}
• (b) (Constant)	{Col. No. 21 – (24 x 20)} x No. of Assessment year
• R normal (Linear regression method) (a x r-normal + b)	Col. No. 24 x normal monsoon rainfall of the Assessment year + Col. No. 25
• R. I. Factor (per cent)	
• Monsoon Recharge by R.I.F. Rrf (normal rifm) (mcm)	Col. No. 2 x Normal monsoon rainfall of the Assessment year x Col. No. 27
• Variation (PD) (per cent)	per cent deviation {(Col. No. 26 – 28)÷(Col. No. 28)} x 100
• Accepted value of normal monsoon rainfall recharge Rrf (normal) (mcm)	If percentage deviation is in between – 20 and + 20 then accept Col. No. 26, If PD is <-20 then accept 80 per cent of Col. No. 28, If PD is > + 20 then accept 120 per cent Col. No. 28
• Total recharge from other sources (mcm)	Sum of Col. No. 6, 7, 8 & 9 of Assessment year
• Total recharge during normal monsoon season R (normal) (mcm)	Sum of Col. No. 30 & 31
• Block	
• Type of Area	Command, Non-command & Saline
• Water bearing formation	
• Potential zone area (sq km)	
• R.I. Factor (per cent)	
• Normal non-monsoon rainfall (m)	Of the Assessment Year
• Normal Annual Rainfall (m)	Of the Assessment Year
• Normal Recharge from Rainfall during non-monsoon season (mcm)	If Col. No. 38 is equal or more than 10 per cent of Col. No. 39 than Col. No. 36 x 37 x 38, otherwise no calculation required
• Recharge from Groundwater irrigation Rgw. (mcm)	75 per cent of gross agriculture draft of particular year multiplied by factor depending on range of water level (25 per cent for <10m, 15 per cent for 10-25m and 5 per cent for > 25m water level)

Contd...

...contd...

Title	*Explanation*
• Non-monsoon recharge from seepage from canal Rc (mcm)	
• Non-monsoon recharge from surface water irrigation Rsw. (mcm)	
• Non-monsoon Recharge from Tank and ponds Rt (mcm)	
• Recharge from other sources (Rgw + Rc + Rsw + Rt) (mcm)	Sum of Col. No. 41, 42, 43 & 44
• Total Recharge in monsoon season (mcm)	Sum of Col. No. 40 & 45
• Total normal Annual groundwater recharge (mcm)	Sum of Col. No. 32 & 46
• Natural discharge during non-monsoon season (mcm)	Allow natural discharge in monsoon season in terms of base flow and sub surface inflow/outflow (5-10 per cent depending on the area)
• Block	
• Area of block (sq. km)	
• Type of area	
• Water bearing formation	
• Potential zone area (sq km)	
• Net groundwater availability (mcm)	Minus Col. No. 48 from Col. No. 47
• Existing gross groundwater draft for irrigation (mcm)	Of the Assessment year
• Existing gross groundwater draft for domestic and industrial use (mcm)	Of the Assessment year
• Existing gross groundwater draft for all uses (mcm)	70 per cent of Col. No. 55 + Col. No. 26
• Present groundwater balance as on Assessment year (mcm)	Minus Col. No. 57 from Col. No. 54
• Water requirement for domestic and industrial as on year 2025 (mcm)	Not required
• Allocation for domestic and industrial requirement as projected for the year 2025 (mcm)	Col. No. 59 – 58
• Stage of groundwater development (per cent)	(Col. No. 57÷ 54) x 100
• Whether significant decline in pre-	Prepare hydrographs and assess monsoon water level (Yes/No)
• Whether significant decline in post-monsoon water level (Yes/No)	Prepare hydrographs and assess

Contd...

...contd...

Title	*Explanation*
• Category	If Col. No. 61 is <70 then categorise the zone as Safe, if between 70-90 categorise it as Semi Critical, if between 90-100 categorise it as Critical, and if>100 then categorise it as Over Exploitation
• Annual Potential Recharge (mcm)	Recharge of areas where water level is <5m. and consider 5m. for fluctuation and then multiply area of the shallow water zone by specific yield and fluc tuation calculated in this column

Source: Dixit, M, and *et al.*, (2001), Reappraisal of Groundwater Resource of Alwar District, Groundwater Department, Government of Rajasthan.

Annexure 2

Estimation of Floral Diversity

Standard methodology was followed to measure floral diversity (Curtis, 1959; Simpson, 1949; Shannon and Weaver, 1963; Odum, 1971; Chapman, 1976). The Quadrate method (10x10 m. and for trees 1x1 m. for grass) was used for sampling of vegetation of both grass and plant species. Several indices were estimated in order to examine the extent to which floral diversity was achieved in the areas where intervention was made. These were as follows

Importance Value Index (IVI) = RF% + RD% + Rdo%

Where,

$$RF\% = \frac{\text{Number of occurrence of a species X 100}}{\text{Number of occurrence of all species}}$$

$$RD\% = \frac{\text{Number of individuals of a species X 100}}{\text{Number of individuals of all species}}$$

$$RDo\% = \frac{\text{Total basal area of a species X 100}}{\text{Total basal area of all species}}$$

Index of Dominance

This shows importance of each species in relation to the community as a whole.

$C = \Sigma(ni/N)^{2,}$

Where,

ni = Importance value of each species or biomass of each species or number of individuals of each species.

N = Total importance value of all species or total biomass or total number of plants.

C = Index of dominance

Index of Diversity

$H = -\Sigma\ (Pi \log Pi)$

where,

$Pi = \frac{ni}{N}$

ni = Importance value of each species.

N= Total importance value of all species.

Species Richness Index

It is defined as the ratio of number of species to the square root of number of total individuals.

$$D = \frac{S}{\sqrt{N}}$$

S= *Number* of species

$\sqrt{N}$ = Root of number of individuals

Similarity Index

It is the measurement of similarity between different stands. The higher the number, higher the similarity between stands and *vice versa.*

$$\text{Similarity index} = \frac{2\ X\ C}{A + B}$$

C= Number of common species in two stands

A= Number of species in A Stand

B= Number of species in B Stand

Annexure 3

Determinants of Probable Alternative Institutional Outcomes

Model

The objective of the model was to estimate probable future scenarios of the renovated village institutions by means of simulating the present one. One may choose the best possible scenario of all in terms of its ramifications on the community initiative, local ecosystem and economy. One may also identify the conditions under which such best possible scenarios emerge. The probable scenarios were modelled on the existing scenario as captured through field surveys. However, the probable scenarios, as estimated by the model, were independent of any time horizon. The analysis was carried out by following the Backward (LR) Multiple Logistic Regression Model (Smith, 1966). This model specifies functional relationships between probable institutional success and the factors responsible for it where institutional success refers to positive institutional outcome consequent to institutional intervention. Indicators of positive institutional outcome of collective action were based on the perceptions of the village community. It may be noted that we are here more concerned with the effect of the independent variables separately ('main effect' model) rather than interaction effects of the variables.

The entire analysis of institutional success or failure was based on the assumption that at least three indicators of the five above must be positive. To convert these positive indicators into a composite unit, the permutation-combination method was followed. The probability of institutional success was estimated by taking combinations of at least three positive outcomes, as mentioned above, into consideration. The combinations of these three positive outcomes generate a composite unit of probable institutional success. Combinations were analysed thoroughly on the basis of these assumptions ($^{5}_{3}c$).

While estimating the probability of success, one must be equally concerned about the probability of failure too at any given point of time. In other words, one has to recognise the dichotomy that exists between two situations related to any institutional outcome. Hence, one has to consider both simultaneously in the statistical analysis. Let us express this situation as dichotomous dependent variable (Y).

Hence probability of Y=1 as P (Y=1)= α_0 + $\Sigma\alpha_i X_I$ (i=1,2,3...n) where α_0 is the constant term, α_1, α_2... α_n are coefficient of respective independent variables in the model. The probability of Y, needless to mention, lies between zero and one. However, the predicted value of Y may be either less than zero or greater than one. In such cases, one has to take a step forward to substitute the probability of Y=1 with its odds. The odds are defined as $\left[\frac{(P(Y=1/X)}{1-P(Y=1/X)}\right]$ (Menard, S.1995). The natural logarithm of odds, which is log, $\left[\frac{(P(Y=1/X)}{1-P(Y=1/X)}\right]$ is called logit of Y. The above expression may be considered as dependent variable. Hence, the relationship between the dependent variable and the independent variables (call 'predictors') may be written as

$$\text{Logit (Y)} = \alpha_0 + \alpha_1 X_1 + \alpha_2 X_2 + \alpha_3 X_3 + \ldots + \alpha_n X_n \text{ -------- (i)}$$

Convert logit (Y) back to the odds by exponentiation

$$\text{Odds (Y=1)} = \exp^{\text{logit (Y)}}$$

$$\text{Odds (Y=1)} = \exp^{\ln (\text{Odds (Y=1)})}. = e^{(\alpha_0 + \alpha_1 X_1 + \alpha_2 X_2 + \alpha_3 X_3 + \ldots + \alpha_n X_n)} \text{ ------------------------(ii)}$$

Equation (ii) ensures P (Y=1) must lie between zero and one. A change in the predictors by one unit should result in change in odds (Y=1) in an exponential rate i.e. exp $^{(a)}$. We can then convert the odds back to probability that Y=1 by $\left[\frac{Odds(Y=1)}{1+Odds(Y=1)}\right]$.

In other words,

$$P(Y=1/X_i) = \frac{Z}{1+Z}$$

$$\text{Where } Z = e^{(\alpha_0 + \alpha_1 X_1 + \alpha_2 X_2 + \alpha_3 X_3 + \ldots + \alpha_n X_n)} \text{ ---------- (iii)}$$

Annexure 4

Estimation of Equivalence of Value of Fuelwood and Grass

Fuelwood	
Average annual requirement of fuelwood per sample household (quintals)	13.69
Number of rural households in the state using fuelwood	54,99,961.00
Total requirement of fuelwood in rural Rajasthan (quintals)	7,53,13,979.80
Quantity of kerosene required equivalent to fuelwood required (quintals)	2,63,61,211.00
Value of kerosene (Rs.)	3163,34,53,171.30

Note: Equivalance of fuelwood is estimated on the basis of following convertion rate 1 kg kerosene = 10000 kcal and 1 kg fuelwood = 3500 kcal value. Hence, 1 litre kerosene = 2.857 kg fuelwood (approx.)

Source: *http://www.indiasolar.com/col-value.htm*

Grass	
Average annual requirement of grass per sample household (quintals)	24.30
Number of rural households in the state	71,56,703.00
Total requirement of grass of all rural households in Rajasthan (quintals)	17,39,07,883.00
Quantity of dry fodder equivalent to grass required by all rural households (quintals)	16,72,19,118.00
Value of dry fodder (Rs.)	4799,18,86,866.00

Note: Equivalence of grass is estimated on the basis of following conversion rate. It may be noted that farmers use both dry and green grass. Moisture content of green grass differs between 70-75 per cent, while for dry grass it is 12 per cent. Similarly, protein content varies between 7-8 per cent in green grass, while it is 4 per cent in dry grass. In this present exercise, average value of both green and dry grass is taken in respect of moisture as well as protein. In the first case, average value is 43.50 per cent while it is 6 per cent in the second case. The corresponding equivalance of moisture content in dry fodder is 12 per cent while for protein it is 4 per cent. On the basis of these equivalance, grass is converted into dry fodder @1kg dry fodder = 1.04 kg grass (with protein content). The price at which dry fodder is bought is collected from the sample households.

Source: Data on total number of rural housholds and the number of rural households using fuelwood in Rajasthan were collected from Population Census of India, 2001. The conversion rate of grass into dry fodder is collected through personal communication with Dr. S.C. Bohra, Central Arid Zone Research, Institute (CAZRI), Jodhpur.

Index

A

B

C

D

E

N

O

P

Q

R

S

T

U

V

W

Y

Z